T0329161

Numerical Simulation of Multiphase Reactors with Continuous Liquid Phase

Numerical Simulation of Multiphase Reactors with Continuous Liquid Phase

Chao Yang and Zai-Sha Mao

Institute of Process Engineering
Chinese Academy of Sciences

 Chemical Industry Press

AMSTERDAM • BOSTON • HEIDELBERG • LONDON
NEW YORK • OXFORD • PARIS • SAN DIEGO
SAN FRANCISCO • SINGAPORE • SYDNEY • TOKYO

Academic Press is an Imprint of Elsevier

ELSEVIER

Academic Press is an imprint of Elsevier
The Boulevard, Langford Lane, Kidlington, Oxford OX5 1GB, UK
225 Wyman Street, Waltham, MA 02451, USA

First edition 2014

British Library Cataloguing in Publication Data
A catalogue record for this book is available from the British Library

Library of Congress Cataloging-in-Publication Data
A catalog record for this book is availabe from the Library of Congress

ISBN–13: 978-0-08-099919-7

For information on all Academic Press publications
visit our web site at store.elsevier.com

Printed and bound in the US
14 15 16 17 18   10 9 8 7 6 5 4 3 2 1

Contents

Preface

Multiphase reactors with continuous liquid phase such as stirred tanks and loop reactors are popularly used equipment for multiphase chemical reactions, crystallization and mixing in chemical and petrochemical, hydrometallurgical and pharmaceutical industries, etc. Much effort has been devoted to numerically resolving the hydrodynamics and transport in multiphase reactors with liquid phases and satisfactory progress has been achieved, especially with respect to single-phase liquid systems. The flow and transport in reactors operating in multiphase systems demand more intensive attention, both numerical and experimental. The computational fluid dynamics and computational transport principles have been developed into reliable and efficient tools to study and optimize the macroscopic performance of unit operations in process equipment. Along with the rapid development of physical and chemical technologies, numerical simulation of multiphase flow and mass transfer in multiphase reactors with continuous liquid phase is now faster than ever before. It is now appropriate to present the state-of-the-art knowledge and research in this very active field. We hope this book is able to provide useful knowledge for our colleagues and to facilitate research and development in the field of multiphase reaction engineering.

To facilitate the exchange of original research results and reviews on the design, scale-up, and optimization of multiphase reactors, we have written this book entitled *Numerical Simulation of Multiphase Reactors with Continuous Liquid Phase* to address many important aspects of multiphase flow and transport fields. This book aims to embrace important interdisciplinary topics in fundamental and applied research of mathematical models, numerical methods, and experimental techniques for multiphase flow and mass transfer in reactors and crystallizers, operating in gas–liquid, liquid–solid, liquid–liquid, gas–liquid–solid, liquid–liquid–solid, and gas–liquid–liquid systems on the macro-scale and meso-scale (namely the scale of particles including solid particles, bubbles, and drops). Thus, important and interesting topics of research frontiers for a wide range of engineering and scientific areas are presented. We believe that this is a good reference book for readers interested in the design and scale-up of multiphase reactors and crystallizers, in particular stirred tanks, loop reactors, and microreactors, using mathematical modeling and numerical simulation as tools.

We express our sincere appreciation to Jie Chen, Yang Wang, Ping Fan, and Zhihui Wang, who contributed to Chapter 2; Xiangyang Li, Xin Feng, Jingcai Cheng, and Guangji Zhang, who contributed to Chapter 3; Qingshan Huang, Weipeng Zhang, and Guangji Zhang, who contributed to Chapter 4; Yumei Yong, Xi Wang, and Yuanyuan Li, who contributed to Chapter 5; and Jingcai Cheng, Xin Feng, and Yuejia Jiang, who contributed to Chapter 6.

We are very grateful to our many students who have contributed to the book. We wish to thank Prof. Jiayong Chen at our institute, for valuable advice and continuous encouragement. We would like to express our gratitude to our families for their great support of our work. This work is partly supported by China Sci-Tech projects including 973 Program (2010CB630904, 2012CB224806), National Science Fund for Distinguished Young Scholars (21025627), National Natural Science Foundation of China (20990224, 21106154, 21306197), and 863 Project (2012AA03A606, 2011AA060704). We also look forward to receiving any comments, criticisms, and suggestions from the readership, which would be of benefit to the book and the authors.

Chao Yang and Zai-Sha Mao

Introduction

To meet the growing need for bulk chemicals in the national economy and in human life, chemical engineers have been trying to develop the best methodology for scaling up all types of reactors for diversified products. Historically, a larger scale reactor was tentatively designed after a series of cold model experiments and hot model tests. Even though these tests were done carefully and the design was backed up with a wealth of valuable intellectual experience, such a scale-up remains quite risky, because a new or large-sized reactor is the result of *extrapolation* based on tests in limited scopes of reactor configuration and experimental conditions. Better extrapolation would result from a basis of scientific laws that have been proved universally true in many industrial tests in addition to numerous natural phenomena. Mathematical models of chemical reactors are believed to be a sound scientific basis for such extrapolations. A practical model of a reactor is very complicated: phenomenologically involved with multiphase flow, macro- and micromixing, heat and mass transfer, and complex chemical reactions; mathematically with algebraic, ordinary and partial differential equations with strong nonlinearity and mutual coupling. Fortunately, we can utilize numerical simulation to solve such models, and tentatively guide the scale-up of chemical reactors to successful commercial operation. This explains why we are advocating the approach of mathematical modeling and numerical simulation so ardently, both in chemical engineering fundamental research and in industrial innovation.

Many interesting methods may be complemented with mathematical modeling and numerical simulation – for example, optimized operation for higher productivity or better product quality, upgrading the performance of reactors already on the process line, etc. There is one further comment here on scaling-up a reactor. Strictly speaking, we are not sure if the present reactor type and configuration are suitable for a larger scale reactor, as judged by our previous experience on extrapolation. A chemist can conduct a reaction successfully in a lab beaker (a small stirred tank), but this does not mean all commercial reactors for the same reaction should be conducted in large beakers. Using the approach of mathematical modeling and numerical simulation, we can conduct many virtual (numerical) tests of several reactor types and configurations on different scales, with the confidence that the capability of such a first-principles-based approach can achieve an optimized extrapolation of reactors. This approach may ultimately resolve the methodology of reactor renovation and

Numerical Simulation of Multiphase Reactors with Continuous Liquid Phase. DOI: 10.1016/B978-0-08-099919-7.00001-9

innovation. The approach relies heavily on an in-depth quantitative understanding of the mechanisms occurring in chemical reactors for building the mathematical models and the various numerical techniques for solving the established models, as itemized and exemplified in this book.

This book is primarily focused on chemical engineering sciences and technologies, and aims to be a reference book for scientists and engineers in the fields of chemical reaction engineering, mass/heat transfer, hydrodynamics, crystallization, etc. The book will provide design, optimization, and scale-up concepts and numerical methods for multiphase reactors and crystallizers such as stirred tanks, loop reactors and microreactors for different application purposes. There are five subsequent chapters on various topics relevant to multiphase reactors with liquid phases.

Chapter 2 deals with the multiphase flow and interphase mass transfer on a particle scale. The mechanism of multiphase flow and mass transfer on the mesoscale is vital to the design and scale-up of reactors and crystallizers. The orthogonal boundary-fitted coordinate system-based simulation and level set method are improved to compute the motion and mass transfer of bubbles and drops, and also the mirror fluid method for motion of solid particles is developed. Thereafter, the modified cell model is proposed to examine the flow and transport behavior of particle swarms. The study on the motion and mass transfer of a solute to/from a single drop with a surfactant adsorbed on the interface and the Marangoni effect is expounded to better understand the liquid extraction and reaction processes. Also, the principal research results for the transport process of a spherical particle in pure extensional and simple shear flows are introduced in this chapter.

Chapter 3 deals with the numerical simulation of multiphase stirred tanks, which are the most used reactors or crystallizers in continuous, batch, or fed-batch modes. Good mixing in stirred tanks is important for minimizing investment and operating costs, providing high yields when mass/heat transfer is limiting, and thus enhancing profitability. Multiphase flow and transport in stirred tanks demand more intensive attention with combined numerical and experimental approaches. In this chapter, we present extensive experimental and numerical simulation results of recent developments for stirred tanks. Multiphase flows (including two- and three-phase flows) are discussed in detail based on numerical methods using the Eulerian multifluid approach and RANS (Reynolds average Navier–Stokes)-based turbulence models (e.g., k–ε model). Novel surface aeration configurations are introduced for better gas dispersion and high pumping capacity, and the hydrodynamic characteristics of multi-impellers and numerical simulation of gas hold-up in surface-aerated stirred tanks are also addressed. Some new advances in numerical simulation are also presented. The algebraic stress model (ASM) and large eddy simulation (LES) are recommended for future research on multiphase flows in stirred reactors.

Chapter 4 deals with the hydrodynamics and transport in loop reactors. Airlift internal loop reactors are commonly used in petrochemical, hydrometallurgical, energy, environmental, and bio-engineering processes due to their excellent advantages of simple structure, high gas–liquid mass and heat transfer rates, good solid suspension, homogeneous shear distribution, and good mixing. Although great achievements have been made on loop reactors, the design and scale-up of these reactors still remain difficult due to the nature of complex multiphase flow. In this chapter, investigations of the flow and mixing characteristics by experiments and computational fluid dynamics simulations are presented on airlift reactors with very high and low height-to-diameter ratios. Also, as an intrinsic element of the new technology of coal liquefaction in China, an internal airlift loop reactor pilot test is introduced on the feasibility of replacing the bubble column reactor on the industrial process line of direct coal liquefaction.

Chapter 5 deals with the preliminary investigation of numerical methods and experiments for flow and mixing in two-phase microreactors. The miniaturization of chemical engineering devices has recently brought significant changes, and the progress in microreactors opens doors to more efficient, economic, and safer process intensification. The selectivity of fast chemical reactions depends on the quality of macro- and micromixing. In this chapter, the flow, pressure drop, mass transfer, and mixing of two-phase flow in microchannels with different wetting properties are investigated for different flow patterns. Immiscible two-phase flows, thermal transfer, and mass diffusion in microchannels are numerically studied by a lattice Boltzmann method based on field mediators.

Chapter 6 deals with the mathematical models and numerical simulation of solid–liquid crystallizers. Crystallizers are widely used to produce fine and bulk chemicals. Most of the theoretical and experimental studies are aimed at understanding important mechanisms in the crystallization process in order to stabilize process control, and ultimately to obtain products with desired crystal size distribution (CSD), morphology, and mean size. In this chapter, numerical simulations towards predicting the full CSD directly in a more practical crystallization reactor are presented in a Eulerian framework, and nucleation, growth, and aggregation are considered. The effects of aggregation, feeding concentration, agitation speed, mean residence time, and the CSDs of different locations are studied numerically. Reaction crystallizations are mixing-sensitive multiphase processes, so macro- and micromixing in crystallizers and some other multiphase reactors are also presented.

Fluid flow and mass transfer on particle scale

2.1 INTRODUCTION

Fluid flow of and mass transfer from/to drops, bubbles, and solid particles are often observed in nature and various areas of engineering. Chemical and metallurgical engineers rely on bubbles and drops for unit operations such as distillation, absorption, flotation and spray drying, while solid particles are used as catalysts or chemical reactants. In these processes, there is relative motion between bubbles, drops or particles on one hand, and a surrounding fluid on the other. In many cases, transfer of mass and/or heat is also of importance. Owing to rapid progress of computer techniques and numerical methods in fluid mechanics and transport phenomena, the application of numerical simulation has recently become increasingly popular in understanding multiphase flow and transport on a particle (a generic term including drops and bubbles) scale.

In this chapter, this topic is discussed in detail in the following six sections. Firstly, the theoretical basis and numerical methods frequently adopted are summarized in Sections 2.2 and 2.3 respectively. We choose to focus mostly on three methods: simulation on orthogonal boundary-fitted coordinates, an improved level set method, and a mirror fluid method. This choice reflects our own background, as well as the fact that these methods are deemed successful and reliable for computing the motion and mass transfer of fluid particles (bubbles and drops) or solid particles. The validity of these methods is demonstrated and compared with the reported experimental data in Section 2.4. Also, considering the trace quantities of surfactants unavoidable in most industrial systems, study of the motion and mass transfer of a solute to/from a single drop with a surfactant adsorbed on the interface is carried out to better understand the liquid extraction processes and for the scientific design of relevant equipment. The Marangoni effect, one of the most sophisticated interphase transport phenomena, interests researchers due to its influence on transport rates and it has been mathematically formulated and numerically simulated to shed light on these mechanisms. Recent studies relating to the Marangoni effect are presented in Section 2.5. In Section 2.6, numerical simulation methods on particle swarms are discussed briefly and modified cell models are introduced to examine the flow and transport behaviors of particle swarms. Section 2.7 incorporates related progress on particle motion controlled by fluid shear or extension.

Numerical Simulation of Multiphase Reactors with Continuous Liquid Phase. DOI: 10.1016/B978-0-08-099919-7.00002-0

2.2 THEORETICAL BASIS

The mathematical formulation of two-phase particle flow may be exemplified using two-fluid systems in which a liquid drop or a gas bubble moves in another continuous liquid as it follows in this section. The fundamental physical laws governing the motion of and mass transfer from/to a single particle immersed in another fluid are Newton's second law, the principle of mass conservation, and Fick's diffusion law. So the flow field and solute transport in both fluid phases must be formulated using the first principles of fluid mechanics and transport phenomena. When a solid particle is involved, the flow in the solid domain is usually not necessary and the particle is tracked mechanically as a rigid body. In this context, two-phase flow with a solid particle is a simplified case of general two-phase systems.

2.2.1 Fluid mechanics

The motion of a small particle (drop, bubble or solid particle) of around 1 mm size under gravity through an immiscible continuous fluid phase can be resolved using the following assumptions: (1) the fluid is viscous and incompressible; (2) the physical properties of the fluid and the particle are constant; (3) the two-phase flow is axisymmetric or two-dimensional; (4) the flow is laminar at low Reynolds numbers.

The flow in each fluid phase is governed by the continuity and Navier–Stokes equations:

$$\nabla \cdot \mathbf{u} = 0 \tag{2.1}$$

$$\rho \left(\frac{\partial \mathbf{u}}{\partial t} + \mathbf{u} \cdot \nabla \mathbf{u} \right) = -\nabla p + \rho \mathbf{g} + \nabla \cdot \boldsymbol{\tau} + \mathbf{S} \tag{2.2}$$

where $\boldsymbol{\tau}$ is the stress tensor defined as

$$\boldsymbol{\tau} = \mu (\nabla \mathbf{u} + (\nabla \mathbf{u})^{\mathrm{T}}) \tag{2.3}$$

and the source term $\mathbf{S}$ is formulated differently in different cases.

Boundary conditions for the governing equations are essential when an interface exists between the two phases. For a bubble or a drop, the normal velocity in each phase is equal at the interface. If the gas in a bubble is taken as inviscid, the bubble surface is mobile and not subject to any shear force. However, if the gas is taken as a viscous fluid, both the velocity vector and shear stress should be continuous across the interface. For a solid particle, both the normal and tangential velocity components of the continuous phase must be zero at the particle surface; that is, the solid surface should satisfy the "no-slip" condition.

For the case with constant physical properties of both fluid phases, including that on the interface, the solution for mass transfer will be decoupled from the problem of fluid flow. Thus, the information of the flow field, required for solution of convective diffusion problems, whether for steady or unsteady mass transfer, can be provided directly from numerical simulation of steady-state fluid flow only once.

2.2.2 Mass transfer

In general, the transient mass transfer to/from a drop (or a bubble) is governed by the convective diffusion equation in vector form:

$$\frac{\partial c}{\partial t} + \mathbf{u} \cdot \nabla c = D \nabla^2 c \tag{2.4}$$

in each phase subject to two interfacial conditions:

$$D_1 \frac{\partial c_1}{\partial n_1} = D_2 \frac{\partial c_2}{\partial n_2} \quad \text{(flux continuity at the interface)} \tag{2.5}$$

$$c_2 = mc_1 \quad \text{(interfacial dissolution equilibrium)} \tag{2.6}$$

In the above equations, subscript 1 indicates the continuous phase and 2 the dispersed phase. The solution of Eq. (2.5) is reliant on the resolved fluid flow both in the dispersed and the continuous phases, as addressed by Li and Mao (2001). In accordance with Fick's first law, for steady external mass transfer the local diffusive flux across the interface is calculated by

$$N_{\text{loc}} = -D_2 \frac{\partial c_2}{\partial n_2} = k_{\text{loc}} (\overline{c}_2 - mc_1^\infty) \tag{2.7}$$

where the remote boundary concentration c_1^∞ and the only available measurement of the bubble/drop concentration $\overline{c}_2$ (averaged over the whole drop, taking a drop as an example) are used to define the driving force and the mass transfer coefficient. The latter may be expressed in terms of dimensional concentration gradient as

$$k_{\text{loc}} = -\frac{D_2}{(\overline{c}_2 - mc_1^\infty)} \frac{\partial c_2}{\partial n_2} \tag{2.8}$$

Then, the local Sherwood number is

$$Sh_{\text{loc}} = \frac{d k_{\text{loc}}}{D_2} = -\frac{d}{(\overline{c}_2 - mc_1^\infty)} \frac{\partial c_2}{\partial n_2} \tag{2.9}$$

and the drop area averaged Sh_{od} is

$$Sh_{\text{od}} = \frac{\oint Sh_{\text{loc}} \, ds}{\oint ds} \tag{2.10}$$

On the other hand, the overall mass transfer coefficient k_{od} may be evaluated from the overall solute conservation based on the drop as follows:

$$k_{\text{od}} (c_2^* - \overline{c}_2) A = V_d \frac{d\overline{c}_2}{dt} \tag{2.11}$$

where $\overline{c}_2$ is the average concentration of the drop at any time instant, which is almost the only available measure of solute concentrations of drops in conventional experiments. If the time interval $t_{out}-t_{in}$ is chosen small enough, k_{od} may be evaluated approximately from integration of the above equation as

$$k_{od} = -\frac{V_d}{A}\frac{1}{t_{out}-t_{in}}\ln\left(\frac{c_2^*-\overline{c}_{2,out}}{c_2^*-\overline{c}_{2,in}}\right)$$

(2.12)

where A and V_d are the volume and the surface area of the drop, and for a spherical drop $V_d/A = d/6$. The corresponding Sherwood number is

$$Sh_{od} = \frac{d}{D}k_{od}$$

(2.13)

2.2.3 Interfacial force balance

When the drop or bubble shape is to be determined, the force balance over the interface must be satisfied. Moreover, the interface that separates two contacting phases is the common boundary of two phases. The interface status must be compatible with the motion of either phase. Thus, the equations governing the momentum and mass balances are often used as the boundary conditions for the governing equations of motion and transport in each phase.

In general macroscopic hydrodynamic formulations, the interface is taken realistically as a weightless layer of zero thickness. Therefore, all forces exerted over an infinitesimal section of the interface have to be summed to be zero, whether the particle is in steady or accelerating motion. As illustrated in Figure 2.1, the pressure and stress tensor in both fluids and the surface force are involved. The overall force balance is as follows:

$$(-p_1\mathbf{I}+\tau_1)\cdot\mathbf{n}+\mathbf{f}_s = (-p_2\mathbf{I}+\tau_2)\cdot\mathbf{n}$$

(2.14)

where the surface force $\mathbf{f}_s$ is the sum of normal and tangential force components:

$$\mathbf{f}_s = -\sigma\kappa\mathbf{n}+\nabla_s\sigma$$

(2.15)

in which $\kappa=-\nabla\cdot\mathbf{n}$ is the mean curvature of the interface, $\mathbf{n}$ is the outward normal unit vector, and ∇_s is the surface gradient operator. The interfacial tension σ is generally a function of temperature, solute concentration at the surface, surfactant

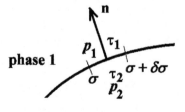

FIGURE 2.1 Interfacial force balance over an interface segment.

adsorption etc., and such constitutive relations need to be known in advance of the numerical simulation of fluid particle motion.

In the case where the surface tension varies only with the solute concentration, the surface force due to the gradient of the above parameters along the interface may be expressed in a more elegant form:

$$\mathbf{f}_s = -\sigma\kappa\mathbf{n} + \frac{d\sigma}{dc}(\mathbf{I} - \mathbf{nn}) \cdot \nabla c \tag{2.16}$$

Equation (2.14) denotes the normal and tangential force balances, and is used as the boundary conditions for solving the hydrodynamic equations of two bulk phases. It will be expanded into diverse forms when simulating various kinds of particles in corresponding reference frames.

2.2.4 Interfacial mass transport

An ordinary solute in a solvent extraction system would not accumulate on the drop surface, and the interface exerts no influence on the interfacial tension. However, in a surfactant-contaminated two-fluid system, or when the surfactant is added intentionally as a manipulating measure, the surfactant will be adsorbed and accumulate on the interface, which is mobile for bubbles and drops. Therefore, the surfactant molecules may be transported within the interface by convection and diffusion mechanisms, while they undergo dynamic exchange with two bulk phases via adsorption and desorption. Since variation of interfacial tension relies on the amount of adsorbed surfactant, its transport in the interface in turn influences the force balance at the interface. The interfacial transport equation is established to account for all the above mechanisms as follows:

$$\frac{\partial\Gamma}{\partial t} + \nabla_s \cdot (\mathbf{u}_s\Gamma) - \nabla_s \cdot (D_s\nabla_s\Gamma) = S \tag{2.17}$$

in which D_s is the surface diffusion coefficient, Γ is the surface adsorption of surfactant, $\mathbf{u}_s$ is the convective velocity at the interface, and ∇_s is the surface counterpart of operator ∇. The source term S consists of the separate net adsorptions from each phase, which accounts for the exchange of surfactant mass with two bulk phases via adsorption and desorption:

$$S = S_1 + S_2 \tag{2.18}$$

$$S_1 = D_1[\mathbf{n}\cdot\nabla c_1]_s = \beta_1 c_{1s}(\Gamma_\infty - \Gamma) - \alpha_1\Gamma \tag{2.19}$$

$$S_2 = -D_2[\mathbf{n}\cdot\nabla c_2]_s = \beta_2 c_{2s}(\Gamma_\infty - \Gamma) - \alpha_2\Gamma \tag{2.20}$$

Here ∇c is the surfactant concentration gradient in the bulk, and subscript s indicates the value immediately at the interface. The surfactant has different adsorption and desorption coefficients β and α in each phase.

Equation (2.17) describes the surface status and its temporal evolution to function as the boundary condition for coupled convective–diffusive mass transfer in two fluid phases, and the possible non-uniform distribution of a solute or a surfactant on the interface will in turn make the hydrodynamic and transport problems coupled and add to the numerical difficulty. Equation (2.17) is itself a differential equation to be solved, and its boundary conditions should be designated appropriately. For example, the following condition:

$$\nabla_s \Gamma = 0 \qquad\qquad (2.21)$$

is suitable for the surfactant loaded interface at the front and the rear stagnant points of a drop.

2.3 NUMERICAL METHODS

The Navier–Stokes equation is the fundamental equation for describing hydrodynamic problems, which may be solved using three types of numerical solution methods: the primitive variables method, the stream function–vorticity formulation method, and the high-order stream function method. The accuracy and efficiency of computational fluid dynamics depends largely on the quality of the computational grid and numerical algorithm. For a specific problem, therefore, an appropriate grid and algorithm should be adopted to meet the requirements of accuracy and computational efficiency. Stream function–vorticity formulation can be adopted in an orthogonal boundary-fitted coordinate system for solving exactly 2D laminar flow with low and medium Reynolds numbers. However, the topological structure of the interface in multiphase flow changes dramatically, for instance, coalescence, breakage, and filamentation of a dispersed phase. In such circumstances, the primitive variables method and interface treatment techniques (such as the level set method) are usually combined to gain greater accuracy and higher efficiency, even in a regular structured grid.

The development of numerical methods for flow containing a sharp interface is currently a hot issue and significant progress has been made by a number of groups. The body-fitted grid technique is an appropriate option to solve a problem with an irregular boundary (Thames et al., 1977; Shyy et al., 1985). The immersed boundary method (Kim et al., 2001; Peskin, 2002) and the mirror fluid method (Yang and Mao, 2005a) can also handle this problem, which is based on mechanical principles and makes a reasonable hypothesis or introduces a variable with well-defined physical meaning to express the effect of a boundary on the fluid flow.

2.3.1 Orthogonal boundary-fitted coordinate system

For the free-boundary problem of a buoyancy-driven particle, it is beneficial to use an orthogonal boundary-fitted coordinate system (OBFCS) so as to enforce the boundary conditions more accurately at the surface of the deformed particle. In order to

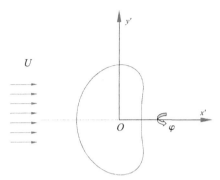

FIGURE 2.2 Sketch of a deformed drop in the cylindrical reference frame.

ensure that the two coordinate grids, inside and outside the particle, match up exactly at the free surface, the strong constraint method is used for the outer domain and the weak constraint method for the inner domain (Ryskin and Leal, 1983). For an axisymmetrical fluid particle in another quiescent fluid (as seen in Figure 2.2), it is convenient to use the common cylindrical frame. When the coordinates are nondimensionalized with the volume-equivalent radius, R, of a particle, the two-dimensional sectional plane (x,y) through the symmetry axis can be mapped orthogonally to the computational plain (ξ,η) by the covariant Laplace equations:

$$\begin{cases} \dfrac{\partial}{\partial \xi}\left(f(\xi,\eta)\dfrac{\partial x}{\partial \xi} \right) + \dfrac{\partial}{\partial \eta}\left(\dfrac{1}{f(\xi,\eta)}\dfrac{\partial x}{\partial \eta} \right) = 0 \\[2ex] \dfrac{\partial}{\partial \xi}\left(f(\xi,\eta)\dfrac{\partial y}{\partial \xi} \right) + \dfrac{\partial}{\partial \eta}\left(\dfrac{1}{f(\xi,\eta)}\dfrac{\partial y}{\partial \eta} \right) = 0 \end{cases} \tag{2.22}$$

where $f(\xi,\eta)$ is called the distortion function, defined as $f(\xi,\eta) = h_\xi/h_\eta$, the ratio of scale factors h_ξ to h_η with

$$h_\xi = \sqrt{\left(\frac{\partial x}{\partial \xi}\right)^2 + \left(\frac{\partial y}{\partial \xi}\right)^2}, \quad h_\eta = \sqrt{\left(\frac{\partial x}{\partial \eta}\right)^2 + \left(\frac{\partial y}{\partial \eta}\right)^2} \tag{2.23}$$

A unit square in the computational plane corresponds physically to either the external or internal domain, as indicated in Figure 2.3 (Li and Mao, 2001).

In choosing the distortion function for implementing orthogonal mapping, Ryskin and Leal (1983, 1984a) and Dandy and Leal (1989) chose an infinite region for the exterior phase and specified the distortion function:

$$f_1(\xi_1,\eta_1) = \pi\xi_1(1 - 0.5\cos\pi\eta_1) \tag{2.24}$$

where subscript 1 denotes the outer continuous phase.

Li et al. (2000) noted that the distortion function became vanishingly small at the far infinity and made it difficult to enforce the remote boundary conditions

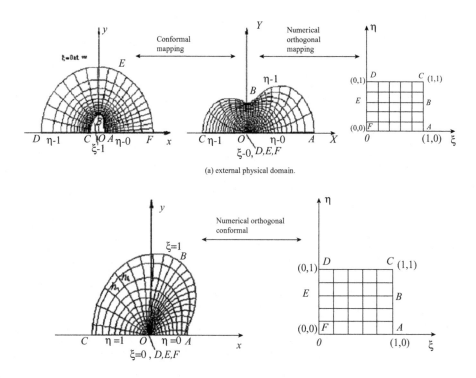

(a) external physical domain.

(b) internal physical domain

FIGURE 2.3 The correspondence of the physical domain (x,y) of a particle to the auxiliary domain (X, Y) and the computational domain (ξ, η).

(a) External physical domain. (b) Internal physical domain.

accurately at $\xi_1 = 0$ (Figure 2.3a). This shortcoming can be overcome by choosing a large enough exterior region around a solid sphere and using a new form of distortion function:

$$f_1(\xi_1, \eta_1) = \frac{\pi}{\beta}(1 - 0.5\cos \pi \eta_1) \qquad (2.25)$$

which is called the finite domain distortion function. Here $\beta = 2.5$ is a parameter for specifying the position of the remote boundary, and f is the same order of unity everywhere in the exterior domain. This choice of the distortion function has been successfully applied in a series of numerical works (Mao and Chen, 1997; Li et al., 2000, 2001; Mao et al., 2001).

For the interior phase, the distortion function provided by Ryskin and Leal (1984a) is used:

$$f_2(\xi_2, \eta_2) = \pi \xi_2 (1 - 0.5\cos \pi \eta_2) \qquad (2.26)$$

2.3.1.1 Stream function–vorticity formulation

The steady-state motion of a single particle in an infinite fluid medium in an axisymmetric orthogonal coordinate system may be described by a set of partial differential equations of stream function ψ and vorticity ω in a sectional plane (x,y) passing through the axis of symmetry. The governing equations are (Li and Mao, 2001)

$$v_1 L_1^2 (y_1 \omega_1) - \frac{1}{h_{\xi_1} h_{\eta_1}} \left[\frac{\partial \psi_1}{\partial \xi_1} \frac{\partial}{\partial \eta_1} \left(\frac{\omega_1}{y_1} \right) - \frac{\partial \psi_1}{\partial \eta_1} \frac{\partial}{\partial \xi_1} \left(\frac{\omega_1}{y_1} \right) \right] = \frac{\partial \omega_1}{\partial t} \tag{2.27}$$

$$L_1^2 \psi_1 + \omega_1 = 0 \tag{2.28}$$

$$v_2 L_2^2 (y_2 \omega_2) + \frac{1}{h_{\xi_2} h_{\eta_2}} \left[\frac{\partial \psi_2}{\partial \xi_2} \frac{\partial}{\partial \eta_2} \left(\frac{\omega_2}{y_2} \right) - \frac{\partial \psi_2}{\partial \eta_2} \frac{\partial}{\partial \xi_1} \left(\frac{\omega_2}{y_2} \right) \right] = \frac{\partial \omega_2}{\partial t} \tag{2.29}$$

$$L_2^2 \psi_2 + \omega_2 = 0 \tag{2.30}$$

where subscript 1 denotes the outer continuous phase, 2 denotes the particle, and the differential operator is

$$L^2 = \frac{1}{h_\xi h_\eta} \left[\frac{\partial}{\partial \xi} \left(\frac{f}{y} \frac{\partial}{\partial \xi} \right) + \frac{\partial}{\partial \eta} \left(\frac{1}{fy} \frac{\partial}{\partial \eta} \right) \right] \tag{2.31}$$

When being transformed in terms of the following nondimensional physical variables (Ω and Ψ) defined by

$$\omega_1 = \frac{U_\mathrm{T}}{R} \Omega_1, \quad \omega_2 = \frac{U_\mathrm{T}}{R} \Omega_2 \tag{2.32}$$

$$\psi_1 = R^2 U_\mathrm{T} \Psi_1, \quad \psi_2 = R^2 U_\mathrm{T} \Psi_2 \tag{2.33}$$

the governing equations become

$$L_1^2 (Y_1 \, \Omega_1) - \frac{Re_1}{2} \frac{1}{H_{\xi_1} H_{\eta_1}} \left[\frac{\partial \Psi_1}{\partial \xi_1} \frac{\partial}{\partial \eta_1} \left(\frac{\Omega_1}{Y_1} \right) - \frac{\partial \Psi_1}{\partial \eta_1} \frac{\partial}{\partial \xi_1} \left(\frac{\Omega_1}{Y_1} \right) \right] = 0 \tag{2.34}$$

$$L_1^2 \Psi_1 + \Omega_1 = 0 \tag{2.35}$$

$$L_2^2 (Y_2 \, \Omega_2) + \frac{Re_2}{2} \frac{1}{H_{\xi_2} H_{\eta_2}} \left[\frac{\partial \Psi_2}{\partial \xi_2} \frac{\partial}{\partial \eta_2} \left(\frac{\Omega_2}{Y_2} \right) - \frac{\partial \Psi_2}{\partial \eta_2} \frac{\partial}{\partial \xi_1} \left(\frac{\Omega_2}{Y_2} \right) \right] = 0 \tag{2.36}$$

$$L_2^2 \Psi_2 + \Omega_2 = 0 \tag{2.37}$$

$$L^2 = \frac{1}{H_\xi H_\eta}\left[\frac{\partial}{\partial \xi}\left(\frac{f}{Y}\frac{\partial}{\partial \xi}\right)+\frac{\partial}{\partial \eta}\left(\frac{1}{fY}\frac{\partial}{\partial \eta}\right)\right] \tag{2.38}$$

where $\theta = tU_T/R$, and the relevant dimensionless parameters are

$$Re_1 = \frac{2RU_T\rho_1}{\mu_1}, \quad Re_2 = \frac{\zeta}{\lambda}Re_1, \quad \lambda = \frac{\mu_2}{\mu_1}, \quad \zeta = \frac{\rho_2}{\rho_1} \tag{2.39}$$

The physical velocity components are related to the stream function by

External domain $\quad U_{\xi_1} = -\dfrac{1}{Y_1 H_{\eta_1}}\dfrac{\partial \Psi_1}{\partial \eta_1}, \quad U_{\eta_1} = -\dfrac{1}{Y_1 H_{\xi_1}}\dfrac{\partial \Psi_1}{\partial \xi_1}$

(left-handed frame) $\hspace{9cm}$ (2.40)

Internal domain $\quad U_{\xi_2} = -\dfrac{1}{Y_2 H_{\eta_2}}\dfrac{\partial \Psi_2}{\partial \eta_2}, \quad U_{\eta_2} = -\dfrac{1}{Y_2 H_{\xi_2}}\dfrac{\partial \Psi_2}{\partial \xi_2}$

(right-handed frame) $\hspace{8.5cm}$ (2.41)

Solution of fluid flow involves the pertinent boundary conditions. In general, non-slip conditions are to be enforced on the solid surface, which is a coordinate line in the orthogonal body-fitted reference system. For a nondeformable fluid particle, kinematical continuity and tangential force balance over the interface need to be satisfied. For a deformable fluid particle, a third condition of normal force balance across the interface is used to adjust the interface position in real time or virtual time so that the force balance is ultimately satisfied. In this course, the orthogonal body-fitted coordinate system has to be refreshed whenever the particle shape is adjusted. More details on this may be found in some recent monographs on multiphase flow and its numerical simulation.

2.3.1.2 Convective transport equation
The expanded form of Eq. (2.4) for the external domain in a two-dimensional orthogonal curvilinear coordinate grid is

$$\frac{\partial c}{\partial t}+\frac{u_\xi}{h_\xi}\frac{\partial c}{\partial \xi}+\frac{u_\eta}{h_\eta}\frac{\partial c}{\partial \eta}=\frac{D}{h_\xi h_\eta y}\left[\frac{\partial}{\partial \xi}\left(\frac{h_\eta y}{h_\xi}\frac{\partial c}{\partial \xi}\right)+\frac{\partial}{\partial \eta}\left(\frac{h_\xi y}{h_\eta}\frac{\partial c}{\partial \eta}\right)\right] \tag{2.42}$$

In a numerical procedure, the governing equations are nondimensionalized and the following nondimensional variables and groups are defined for this purpose:

$$C_1 = \frac{c_1}{c_1^\infty}, \quad H_{\xi_1} = \frac{h_{\xi_1}}{R}, \quad H_{\eta_1} = \frac{h_{\eta_1}}{R}, \quad Y_1 = \frac{y_1}{R}, \quad Pe_1 = \frac{2RU}{D_1} \tag{2.43}$$

Here Pe is the Peclet number, symbolizing the relative strength of convection to molecular diffusion. With subscript 1 inserted into Eq. (2.42) to index the external domain and Eq. (2.40) incorporated, the nondimensional convective diffusion equation for the external region is

$$\left(\frac{Pe_1}{2}H_{\xi_1}H_{\eta_1}Y_1\right)\frac{\partial C_1}{\partial\theta}+\frac{Pe_1}{2}\left[\frac{\partial}{\partial\xi_1}\left(-\frac{\partial\Psi_1}{\partial\eta_1}C_1\right)+\frac{\partial}{\partial\eta_1}\left(\frac{\partial\Psi_1}{\partial\xi_1}C_1\right)\right]$$
$$=\left[\frac{\partial}{\partial\xi_1}\left(f_1Y_1\frac{\partial C_1}{\partial\xi_1}\right)+\frac{\partial}{\partial\eta_1}\left(\frac{Y_1}{f_1}\frac{\partial C_1}{\partial\eta_1}\right)\right]$$

(2.44)

The governing equation for the interior of the drop is obtained with subscript 1 replaced by 2 and Eq. (2.41) incorporated. The nondimensional equation for the drop phase in the reference frame as in Figure 2.3a is similar:

$$\left(\frac{Pe_2}{2}H_{\xi_2}H_{\eta_2}Y_2\right)\frac{\partial C_2}{\partial\theta}+\frac{Pe_2}{2}\left[\frac{\partial}{\partial\xi_2}\left(-\frac{\partial\Psi_2}{\partial\eta_2}C_2\right)+\frac{\partial}{\partial\eta_2}\left(\frac{\partial\Psi_2}{\partial\xi_2}C_2\right)\right]$$
$$=\left[\frac{\partial}{\partial\xi_2}\left(f_2Y_2\frac{\partial C_2}{\partial\xi_2}\right)+\frac{\partial}{\partial\eta_2}\left(\frac{Y_2}{f_2}\frac{\partial C_2}{\partial\eta_2}\right)\right]$$

(2.45)

It is difficult to construct the orthogonal curvilinear coordinates for complicated and seriously deformed interfaces and the orthogonal boundary-fitted coordinate system is usually used to calculate a buoyancy-driven drop with slightly deformed interface at low Reynolds numbers. In order to calculate the seriously deformed interface at high Reynolds numbers, an improved level set method is proposed (Yang and Mao, 2002, 2005b; Li et al., 2008).

2.3.1.3 Numerical solution procedure

The numerical simulation proceeds via the following steps: (1) Numerical simulation of steady fluid flow inside and outside the particle is carried out by solving Eqs. (2.34)–(2.37) with pertinent boundary conditions. This can be done either by numerical solution of steady Navier–Stokes equations with necessary under-relaxation, or by solution of time-dependent Navier–Stokes equations for sufficient time so that the flow approaches the steady state. (2) Utilizing the solved velocity field (or the stream function) as the starting initial flow conditions, transient mass transfer of a solute is simultaneously solved based on Eqs. (2.44) and (2.45). This stage of simulation is carried out in the real-time domain: Exact solution of the flow and concentration fields must be achieved for each time step, so that the correct mass transfer rate in terms of Sh_{od} can be evaluated using Eq. (2.10). In each time step, it is necessary to solve the stream function Ψ, the vorticity Ω and the concentration of solute C, and to refresh the surface parameters involved in boundary conditions in a few iterations so that these fields become compatible with one another at the end of a time step.

2.3.2 Level set method

Since Osher and Sethian (1988) introduced the level set method for modeling the advected phase front, this method has been applied in many fields with particular advantages being easy and accurate capture of the interface in a fixed Euler grid. Using this method, the topology of an interface and its changes are fully described by the zero set of the level set function, which is initially defined in the whole flow field and then advanced along with fluid flow. The method is stable numerically, the surface geometric parameters such as curvature are easy to be calculated, and three-dimensional calculation problems are also easy to be implemented. Unfortunately, the algorithm for the level set function is in a nonconservative form and requires a reinitialization procedure to guarantee mass conservation.

In the following sections the numerical simulation of a single drop in a liquid–liquid extraction system has been performed in an axisymmetric cylindrical reference frame fixed on the drop. The motion of a drop with a finite degree of deformation is coupled with simultaneous mass transfer to consider the unsteady-state motion during the drop formation stage, etc.

2.3.2.1 Level set method for fluid flow

In a two-dimensional coordinate system, mass and momentum conservation with the level set method incorporated (Yang and Mao, 2002) are written in terms of dimensionless variables as

$$\frac{\partial u}{\partial x} + \frac{1}{r}\frac{\partial}{\partial y}(rv) = 0 \tag{2.46}$$

$$
\begin{aligned}
&\frac{\partial}{\partial \theta}(\rho u) + \frac{\partial}{\partial x}\left(\rho uu - \frac{\mu}{Re}\frac{\partial u}{\partial x}\right) + \frac{1}{r}\frac{\partial}{\partial y}\left(r\rho vu - r\frac{\mu}{Re}\frac{\partial u}{\partial y}\right) \\
&\quad = -\frac{\partial p}{\partial x} + \frac{1}{Fr}\rho g_x - [\rho_r a] - \frac{1}{We}\kappa(\phi)\delta_\varepsilon(\phi)\frac{\partial \phi}{\partial x} \\
&\quad + \frac{1}{Re}\frac{\partial}{\partial x}\left(\mu\frac{\partial u}{\partial x}\right) + \frac{1}{Re}\frac{1}{r}\frac{\partial}{\partial y}\left(r\mu\frac{\partial v}{\partial x}\right)
\end{aligned} \tag{2.47}
$$

$$
\begin{aligned}
&\frac{\partial}{\partial \theta}(\rho v) + \frac{\partial}{\partial x}\left(\rho uv - \frac{\mu}{Re}\frac{\partial v}{\partial x}\right) + \frac{1}{r}\frac{\partial}{\partial y}\left(r\rho vv - r\frac{\mu}{Re}\frac{\partial v}{\partial y}\right) \\
&\quad = -\frac{\partial p}{\partial y} + \frac{1}{Fr}\rho g_y - \frac{1}{We}\kappa(\phi)\delta_\varepsilon(\phi)\frac{\partial \phi}{\partial y} \\
&\quad + \frac{1}{Re}\frac{\partial}{\partial x}\left(\mu\frac{\partial u}{\partial y}\right) + \frac{1}{Re}\frac{1}{r}\frac{\partial}{\partial x}\left(r\mu\frac{\partial v}{\partial y}\right) - \left\{\frac{2}{Re}\frac{\mu v}{r^2}\right\}
\end{aligned} \tag{2.48}
$$

where $r \equiv 1$ for Cartesian coordinates, $r \equiv y$ for cylindrical coordinates, and curly brackets indicate that the term is presented only in cylindrical coordinates. The acceleration of a drop is formulated in the moving reference coordinate via the term marked with the square brackets, which is absent in the fixed inertial reference coordinate, $\rho_r = \rho_2/\rho_1$, and a is the acceleration of the drop. The dimensionless groups Re, Fr, and We are the Reynolds, Froude, and Weber numbers respectively:

$$Re \equiv \frac{\rho dV}{\mu}, \quad Fr \equiv \frac{V^2}{dg}, \quad We \equiv \frac{d\rho V^2}{\sigma} \tag{2.49}$$

where the corresponding reference velocity is defined as $V = \sqrt{dg}$.

The level set function ϕ is introduced into the formulation of multiphase flow and mass transfer systems to define and capture the interface between two fluids. The interface is defined as the zero set of ϕ, which is defined as the signed algebraic distance of a node to the interface, being positive in the continuous fluid phase and negative in the drop.

The following equation is used to advance the level set function exactly as the drop moves:

$$\frac{\partial\phi}{\partial\theta} + \frac{\partial(u\phi)}{\partial x} + \frac{1}{r}\frac{\partial}{\partial y}(rv\phi) = 0 \tag{2.50}$$

$\kappa(\phi)$ is the curvature of the drop surface, defined as

$$\kappa(\phi) = -\nabla \cdot \mathbf{n} = -\nabla \cdot \left(\frac{\nabla\phi}{|\nabla\phi|}\right) \tag{2.51}$$

where $\mathbf{n}$ is the unit vector normal to the interface pointing towards the continuous phase and $\delta_\varepsilon(\phi)$ is the regularized delta function, defined as

$$\delta_\varepsilon(\phi) = \begin{cases} \frac{1}{2\varepsilon}\left(1+\cos\left(\frac{\pi\phi}{\varepsilon}\right)\right) & \text{if } |\phi| < \varepsilon \\ 0 & \text{otherwise} \end{cases} \tag{2.52}$$

where ε prescribes the finite "half thickness" of the interface. The interface is supposed to have the finite "thickness" to ease the convergence and numerical stability of the numerical algorithm, as seen in Figure 2.4. Usually $\varepsilon = 1.5\Delta x$ is taken, where Δx is the dimensionless uniform mesh size near the interface. $H_\varepsilon(\phi)$ is the regularized Heaviside function, expressed as

$$H_\varepsilon(\phi) = \begin{cases} 0 & \text{if } \phi < -\varepsilon \\ \frac{1}{2}\left(1+\frac{\phi}{\varepsilon}+\frac{\sin\left(\frac{\pi\phi}{\varepsilon}\right)}{\pi}\right) & \text{if } |\phi| \leq \varepsilon \\ 1 & \text{if } \phi > \varepsilon \end{cases} \tag{2.53}$$

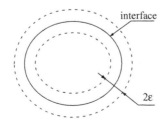

FIGURE 2.4 Schematic diagram of the interface for the level set approach.

The corresponding regularized (smoothed) density function ρ and the regularized viscosity μ are then defined as

$$\rho_\varepsilon(\phi) = \rho_2/\rho_1 + (1 - \rho_2/\rho_1)H_\varepsilon(\phi) \tag{2.54}$$

$$\mu_\varepsilon(\phi) = \mu_2/\mu_1 + (1 - \mu_2/\mu_1)H_\varepsilon(\phi) \tag{2.55}$$

The subscripts 1 and 2 denote the continuous phase and the drop respectively.

Generally, ϕ will no longer be a distance function (i.e., $|\nabla\phi| \neq 1$) after some iterations, even if Eq. (2.50) moves the interface ($\phi = 0$) at correct velocities. Maintaining ϕ as a distance function is essential to provide the interface with an invariant width and a sound basis for estimating the surface curvature. This problem can be solved by adopting the reinitialization method proposed by Sussman et al. (1994) to solve the following initial problem to steady state in a virtual time domain:

$$\frac{\partial\phi}{\partial\tau} = \text{sgn}(\phi_0)\left(1 - |\nabla\phi|\right) \tag{2.56}$$

$$\phi(x, 0) = \phi_0(x) \tag{2.57}$$

where τ is the virtual time for reinitialization, $\phi_0(x)$ is the level set function at any computational instant, and $\text{sgn}(\phi_0)$ is the sign function needed for enforcing $\nabla\phi = 1$. Equation (2.56) has the property that ϕ remains unchanged at the interface; therefore, the zero level set of ϕ_0 and ϕ is the same. Away from the interface ϕ will converge to $\nabla\phi = 1$, i.e., the actual distance function.

The following perturbed Hamilton–Jacobi equation was introduced by Chang et al. (1996) to guarantee mass conservation:

$$\frac{\partial\phi}{\partial\tau} + (A_0 - A(\tau))(-Q + \kappa(\phi))|\nabla\phi| = 0 \tag{2.58}$$

where A_0 denotes the initial total mass of both fluids at $\tau = 0$ and $A(\tau)$ the total mass corresponding to the level set function $\phi(\tau)$. The parameter Q is a positive constant and is usually set to be 1.

Yang and Mao (2002) found that a considerable amount of mass in the fluid particle vanished gradually, though an excellent conservation of total mass with time was observed with the reinitialization using Eq. (2.58). Therefore, the definition of $A(\tau)$ was modified to

$$A(\tau) = \sum_{\phi_{ij} \leq \varepsilon} \rho_\varepsilon(\phi_{ij})R_j\Delta x\Delta y \tag{2.59}$$

where $\phi_{ij} \leq \varepsilon$ denotes the nodes in the fluid particle and the interface has a virtual thickness of 2ε, i.e., $A(\tau)$ in Eq. (2.59) is taken as the mass of a drop/bubble instead of the total mass as in Eq. (2.58). The improved reinitialization procedure can maintain the level set function as a distance function and guarantee the drop mass conservation.

Since the fluids are incompressible, mass conservation is equivalent to volume conservation. In light of this, a volume-amending method (Li et al., 2008) was proposed, in which the level set function ϕ to an exact signed distance function was reinitialized by solving Eq. (2.56) firstly and then the position of the interface was corrected according to the loss or gain of the particle (bubble/drop) volume ($V = \Sigma_{\phi \leq 0} H(\phi) r \Delta x \Delta y$):

$$\Delta V = \frac{V(t) - V_0}{V_0} \tag{2.60}$$

Assuming the bubble/drop is spherical, the increment of the particle radius may be expressed as

$$\Delta R = R - R_0 = ((1 + \Delta V)^{1/\alpha} - 1) R_0 \tag{2.61}$$

where α is the dimension of the system simulated: $\alpha = 2$ for an axisymmetric fluid particle and $\alpha = 3$ for a three-dimensional case. Since the volume increase of the particle corresponds to that of the distance from the interface to the particle center, the correction to the level set function ϕ may be taken in proportion to ΔR. Thus, a relationship is established between the increment of the level set function ϕ and the loss or gain of the particle (bubble/drop) volume, so that the correction to the level set function ϕ may be expressed as

$$\delta\phi = \beta((1 + \Delta V)^{1/\alpha} - 1) \tag{2.62}$$

where β is a relaxation coefficient, and a large β may cause the reinitialization to diverge and a small value leads to low computational efficiency. β is usually chosen empirically between 0.01 and 0.1. Thus, the volume-amending equation can be written as

$$\phi = \phi_0 + \delta\phi = \phi_0 + \beta((1 + \Delta V)^{1/\alpha} - 1) \tag{2.63}$$

After the volume-amending reinitialization procedure, mass conservation is well satisfied (Li et al., 2008).

2.3.2.2 Level set method for mass transfer

In the following, a "one-fluid" formulation coupled with the level set function for interphase mass transfer is derived. When the distribution coefficient of a solute, m, is unity, the solute concentration across the interface is continuous. The only difficulty is the discontinuity of molecular diffusion coefficients, which can easily be smoothed with the same Heaviside function as used in Eqs. (2.54) and (2.55). Thus, Eq. (2.4) for two phases may be expediently solved in a single domain by the level set method similar to the solution of multiphase flow. In more general cases, m is not equal to unity, and some measures must be taken to make the concentration field become continuous across the interface, in the same manner as the continuity of fluid velocity at the interface was handled. For this purpose, a concentration transformation method, such as $\hat{c}_1 = c_1 \sqrt{m}$ and $\hat{c}_2 = c_2 / \sqrt{m}$, is proposed (Yang and Mao, 2005b).

Equation (2.6) becomes $\hat{c}_2 = \hat{c}_1$ at the interface, and the concentration field becomes continuous in the whole domain. Using these definitions, the transformed Eq. (2.4) remains in the following form:

$$\frac{\partial \hat{c}_i}{\partial t} + \mathbf{u} \cdot \nabla \hat{c}_i = D_i \nabla^2 \hat{c}_i, \; i = 1, \; 2 \tag{2.64}$$

where $\hat{c}_i$ denotes $\hat{c}_1$ or $\hat{c}_2$, and D_i represents the corresponding D_1 or D_2. When the transformations are utilized, Eq. (2.5) becomes

$$\frac{D_1}{\sqrt{m}} \frac{\partial \hat{c}_1}{\partial n_1} = \sqrt{m} D_2 \frac{\partial \hat{c}_2}{\partial n_2} \tag{2.65}$$

Therefore, at the interface, which is of finite thickness, the diffusion coefficients in both phases should be locally replaced by $D_1/\sqrt{m}$ and $D_2\sqrt{m}$ to satisfy the original condition of mass flux continuity. This would make the diffusivity in the interface region different from that in the bulk domain and may result in unacceptable errors. To mitigate this problem, a simple transformation is applied to make the diffusivity equal throughout a fluid phase. For the continuous and dispersed phases, Eq. (2.64) can be rewritten separately as

$$\frac{\partial \hat{c}_1}{\partial \left(\sqrt{m} t \right)} + \frac{1}{\sqrt{m}} \mathbf{u} \cdot \nabla \hat{c}_1 = \frac{1}{\sqrt{m}} D_1 \nabla^2 \hat{c}_1 \tag{2.66}$$

$$\frac{\partial \hat{c}_2}{\partial \left(\dfrac{1}{\sqrt{m}} t \right)} + \sqrt{m} \mathbf{u} \cdot \nabla \hat{c}_2 = \sqrt{m} D_2 \nabla^2 \hat{c}_2 \tag{2.67}$$

and then put into a unified equation over the whole domain in a form analogous to the momentum equation:

$$\frac{\partial \hat{c}}{\partial \hat{t}} + \hat{\mathbf{u}} \cdot \nabla \hat{c} = \nabla \cdot (\hat{D} \nabla \hat{c}) \tag{2.68}$$

where $\hat{t}$, $\hat{D}$, and $\hat{\mathbf{u}}$ are defined using the regularized Heaviside function $H_\varepsilon(\phi)$:

$$\hat{t}(\phi) = \begin{cases} \sqrt{m} t, & \text{if } \phi \geq 0 \\ \dfrac{1}{\sqrt{m}} t, & \text{if } \phi < 0 \end{cases} \tag{2.69}$$

$$\hat{D}(\phi) = \sqrt{m} D_2 + \left(\frac{1}{\sqrt{m}} D_1 - \sqrt{m} D_2 \right) H_\varepsilon(\phi) \tag{2.70}$$

$$\hat{\mathbf{u}}_\varepsilon(\phi) = \sqrt{m} \mathbf{u} + \left(\frac{1}{\sqrt{m}} \mathbf{u} - \sqrt{m} \mathbf{u} \right) H_\varepsilon(\phi) \tag{2.71}$$

where $\hat{\mathbf{u}}$ should be the velocity field in a frame of reference moving with the drop.

The governing equation for interphase mass transfer is nondimensionalized and the expanded expression in a two-dimensional axisymmetric coordinate system becomes

$$\frac{\partial C}{\partial \theta} + u'\frac{\partial C}{\partial X} + v'\frac{\partial C}{\partial Y} = \frac{1}{Pe_1}\left[\frac{\partial}{\partial X}\left(D\frac{\partial C}{\partial X}\right) + \frac{1}{Y}\frac{\partial}{\partial Y}\left(rD\frac{\partial C}{\partial Y}\right)\right] \qquad (2.72)$$

where C is the dimensionless concentration based on the reference concentration, using c_1^{∞} for solute transportation from the continuous phase to the drop or c_2 for the opposite transport direction, and the dimensionless group Pe_1 is the Peclet number, denoted as

$$Pe_1 \equiv \frac{dV}{D_1} \qquad (2.73)$$

where D_1 is chosen as the characteristic molecular diffusivity.

Other transformation forms of concentration, such as $\hat{c}_1 = mc_1$ and $\hat{c}_2 = c_2$, or $\hat{c}_1 = c_1$ and $\hat{c}_2 = c_2/m$, were also tested. The interphase mass transfer can be solved in a similar way, and identical results were obtained. Therefore, the transformation form has no effect on the numerical solutions of interphase mass transfer.

2.3.2.3 Numerical solution procedure

The main calculation steps of the level set algorithm are outlined as follows: (1) Initialize the flow field ($\mathbf{u}$ and p) with a quiescent drop/bubble, physical parameters (density, viscosity, surface tension, and molecular diffusivity of solute in each phase), concentration, and ϕ as the signed normal distance to the interface. (2) Compute velocity and pressure to be convergent. If unsteady motion is simulated in moving reference coordinates, Eq. (2.47) should be iterated until the acceleration a updated from the overall force balance over the particle also converges at the current time step. (3) Update the level set function. (4) Compute the unsteady concentration field over the whole domain with a steady-state flow field, and calculate the corresponding overall mass transfer coefficients and Sherwood numbers.

2.3.3 Mirror fluid method

The mirror fluid method (Yang and Mao, 2005a) was proposed to simulate solid–fluid two-phase flows. The mirror fluid method has been implemented to compute the motion of a rigid spherical or elliptic particle in a Newtonian fluid (Yang and Mao, 2005a) and has been extended to compute a stirred tank with a pitched blade stirrer (Wang, T. et al., 2013). The basic idea of the mirror fluid method is to take the whole domain as an Eulerian one for the fluid with a Lagrangian subdomain embedded; i.e., a solid–fluid flow problem with geometric complexity is resolved on a single domain, including a mirror fluid domain that is originally occupied by the solid. In this aspect, the mirror fluid method and the fictitious-domain method (Glowinski et al., 2001) have the same advantages that a fixed and regular mesh can be applied for the entire

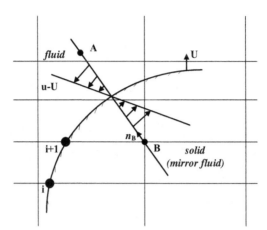

FIGURE 2.5 Schematic diagram of the solid–fluid interface for the mirror fluid method (Yang and Mao, 2005a).

computation without repeated remeshing or using boundary-fitted coordinates, and a simple finite-difference method can be used in the mirror fluid method instead of the finite-element method and a complicated weak formulation (Glowinski et al., 1999, 2001). The no-slip boundary condition is enforced implicitly on solid–fluid surface segments by mirror relations. Therefore, to make a surface segment subjected to a zero net force, the key is to guarantee the shear rates across the surface have the same magnitude but in opposite directions. This may be implemented by taking the inside flow in the subdomain occupied by the solid particle (i.e., the mirror fluid domain) as the flipped mirror image of the outside flow in the real fluid phase at the same surface segment or, in other words, by rotating the outside flow field and pressure field by 180° around the surface segment, as shown in Figure 2.5 (Yang and Mao, 2005a).

The key step of the mirror fluid method is to accurately specify the flow parameters inside the mirror fluid domain, in which the flow variables ($\mathbf{u}$ and p) at each node correspond to that of a point in a real fluid by the mirror relations. The following formulations are derived only in two-dimensional Cartesian or cylindrical coordinates for a single solid particle moving in a fluid as a typical example, but it is straightforward to extend this to a three-dimensional space and to multiple-particle–fluid systems.

As depicted in Figure 2.5, we can find the mirror location of A (x_A, y_A) in the real fluid region, corresponding to node B (x_B, y_B) in the mirror fluid domain, in terms of a distance function defined similarly as in the level set method (Sussman et al., 1994; Yang and Mao, 2002). The signed algebraic distance function denoted as ϕ, being positive in the continuous fluid phase, negative in the solid phase, and zero at the solid–fluid interface, facilitates the mirror calculations. After a particle is advanced by a time step, ϕ can be redesignated based on the new position of the particle. The unit normal vector to the interface is denoted as $\mathbf{n} = \nabla\phi / |\nabla\phi|$ and is expanded as

$$\mathbf{n} = \begin{pmatrix} n_x \\ n_y \end{pmatrix} = \begin{pmatrix} \dfrac{\partial\phi/\partial x}{\sqrt{(\partial\phi/\partial x)^2 + (\partial\phi/\partial y)^2}} \\ \dfrac{\partial\phi/\partial y}{\sqrt{(\partial\phi/\partial x)^2 + (\partial\phi/\partial y)^2}} \end{pmatrix} \tag{2.74}$$

It can be calculated at the nodes immediately close to the solid–fluid interface. So the unit normal vector passing through B to the interface is thus calculated:

$$\mathbf{n}_B = \begin{pmatrix} n_{x_B} \\ n_{y_B} \end{pmatrix} = \left(\frac{\nabla\phi}{|\nabla\phi|} \right)_B \tag{2.75}$$

In Figure 2.6, the straight line along $\mathbf{n}_B$ passing through B and its mirror image A is

$$\frac{x_A - x_B}{n_{x_B}} = \frac{y_A - y_B}{n_{y_B}} \tag{2.76}$$

The equal distance of either A or B to the interface requires

$$(x_A - x_B)^2 + (y_A - y_B)^2 = (2\phi_B)^2 \tag{2.77}$$

$$\phi_A \phi_B \leq 0 \quad (\text{if } \phi_B \leq 0, \text{ then } \phi_A \geq 0) \tag{2.78}$$

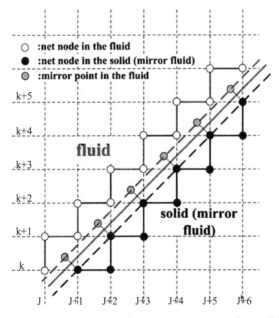

FIGURE 2.6 Sketch of a typical mirror relation between the nodes in a solid particle and the corresponding mirror nodes in a structured grid.

So the coordinates of A mirrored with B can be found by solving the equation set of Eqs. (2.76) and (2.77) coupled with the constraint in Eq. (2.78).

As shown in Figure 2.5, the points denoted by solid circle symbols (A) are one-to-one mirror images of the nodes in a solid particle denoted by open circle symbols (B). Then the fictitious velocity and pressure of node B in the mirror fluid are easily obtained:

$$\mathbf{u}_B = -(\mathbf{u}_A - \mathbf{U}) + \mathbf{U} = 2\mathbf{U} - \mathbf{u}_A \tag{2.79}$$

$$p_B = p_A \tag{2.80}$$

Be aware that for a rotating solid particle, the velocity at the particle–fluid interface (i.e., $\mathbf{U}$) should be replaced by $\mathbf{U} + \omega \times (\mathbf{x} - \mathbf{x}_p)$. Such specification ensures that the shear and normal stresses on the two sides of the solid–fluid interface have the same magnitude but opposite directions. The density and viscosity of the mirror fluid are designated as simply equal to those of the real fluid.

Using the mirror fluid method, the main calculation steps of solid–fluid flows are summarized as follows: (1) Initialize the flow field ($\mathbf{u}$, $\mathbf{U}$, p), $\mathbf{a}$, ω, physical parameters (density and viscosity), and ϕ as the signed normal distance to the interface. (2) Assume the velocity of the solid particle is $\mathbf{U}$ at the end of a time increment Δt: $\mathbf{U} = \mathbf{U}_0 + \mathbf{a}_0 \Delta t$, where $\mathbf{U}_0$ and $\mathbf{a}_0$ are the velocity and acceleration of the solid particle at the end of the previous time step. (3) Calculate the location A in the real fluid mirrored with any node B in the subdomain occupied by the solid particle by solving Eqs. (2.76) and (2.77). (4) Solve the mass and momentum conservation equations of fluid in the whole domain using the SIMPLE algorithm with the interface boundary condition implicitly enforced by the mirror fluid method. (5) Calculate the total drag force exerted by the liquid on the particle and the body force by the external field, and then calculate the acceleration $\mathbf{a}$ using Newton's second law. (6) Estimate the tentative particle velocity $\tilde{\mathbf{U}}$ at the end of current time step according to the new acceleration value: $\mathbf{U} = \mathbf{U}_0 + (\mathbf{a}_0 + \mathbf{a})\Delta t/2$. (7) Compare $\tilde{\mathbf{U}}$ and $\mathbf{U}$. If $\tilde{\mathbf{U}} = \mathbf{U}$ (meaning $\mathbf{a}_0 = \mathbf{a}$), go on to the next time step of the simulation. If not, go back to step 2 with suitably adjusted $\mathbf{a}_0$. (8) Move the particle center to a new position with the converged surface velocities and respecify ϕ of the fluid and the solid particle. Keep the $\mathbf{u}$ and p values for the nodes in the fluid and discard these in the particle according to the renewed signed distance function ϕ. (9) Let $\mathbf{U}_0 = \mathbf{U}$, $\mathbf{u}_0 = \mathbf{u}$, $\mathbf{a}_0 = \mathbf{a}$, and $\omega_0 = \omega$, and repeat steps 2–8 for the next time step.

2.4 BUOYANCY-DRIVEN MOTION AND MASS TRANSFER OF A SINGLE PARTICLE

Drop and bubble motion is a classical unsteady flow with a moving interface in multiphase flow, and in past decades many experimental studies have focused on the motion of bubbles and drops (Bhaga and Webber, 1981; Ryskin and Leal, 1984b, c; Tomiyama et al., 1993; Lin et al., 1996; Oka and Ishii, 1999; Sankaranarayanan

et al., 1999; Yang and Mao, 2002). Due to experimental restrictions, the flow characteristics are limited to apparent records of multiphase flow phenomena. Fortunately,
numerical simulation provides an effective way to profile the flow details and probe
the underlying mechanism, particularly the interactions between phases.

2.4.1 Drop, bubble and solid particle motion

Researchers have simulated various behaviors of a single particle or several particles,
with respect to their formation, unsteady motion, coalescence, and breakage. Most
earlier studies simplified the fluid's physical properties. Ryskin and Leal (1984a–c)
took only the liquid phase into account, neglected the gas-phase physics, and computed the shape of a single free-rising bubble under equilibrium conditions using a
finite difference method. Unverdi and Tryggvason (1992) simulated the deformation
of a single moving bubble and the coalescence of two bubbles using a VIC (vortex-
in-cell) method. The VIC method is a hybrid with distinct Lagrangian particles based
on a stationary Eulerian grid and has less calculating load than the Lagrangian discrete vortex method.

2.4.1.1 Bubble/drop formation

The hydrodynamics of the growth and detachment of bubbles/drops from a nozzle,
a capillary or an orifice plate into another immiscible fluid is of practical importance
in many applications, including extraction operation, inkjet printing, flight vehicle
protection, and measurement of dynamic surface tension (Lu et al., 2010; Wang,
Z. H. et al., 2013). Bubble/drop formation at an orifice is a complex phenomenon
described by various parameters, such as the time-mean flow rate, the fluid physical properties, the geometric parameter of orifices or orifice plates, etc. Chen et al.
(2009) simulated the formation of axisymmetric bubbles on an orifice plate using a
level set method for tracking the two-phase interface and demonstrated that bubble
formation was largely controlled by the wetting properties of the orifice plate, as seen
in Figure 2.7 (Chen et al., 2009). As the bubble grows, both the apparent contact angle and the contact diameter vary in a complex way. It is known that the mass transfer during the bubble/drop formation stage is very important in solvent extraction,
which is also very complex, with surface area creation, interface renewal, interfacial
convection, drop growth, and detachment proceeding concurrently or successively.
So the computation of bubble/drop formation processes has real significance and
provides the foundation to numerically investigate the mass transfer mechanism of
the growth stages.

  The formation of a drop at the tip of a vertical capillary into an immiscible liquid
was simulated using the level set method (LSM) in an axisymmetric system and
the results are shown in Figure 2.8. The predicted shapes are similar to the reported
results of VOF simulations and experiments. The velocity field and stream function contours around a growing drop at several typical time instants are depicted in
Figures 2.8 and 2.9 respectively. When the volume of the drop exceeds a critical
value at which the surface tension force cannot hold the drop steady at the nozzle

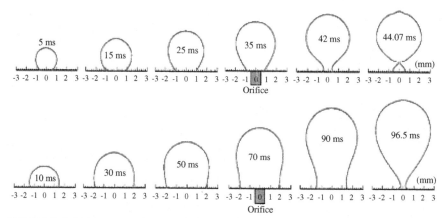

FIGURE 2.7 Bubble contours for a wetting surface ($\theta_{ad} = 50°$, $\theta_{re} = 40°$) (upper) and a less wetting surface ($\theta_{ad} = 110°$, $\theta_{re} = 70°$) (lower) with $Do = 1$ mm and $Q = 1$ cm³/s (Chen et al., 2009).

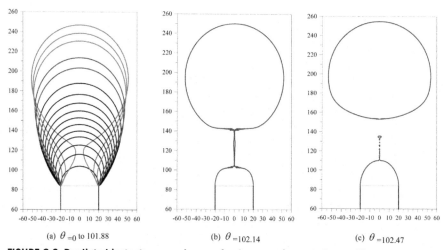

(a) $\theta_{=0}$ to 101.88   (b) $\theta_{=102.14}$ (c) $\theta_{=102.47}$

FIGURE 2.8 Predicted instantaneous shapes of a drop growing out of a capillary ($R = 1.13$ mm) (Yang et al., 2005).

(a) $\theta = 0$–101.88. (b) $\theta = 102.14$. (c) $\theta = 102.47$.

tip against the buoyancy force, the drop detaches and rises. Just after the drop has detached from the nozzle, some satellite droplets are generated from the shrinking liquid neck, and the bottom surface of the main drop formed is rather flat at this time.

2.4.1.2 Unsteady and steady motion

The fluid particle shape and terminal velocity are the key parameters in unsteady and steady motions. An axisymmetric single air bubble simulated with LSM when rising in an aqueous solution of sugar with high and low viscosities respectively is shown in

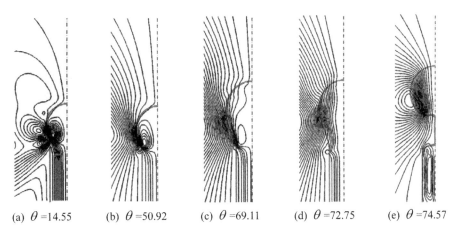

(a) θ =14.55   (b) θ =50.92   (c) θ =69.11   (d) θ =72.75   (e) θ =74.57

FIGURE 2.9 Stream function contours with time of a drop into a liquid (R = 1.98 mm) (Yang et al., 2005).

(a) θ = 14.55. (b) θ = 50.92. (c) θ = 69.11. (d) θ = 72.75. (e) θ = 74.57.

Figures 2.10 and 2.11 (Li et al., 2008). The predicted bubble terminal velocities are 0.181 and 0.317 m/s, which agree with the experimental data of 0.190 and 0.306 m/s respectively (Wang et al., 2004). The numerical results show a spherical cap bubble in the higher viscosity liquid and a cap bubble in the lower viscosity liquid, which agrees well with experimental observations (Bhaga and Webber, 1981).

A single bubble moving in a continuous Newtonian liquid with different *Mo* and *Eo* numbers is shown in Figure 2.12, which is detailed with stream lines. The simulation was performed using the VIC method and the interface was tracked explicitly using a front tracking method incorporating the features of both the volume-tracking

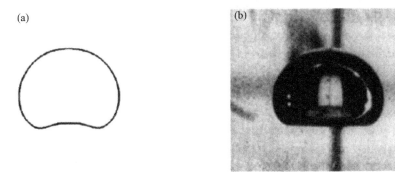

(a)   (b)

(a) Calculated shape   (b) Measured shape

FIGURE 2.10 Comparison of the shape of a 9.3 cm³ bubble in a sugar solution with high viscosity at steady state (ρ_c = 1390 kg/m³, ρ_b = 1.226 kg/m³, μ_c = 2.786 Pa s, μ_b = 1.78 ×10⁻⁵ Pa s, σ = 0.0794 N/m, R = 0.013 m) (Li et al., 2008).

(a) Calculated shape. (b) Measured shape (Bhaga & Webber, 1981).

(a)

(b)

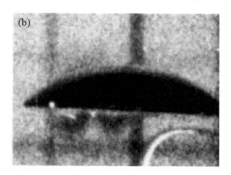

(a) Calculated shape   (b) Measured shape

FIGURE 2.11 Comparison of the shape of a 9.3 cm³ bubble in a sugar solution with low viscosity at steady state (ρ_c = 1320 kg/m³, ρ_b = 1.226 kg/m³, μ_c = 0.525 Pa s, μ_b = 1.78 × 10⁻⁵ Pa s, σ = 0.0754 N/m, R = 0.013 m) (Li et al., 2008).

(a) Calculated shape. (b) Measured shape (Bhaga & Webber, 1981).

and shock-tracking schemes (Unverdi and Tryggvason, 1992). The flow around the bubble is nearly a Stokes flow with high Morton numbers and represents a classical internal circulation. With decreasing Morton number, a wake appears progressively at the rear of the bubble (Figure 2.12a, f, g). The bubble deformation depends critically on the Eo number, and emerges as a spherical, ellipsoidal, spherical cap, or cap shape successively with increasing Eo number. In the meantime the wake size expands and a secondary vortex emerges between the larger one and the bubble rear, which is actually much weaker than the primary one, as magnified in Figure 2.12e and f.

2.4.1.3 Coalescence

Two spherical bubbles with radius ratio R_1/R_2 = 1.5 initially at rest but not aligned vertically are simulated in fixed reference coordinates based on a 3D LSM framework. The most important conditions, such as $We \approx 50$ and $Re \approx 5 \times 10^{-3}$, are set as in Manga and Stone's (1993) experiments, except for the initial distance between two bubbles. Some of the important stages of the interactive behavior are predicted as in Figure 2.13 (Li et al., 2008). The two bubbles deform considerably while rising, but in quite different ways, which resemble the experimental observations. The upper bubble deforms first towards the steady-state shape of a single bubble in free rise, the trailing bubble deforms a little, and the lower part of the upper bubble becomes concave while that of the trailing one becomes protuberant. Subsequently, the distance between the two bubbles becomes shorter and shorter. The trailing bubble is finally entrained into the wake of the preceding one and they then merge into a single bubble. The numerical results are in qualitative agreement with the experiment (Manga and Stone, 1993).

2.4.1.4 Bubbles and drops in a non-Newtonian fluid

It is very important to understand bubble behavior not only in a Newtonian continuous phase but also in non-Newtonian conditions, because knowledge of bubble motion provides a useful and essential basis for optimal process design and

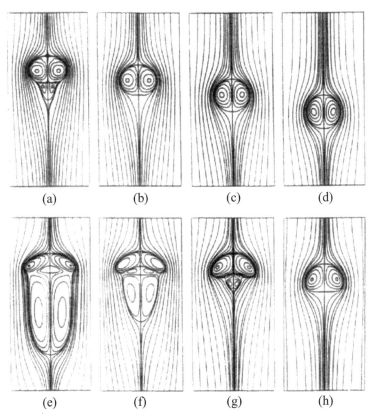

FIGURE 2.12 Streamlines for a steady-state rising bubble for various *Eo* and *Mo* with $\rho_d/\rho_b = 40$ (Unverdi and Tryggvason, 1992).

(a) $Eo = 1$, $Mo = 10^{-7}$, $\mu_d/\mu_b = 88$. (b) $Eo = 1$, $Mo = 10^{-6}$, $\mu_d/\mu_b = 156$. (c) $Eo = 1$, $Mo = 10^{-5}$, $\mu_d/\mu_b = 277$. (d) $Eo = 1$, $Mo = 10^{-4}$, $\mu_d/\mu_b = 493$. (e) $Eo = 10$, $Mo = 10^{-4}$, $\mu_d/\mu_b = 88$. (f) $Eo = 10$, $Mo = 10^{-3}$, $\mu_d/\mu_b = 156$. (g) $Eo = 10$, $Mo = 10^{-2}$, $\mu_d/\mu_b = 277$. (h) $Eo = 10$, $Mo = 10^{-1}$, $\mu_d/\mu_b = 493$.

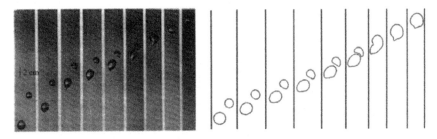

(a) Experimental results of Manga and Stone (1993)   (b) Numerical result

FIGURE 2.13 Experimental and numerical results for interactive behavior of two bubbles rising in corn syrup (photographs and simulated results are shown at 10 s intervals) (Li et al., 2008).

(a) Experimental results of Manga and Stone (1993). (b) Numerical results.

operation. Many investigations concerning various aspects of bubble or drop motion in non-Newtonian fluids have been reported theoretically and experimentally. Many have been well summarized and reviewed by Chhabra (1993) and Kulkarni and Joshi (2005).

The motion of a single bubble rising freely through shear thinning fluids represented by the Carreau model was computed numerically using VOF and LSM for tracking the bubble interface in a few reports (Kulkarni and Joshi, 2005; Ohta et al., 2005; Zhang et al., 2010). The local viscosity field around a bubble rising in shear thinning non-Newtonian fluids was examined numerically with real density and viscosity ratios of gas and liquid adopted.

Ohta et al. (2005) simulated a Newtonian liquid drop moving in a non-Newtonian liquid using a VOF method. Some of their results are shown in Figure 2.14. As *Eo* number is increased, the drop deforms, showing similar shapes close to the experimental ones. The viscosity varies drastically due to the strong shear-thinning property, especially in the area adjacent to the interface. The viscosity distribution is presented in Figure 2.14, showing that a much higher viscosity region is formed gradually in the bubble wake region with increasing *Eo* number. This high-viscosity block remains detached from the drop's rear surface. Besides these macroscopic phenomena, the local details of the shear-thinning behavior are revealed in the numerical results. These details are difficult to be discerned from experiments. From the legend of Figure 2.14, the lower limit viscosity, 0.0013, 0.0015 and 0.0016 Pa s, can be observed respectively for the three cases. The case with the lowest lower limit viscosity instead turns out to have the lowest velocity (the lowest Reynolds number). This is a good reminder that although the shear-thinning property has a significant influence on drop motion,

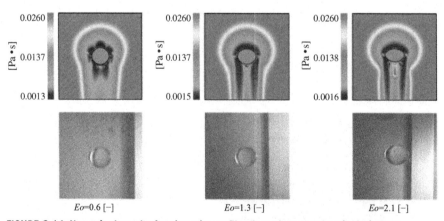

FIGURE 2.14 Numerical results for viscosity profile, drop shape, and typical pictures of a rising silicone oil drop in a Carreau–Yasuda shear-thinning solution of SAP (sodium acrylate polymer).

(a) $d = 5.90$ mm, $Eo = 0.6$, $Re = 12.7$. (b) $d = 9.14$ mm, $Eo = 1.3$, $Re = 35.2$.
(c) $d = 11.6$ mm, $Eo = 2.1$, $Re = 47.7$.

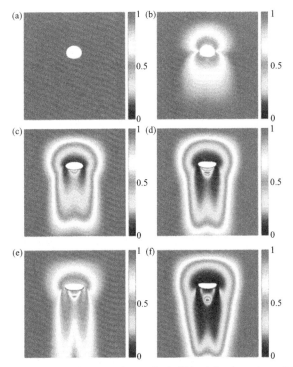

FIGURE 2.15 Viscosity distribution and shape of a bubble rising in a shear-thinning solution (system parameters are listed in Table 2.1) (Zhang et al., 2010).

the viscosity variation and the magnitude of velocity are correlated in a complex way rather than in having a simple linear relationship.

Figure 2.15 shows the viscosity distribution (nondimensionalized by μ_0) around a bubble and the real bubble shapes for systems 1–6 (Table 2.1) as simulated using LSM (Zhang et al., 2010). The bar shown at the right of the figures depicts the range of reduced apparent viscosity around the bubble. System 1 corresponds to the bubble

Table 2.1 Carreau Model Parameters and Numerical Results for Bubbles with $d_e = 1$ cm

System	μ_0 (Pa s)	λ (s)	n	U_T (cm/s)	$2\lambda U_T/d_e$	Re_M
1	0.5	–	1.0	11.2	–	2.24
2	0.5	1.00	0.8	16.5	33.0	6.65
3	0.5	1.00	0.5	22.7	45.4	30.69
4	0.5	1.00	0.4	23.9	47.8	48.8
5	0.1	1.00	0.8	25.3	50.6	55.5
6	0.5	1.00	0.3	25.6	51.2	80.85

rising in a Newtonian fluid with the same zero shear rate viscosity and the viscosity around the bubble is constant. As for systems 2–6, the viscosity around the bubble varies according to the bubble shape and the shear thinning property of the liquid, and the degree of decrease in viscosity is largest in close vicinity to the bubble. The change in viscosity becomes gradually more drastic from system 3 to system 6, showing that a confined region with high viscosity exists in the wake of these bubbles and it finally becomes detached from the bubble's rear surface. As seen in Table 2.1, it is obvious that U_T in terms of

$$Re_M = \frac{\rho_1 d_e U_T}{\eta_0 / [1 + (\lambda(2U_T/d_e))^2]^{(1-n)/2}} \tag{2.81}$$

increases gradually due to the stronger shear thinning effect. The bubble can rise faster due to the large decrease in viscosity around the bubble. Obviously, the bubble shape in the shear-thinning non-Newtonian fluid differs considerably from that in the Newtonian case. As the shear-thinning effect gradually becomes intensive, the bubble takes a more oblate shape, with the front surface flatter than the rear part.

For non-Newtonian fluids, the apparent viscosity of the shear-thinning fluid is a function of shear rate $\dot{\gamma}$. Consider the Carreau–Yasuda model, for instance:

$$\mu(\dot{\gamma}) = \mu_\infty + (\mu_0 + \mu_\infty)[1 + (\lambda\dot{\gamma})^\beta]^{s-1/\beta} \tag{2.82}$$

where μ_0 and μ_∞ are the viscosities corresponding to zero shear rate and infinite shear rate respectively, λ is the inelastic time constant, n is the power-law index, β is a dimensionless parameter describing the transition region between the zero shear-rate region and the power-law region, and shear rate $\dot{\gamma} = (1/2(\dot{\gamma}:\dot{\gamma}))^{1/2}$ can be calculated from the velocity field:

$$\dot{\gamma} = \nabla\mathbf{u} + (\nabla\mathbf{u})^T \tag{2.83}$$

As shown in Figure 2.16, the shear rate distribution around a bubble in Newtonian and non-Newtonian liquids is in accordance with the viscosity distribution. A larger area of higher local shear rates at the bubble nose in Figure 2.16b can be observed in the shear-thinning liquid than that in Figure 2.16a, as a joint effect of shear-thinning behavior and larger Re_M in Figure 2.16b. As demonstrated in Figure 2.16c–f, a region of much lower shear rates is at the rear of the bubble corresponding to a much higher viscosity region in the bubble wake. As the shear-thinning effect becomes stronger (Figure 2.16e, f), the low shear region even becomes isolated and detached from the bubble rear. It is clearly seen that there exists a higher shear-rate region in front of the bubble and a streaming toroidal one at the outer region in the wake, where the viscosity decreases drastically.

2.4.1.5 Simulation of solid particle motion by the mirror fluid method
The mirror fluid method is proposed for simulating solid–fluid two-phase flows (Yang and Mao, 2005a). In this method the entire computational domain is taken as a fixed Eulerian one, and suitable flow parameters are assigned to the solid particle domain

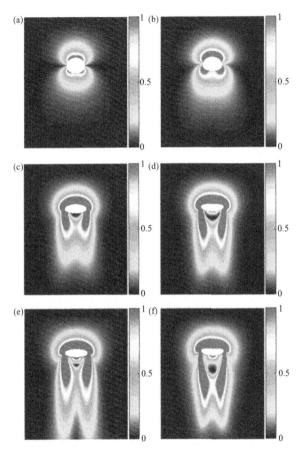

FIGURE 2.16 Shear rate distribution and shape of a bubble rising in a shear-thinning solution (system parameters are listed in Table 2.1) (Zhang et al., 2010).

by mirror relations to ensure the fluid–solid boundary conditions are enforced implicitly. Meanwhile the motion of solid particles is tracked using a Lagrangian scheme.

The Reynolds number, drag coefficient, and wake length of single spherical particles are well predicted as verified against the acknowledged experimental data shown in Figures 2.17 and 2.18 (Yang and Mao, 2005a). The oscillating trajectory and orientation angle of an ellipse settling in a Newtonian fluid are also well predicted as compared with the reported simulation results, as shown in Figure 2.19 (Yang and Mao, 2005a). These numerical tests indicate that the mirror fluid method is simple, robust, and effective in simulating real solid–fluid flows.

Numerical simulation for a single particle is fundamental research, but bears much significance for real processes. From these results, OBFCS and LSM agree well with the experimental data, but OBFCS can satisfy mass conservation commendably while the level set method is superior in dealing with deformable interfaces. Also, the mirror fluid

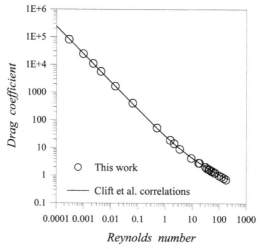

FIGURE 2.17 Plot of drag coefficient against Reynolds number for a falling spherical particle ($\rho_c = 1000$ kg/m³, $\rho_s = 2000$ kg/m³, $\mu_c = 1.0 \times 10^{-3}$ Pa s, $R = 0.21$ mm).

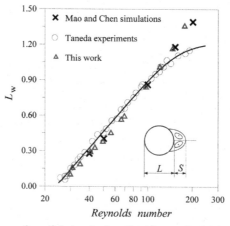

FIGURE 2.18 Comparison of predicted wake length with experimental data ($L_w = S/L$).

method is effective in simulating solid–fluid flows and can be extended to other, more challenging particle flow problems (e.g., multiphase flow in stirred tanks) with fewer restrictions, especially when complicated and irregular solid–fluid interfaces are involved (e.g., chemical reactors with internals of diversified geometry).

2.4.2 Mass transfer to/from a drop

The main difficulty in numerical simulation is that the motion of a deformed drop with simultaneous mass transfer must be solved with the shape of the free surface unknown (Petera and Weatherley, 2001). Numerical methods must ensure accurate

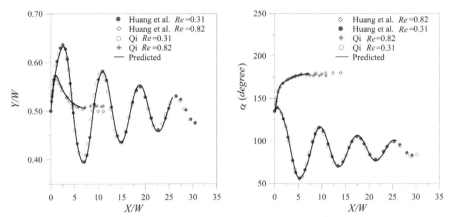

FIGURE 2.19 Comparison of the predicted trajectory *Y/W* and orientation angle α of an ellipse with the results of Huang et al. (1998) and Qi (1999) (*X* is the displacement at the gravity or vertical direction and *Y* that at the horizontal direction).

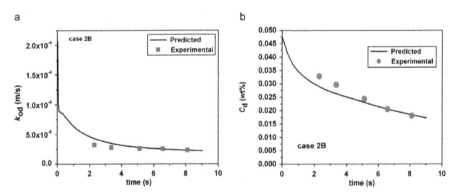

FIGURE 2.20 Predicted and experimental concentrations and overall mass transfer coefficients at various times (Wang et al., 2008a).

(a) Overall mass transfer coefficient. (b) Average drop concentration. $\rho_b = 996.1$ kg/m³, $\rho_c = 866.2$ kg/m³, $\mu_b = 1.185 \times 10^{-3}$ Pa s, $\mu_c = 2.837 \times 10^{-3}$ Pa s, $D_b = 7.2 \times 10^{-10}$ m²/s, $D_c = 3.1 \times 10^{-10}$ m²/s, $c_{b,0} = 4.79$ wt%, $c_{c,0} = 0$ wt%, $\sigma = 1.0 \times 10^{-3}$ N/m, $m = 0.88$, $R = 0.59$ mm.

resolution of the interfacial position and the concentration field so that mass transfer can be accurately evaluated. The level set method was used to simulate the interphase mass transfer to/from a drop in the stage of unsteady motion (Wang et al., 2008a). Figure 2.20 shows the evolution of the average drop concentration and overall mass transfer coefficients for mass transfer from the continuous phase to the drop or in the inverse direction. The predictions are in good agreement with the corresponding experimental data (Wang et al., 2008a). Figure 2.21 shows the predicted solute

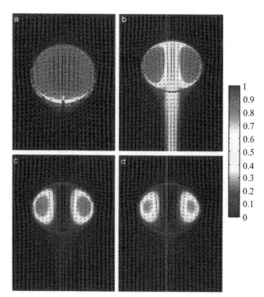

FIGURE 2.21 Predicted fractional solute concentration distribution and velocity vector field relative to the motion of the drop (the same conditions as in Figure 2.20) at various dimensionless times (Wang et al., 2008a).

(a) $t = 0.04$ s. (b) $t = 0.14$ s. (c) $t = 1.39$ s. (d) $t = 2.10$ s.

distribution in both phases and the velocity vector field relative to the drop motion at different dimensionless time for the same system as in Figure 2.20. The results show that in the drop center the solute concentration decreases gradually because the recirculation sweeps the solute towards the drop surface, where it is transferred into the continuous phase as a plume. The plume is more distinct early when the concentration in the drop is highest.

In that work, the effect of the formation stage of the drop on mass transfer is not considered in the simulation as the initial solute concentration in the drop is assigned as zero. The test is performed on the effect of the initial extraction fraction of the drop being assigned as 5% of the initial condition in the simulation (Wang et al., 2008a). The results with and without the guessed initial extraction fraction are compared in Figure 2.22. The effect of mass transfer in the drop formation stage seems to have only moderate effect on the average mass transfer coefficient of the drop. The predicted overall transfer coefficients k_{od} with guessed initial drop concentrations are in even better agreement with the experimental data, but this effect does not seem to be very significant in the later stages. It seems that the mass transfer and internal liquid circulation in the stage of drop formation should be accounted for more accurately when the accelerating motion and mass transfer of single drops are the major concern.

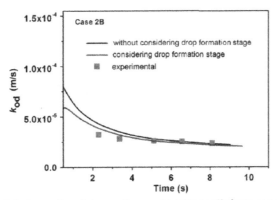

FIGURE 2.22 Predicted experimental overall mass transfer coefficients at various times (the same conditions as in Figure 2.20) (Wang et al., 2008a).

2.5 MASS TRANSFER-INDUCED MARANGONI EFFECT

The well-known Marangoni effect on the sub-particle scale is an important factor in many practical cases in interpreting the interphase mass transfer and interfacial phenomena. With local vortices induced at the interface, the convective currents normal to the interface would add to the interphase mass transfer otherwise dominated by molecular diffusion. A numerical approach is now available to simulate the surface tension gradient-induced Marangoni convection at the interface and the thereby enhanced interphase mass transfer. The resultant information can shed light on the mechanisms in which the interphase interaction is to be either promoted or suppressed.

2.5.1 Solute-induced Marangoni effect

Mao and Chen (2004) adopted an axisymmetric (orthogonal) polar coordinate frame to numerically simulate the influence of the Marangoni effect on unsteady mass transfer to a drop rising at a steady low velocity in liquid–liquid systems. In their work, the drop deformation was not taken into account so that the momentum equation and the convective diffusion equation were decoupled. However, in realistic conditions, the mass transfer in the unsteady motion stage cannot be ignored.

Based on the CSF (continuum surface force) model, Wang et al. (2006) suggested a simple weighted integration method to calculate the surface tension force to suppress parasitic flow. The Marangoni effect induced by interphase mass transfer in unsteady deformable single drop extraction was performed in an Eulerian axisymmetric moving reference frame (Wang et al., 2008b), and the results were compared with classical Sternling and Scriven (S&S) theory (1959). Figure 2.23 (Wang et al., 2008b) shows that the mass transfer rate with the Marangoni effect is

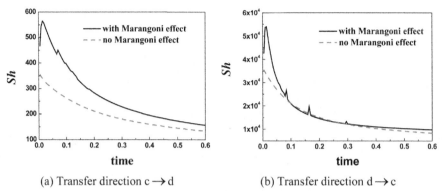

(a) Transfer direction c → d   (b) Transfer direction d → c

FIGURE 2.23 Influence of Marangoni effect on Sherwood number for mass transfer to/from a deformable drop (Wang et al., 2008b).

(a) Transfer direction c → d. (b) Transfer direction d → c.

apparently higher than that without the Marangoni effect. Figure 2.23a shows that the simulation agrees with S&S theory. However, Figure 2.23b displays discrepancies with S&S theory. A reasonable explanation may be that S&S theory analyzed mass transfer after the drop achieved steady motion while the object of this simulation was unsteady rise and mass transfer.

Wang, J. F. et al. (2011) determined experimentally the mass transfer rate in MIBK (methyl isobutyl ketone)–acetic acid–water systems. The larger the concentration of the drop, the more evident are Marangoni convection and the oscillation of rising velocities. The oscillation is also more evident in the initial rising stage when the Marangoni convection is stronger due to larger mass-transfer driving force. In the final stage of steady rising, the oscillation of the rising drop diminishes gradually as the Marangoni convection dies away with elapsed time. To understand the influence of the Marangoni effect on the interphase mass transfer, this process was also investigated numerically. The predicted transient evolutions of streamlines and concentration contours at different instants are shown in Figure 2.24. The transient streamline map changes constantly due to the Marangoni effect. The drop deforms gradually from initially spherical to finally ellipsoidal as an overall trend. Wegener et al. (2009) also observed similar large irregular deformation of the drop in the toluene–water system without mass transfer and Marangoni convection. Some eddies of sub-drop dimensions near the interface make the interfacial tension change clearly along the interface to bring forth strong local interfacial convection with velocity components perpendicular to the interface. The deformation of the interface is irregular due to the transient nature of the Marangoni effect. The shape of the drop is gradually deformed to ellipsoidal that is characterized by steady-state drop motion. Marangoni convection demonstrates its complexity both on the spatial scale and the temporal scale. Although it can be predicted according to the system's properties (such as *Ma* number and initial concentration) and determined whether the Marangoni effect occurs or not based on the occurrence of interface phenomena and the extraordinary

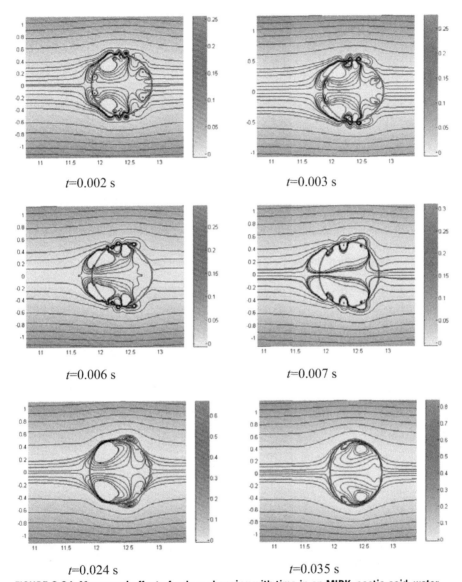

t=0.002 s

t=0.003 s

t=0.006 s

t=0.007 s

t=0.024 s

t=0.035 s

FIGURE 2.24 Marangoni effect of a drop changing with time in an MIBK–acetic acid–water system (Wang, J. F. et al., 2011).

enhancement of mass transfer rate, it is still far from being able to predict it with a universal criterion, as reliably as laminar flow and turbulent flow are distinguished with the Reynolds number.

Marangoni convection is an inherently three-dimensional hydrodynamic phenomenon. Wegener et al. (2009) simulated single drop extraction in a moving 3D reference coordinate system using the commercial CFD code STAR-CD and solved

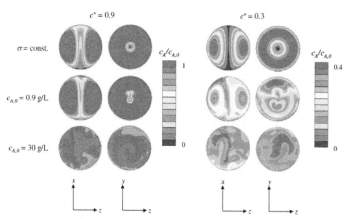

FIGURE 2.25 Simulated concentration plots (drop only) for three different initial concentrations at $c^* = 0.9$ g/L (right) and at $c^* = 0.3$ g/L (left) in two different views (x–z plane: view parallel to the flow; y–z plane: view perpendicular to the flow) (Wegener et al., 2009).

simultaneously the coupled flow field and concentration field by the concentration-dependent surface tension. The simulated results show that, compared with the results without the Marangoni effect, mass transfer was always enhanced. The mass transfer enhancement factor ranges from 2 to 3. Figure 2.25 shows sections of the concentration plots in different views for three different initial concentrations at $c^* = 0.9$ g/L and $c^* = 0.3$ g/L. The typical toroidal axis-symmetric flow pattern is developed without Marangoni convection, and until Marangoni convection occurs, the typical toroidal axis-symmetric flow pattern gradually becomes disturbed over time. Also, the concentration distribution is much more chaotic with higher initial concentrations. Therefore, it is confirmed that the magnitude of Marangoni convection is related to the initial concentration.

2.5.2 Effect of surfactant on drop motion and mass transfer

In most industrial extraction equipment, unavoidable trace quantities of surface-active contaminants may have a profound effect on the behavior of drops and bubbles. Contaminant can eliminate internal circulation, significantly increase the drag, and reduce mass transfer rates. A more accurate account of the effect of surfactants can be resolved using the principles of computational fluid dynamics and transport. This numerical approach is applied to the motion and mass transfer of a single liquid drop and the drop behavior is interpreted with reasonable precision (Li et al., 2001, 2003). A numerical investigation of the flow field inside and around a deformed drop translating in another quiescent liquid contaminated by a surfactant that is soluble in the continuous phase but insoluble in the dispersed phase is presented in this section.

2.5.2.1 *Formulation*

It is interesting to investigate the effects of various dimensionless groups on the flow field inside and outside the drop of an axisymmetric deformed drop rising at steady state and to observe the surfactant concentration profile at the interface of a single drop. In dimensional parameters, the interfacial transport equation of a surfactant can be written as

$$\nabla_S \cdot (\mathbf{u}_S \Gamma) - D_S \nabla_S^2 \Gamma = -D[\mathbf{n} \cdot \nabla c]_S \tag{2.84}$$

in the case where the surfactant is dissolved in the continuous phase only. In this equation, the left-hand quantity is the source term to the transport of the surfactant along the interface. D_S is the surface diffusion coefficient, Γ is the surface concentration of the surfactant, ∇_S is the surface counterpart of operator ∇, $\mathbf{n}$ is the normal unit vector pointing into the droplet, $\mathbf{u}_S$ is the convective velocity at the interface, and ∇c is the concentration gradient of the surfactant in the continuous phase.

Equation (2.84) is actually the boundary condition for the governing equation for the convective diffusion of the surfactant in the bulk phase. It may be nondimensionalized to

$$\nabla_S \cdot (\mathbf{U}_S \overline{\Gamma}) - \frac{1}{Pe_S} \nabla_S^2 \overline{\Gamma} = -\left[\mathbf{n} \cdot \nabla \left(\frac{C}{Pe^*} \right) \right]_S \tag{2.85}$$

and the following nondimensional variables and groups are defined for this purpose:

$$U_S = \frac{u_S}{U}, \quad \overline{\Gamma} = \frac{\Gamma}{\Gamma_\infty}, \quad Pe_S = \frac{2UR}{D_S}, \quad Pe^* = \frac{U\Gamma_\infty}{Dc_\infty} \tag{2.86}$$

The expanded form of Eq. (2.85) in the present orthogonal curvilinear coordinate system as in Figure 2.3 is

$$-\frac{Pe_S}{Pe^*} f_1 y_1 \frac{\partial C}{\partial \xi_1}\bigg|_{\xi_1=1} + \frac{Pe_S}{2} \frac{\partial (U_S y_1 \overline{\Gamma})}{\partial \eta_1} = \frac{\partial}{\partial \eta_1} \left(\frac{y_1}{h_{\eta_1}} \frac{\partial \overline{\Gamma}}{\partial \eta_1} \right) \tag{2.87}$$

Considering the case where the adsorption of the surfactant at the interface reaches steady state and obeys Langmuir's kinetic law, the boundary condition at the interface can also be formulated in another form as

$$-D[\mathbf{n} \cdot \nabla c]_S = \beta c_S (\Gamma_\infty - \Gamma) - \alpha \Gamma \tag{2.88}$$

in which α and β are the adsorption and desorption rate constants respectively, and c_S is the sublayer concentration of the surfactant immediately at the interface in the continuous phase. Using the following nondimensional variables and groups:

$$Bi = \frac{2R\alpha}{U}, \quad K = \frac{\beta c_\infty}{\alpha}, \quad C_S = \frac{c_S}{c_\infty} \tag{2.89}$$

K is termed as the dimensionless bulk concentration, and Eq. (2.88) can be nondimensionalized to

$$-\left[\mathbf{n}\cdot\nabla\left(\frac{C}{Pe^*}\right)\right]_s = Bi(KC_s(1-\overline{\Gamma})-\overline{\Gamma}) \tag{2.90}$$

Similarly, in the external orthogonal grid, the expanded form of Eq. (2.85) is

$$-\frac{1}{Pe^*h_{\xi_1}}\frac{\partial C}{\partial\xi_1}\Big|_{\xi_1=1} = Bi(KC_s(1-\overline{\Gamma})-\overline{\Gamma}) \tag{2.91}$$

Equation (2.90) is itself a differential equation, and its boundary condition is

$$\frac{\partial\overline{\Gamma}}{\partial\eta_1} = 0 \quad \text{(at the front and rear stagnation points)} \tag{2.92}$$

2.5.2.2 Effect of surfactant on drop motion

A series of parametric studies were conducted with only one variable changed from the base set of parameters ($Re = 100$, $We = 0.5$, $Ma = 100$, $Bi = 0.1$, $\lambda = 2.0$, $\zeta = 2.0$, $K = 1 \times 10^{-3}$, $Pe = 10^3$, $Pe_s = 10^5$, $Pe^* = 10^3$). The influential nondimensional groups include Re, We, Ma, λ, and ζ, which arise in the Navier–Stokes equations and boundary conditions, and Bi, K, Pe, Pe_s, and Pe^*, which arise in the convection–diffusion equation and boundary conditions. Among them, the density ratio and viscosity ratio were chosen to be somewhat arbitrarily fixed at representative values of $\lambda = 2.0$ and $\zeta = 2.0$ respectively. The existing simulation of drop flow in pure systems illustrates that the flow structure is relatively insensitive to λ and ζ. Moreover, Pe, Pe^* and Pe_s are usually very large because of the small diffusion coefficients of a surfactant in the liquid, and the distribution of a surfactant is dominated solely by surface convection and adsorption kinetics. This led to the stagnant-cap regime being recognized as the most common physical reality in practical drop flow (He et al., 1991). Thus, only the effects of Biot number and dimensionless concentration on the motion of a drop are expected to be significant and worthy of numerical investigation. According to the above analysis, the above reference state is selected. Since the Weber number is 0.5, the deformation of the drop is negligible and the emphasis is focused on the effect of the physicochemical properties of surfactants on the flow field of a nearly spherical drop.

The dimensionless concentration, K, characterizes the ratio of desorption to adsorption rates. The effect of K on the motion of the drop is examined. When the bulk concentration is large, the surface adsorption at the rear of the droplet should accumulate to a high level due to adsorption flux convected from the droplet front, Γ reaches a very high value on the droplet rear, and a large $\nabla_s\sigma$ builds up. In this case, the surfactant has great effect on the motion of the droplet. In contrast, when $K << 1$, the surface adsorption over the whole interface is low and nearly in equilibrium with the bulk phase. In this case, the surface tension gradient is small and has little influence on the motion of droplets.

Figure 2.26 (Li et al., 2001) illustrates the variation of the flow field with K ranging from 0 to 0.1. In a pure system ($K = 0$), no recirculating wake is formed behind the drop, as shown in Figure 2.26a. However, when the drop is slightly contaminated ($K = 0.005$), the detached recirculating wake shown in Figure 2.26b appears. From a comparison among Figure 2.26b–e, it is noted that detached wakes become progressively closer to the drop's rear surface when K increases. Eventually, when $K = 0.1$, it becomes difficult to tell whether the recirculating wake is marginally detached or simply attached. The recirculating wake behind a drop also increases its size. However, the interfacial profiles of relevant variables (for example, adsorption, vorticity, velocity, interfacial tension) provide evidence contrary to the basic assumptions of the stagnant-cap model. The stagnant-cap model seems to lack the basis of a predictive and quantitative tool for modeling drop flow under the influence of surfactants (Li et al., 2001).

Figure 2.26 also demonstrates that the size of the recirculating wake is largely determined by K. According to Leal (1989), the volume of the recirculating wake depends on the relative rate of production of vorticity generated at the drop nose

(a) $K=0$

(Pure system)

$C_D=0.796$

(b) $K=0.005$

$C_D=0.834$

(c) $K=0.01$

$C_D=0.855$

(d) $K=0.04$

$C_D=0.917$

(e) $K=0.1$

$C_D=0.956$

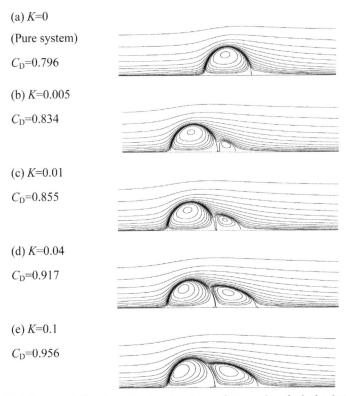

FIGURE 2.26 Influence of K on flow structure (contours of stream function) of a drop at the reference state except K in variation (Li et al., 2001).

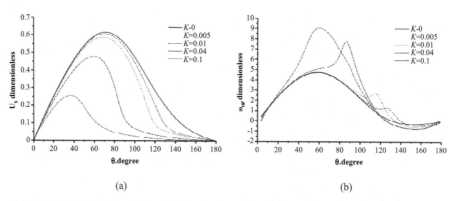

(a)   (b)

FIGURE 2.27 Distribution of tangential velocity and vorticity at the surface for the cases in Figure 2.26 (Li et al., 2001).

and its transport downstream to the wake region by convection and lateral diffusion. From the interfacial vorticity distribution around the drop surface in Figure 2.27b, it is observed that the vorticity at the drop surface is increased, apparently due to reduced surface mobility, as indicated in Figure 2.27a. As K is increased, the vorticity becomes negative at the rear surface, resulting naturally in the accumulation of vorticity in the wake. Therefore, the fact that the recirculating wake expands in volume as K increases seems a natural consequence from the theory of vorticity accumulation.

In a practical system, the numerical drag coefficients of single MIBK (methyl isobutyl ketone) drops either contaminated by sodium dodecyl sulfate (SDS) or not are compared with the experimental data in Figure 2.28 (Li et al., 2003). For the same

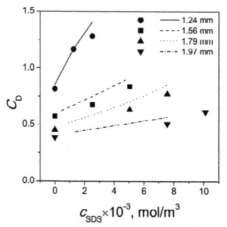

FIGURE 2.28 Effect of SDS concentration in the bulk phase on the drag coefficient of an MIBK drop with an equivalent diameter varying from 1.24 to 1.97 mm (solid symbols denote experimental data and lines denote simulated results) (Li et al., 2003).

diameter drop, it is obvious that the drag coefficient increases gradually with SDS concentration as the drop progressively behaves more like a rigid sphere. It is also seen from Figure 2.28 that a much higher SDS concentration is needed for a larger drop to reach such an asymptotic state.

2.5.2.3 Effect of surfactant on mass transfer

The study of the mass transfer of solutes to/from a single drop with adsorbed surfactants on the interface provides an essential basis for better understanding of practical liquid extraction processes and for scientific design of industrial equipments. A numerical approach on the effect of a surfactant on mass transfer of a buoyancy-driven drop at intermediate Reynolds numbers is necessary (Li et al., 2003). If there is any interfacial resistance (or so-called energy barrier) caused by added surface active agents in a liquid–liquid extraction system, the solute concentration difference, $mc_1^s - c_2^s$, will be greater than zero along the interface. The resistance can be evaluated from the concentrations in the vicinity of the interface during mass transfer, which is a direct indication of the barrier mechanism. On the other hand, when the drop surface is covered by the adsorbed surfactant, the mass transfer area is simply assumed to be $(1 - \bar{\Gamma})$ of interfacial area. Then the total instant diffusion flux over the interface can be formulated as

$$N = k_s (1 - \bar{\Gamma})(mc_1^s - c_2^s)$$

(2.93)

In this equation, the interfacial mass transfer coefficient, k_s, is introduced, which can be evaluated later by least squares fitting. The following boundary conditions can be obtained on the local flux conservation of the solute through the surfactant-covered surface:

$$\frac{D_1}{h_{\xi_1}} \frac{\partial c_1}{\partial \xi_1} = -\frac{D_2}{h_{\xi_2}} \frac{\partial c_2}{\partial \xi_2} = k_s (1 - \bar{\Gamma})(mc_1^s - c_2^s)$$

(2.94)

Thus, Eq. (2.94) can be used as one of the boundary conditions to solve the convective diffusion equations, Eq. (2.42), and the joint effect of hydrodynamics and energy barrier on mass transfer can be evaluated numerically. It can be shown by variable transformations that the characteristic value of Sh_{od} from the simulation is the same, whether the direction of mass transfer is c $\rightarrow$ d or d $\rightarrow$ c.

2.5.2.3.1 Hydrodynamic effect on transient mass transfer

The transient mass transfer to an MIBK drop contaminated by SDS is simulated. The comparison between the numerical simulation and the experimental data is shown in Figure 2.29 (Li et al., 2003). It is observed that the agreement is quite satisfactory, and both the extraction fraction and the overall mass transfer coefficient decrease with the increase of SDS bulk concentration. When the SDS bulk concentration reaches 5.04×10^{-3} mol/m^3, the value of Sh_{od} is only about one-third of that in the pure system. This can be explained by the distribution of interfacial velocities and the local mass transfer coefficient at $t = 0$ along the drop surface, which are plotted against θ in Figures 2.30 and 2.31 (Li et al., 2003). The local Sherwood number at

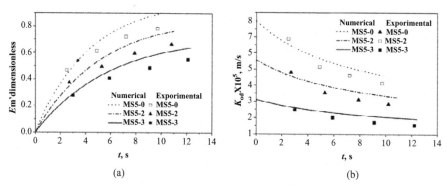

FIGURE 2.29 Contrast of simulation with experiment on the relationships of: (a) extraction fraction E_m, and (b) overall mass transfer coefficient k_{od} with mass transfer time ($d = 1.56$ mm) (serial No. MS5-0: $c_{SDS} = 0$ mol/m³; MS5-2: $c_{SDS} = 2.52 \times 10^{-3}$ mol/m³; MS5-3: $c_{SDS} = 5.04 \times 10^{-3}$ mol/m³).

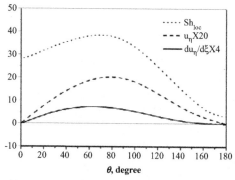

FIGURE 2.30 Profiles of interfacial shear, surface velocity, and local Sherwood numbers ($t = 0$) along the clean drop surface ($d = 1.56$ mm).

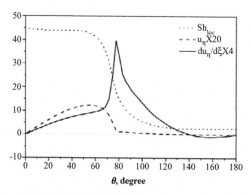

FIGURE 2.31 Profiles of interfacial shear, velocity, and local Sherwood numbers ($t = 0$) along the heavily contaminated drop surface ($d = 1.56$ cm).

the nose ($\theta = 0°$) is, in general, much higher than that in the rear or wake region (θ near 180°). When MIBK is contaminated heavily, the profiles of tangential velocities u_η in Figure 2.31 indicate that the interfacial surface becomes stagnant at the rear region and the surface mobility is much lower than that of a clean drop in Figure 2.30. Correspondingly, the value of Sh_{loc} decreases abruptly along the stagnant surface (Figure 2.31) and the overall k_{od} assumes much lower values than in the pure system (Figure 2.29).

2.5.2.3.2 Effect of interfacial resistance on transient mass transfer

To identify possible interfacial resistance to mass transfer by an adsorbed surfactant layer, the mass transfer to a Tween 80 contaminated MIBK drop is simulated numerically, and the hydrodynamic and energy barrier contribution to mass transfer are taken into account altogether by adopting Eq. (2.94). It is found for the clean extraction system that the numerical results are about 12% higher than the experimental data (Figure 2.32) (Li et al., 2003), whereas they are about 30% higher for the Tween 80 contaminated system (Figure 2.33) (Li et al., 2003). Compared with the simulation results of the SDS contaminated system, it is deduced that Tween 80 molecules adsorbed on the interface have evident mass transfer resistance ($1/k_s$) on acetic acid, which is estimated by least squares fitting to the experimental data. It is noted that interfacial resistance in a surfactant monolayer film increases 10 times higher than that in a clean film, due to the physicochemical interaction between surface active agent and solute molecules.

To resolve the reliable prediction of motion and mass transfer in swarms of bubbles or drops, the study of hydrodynamic behaviors of single bubbles and contaminated drops is of fundamental importance, as it provides a useful starting point toward understanding more complicated and practical multiphase systems. With the help of numerical simulation, it is demonstrated that a surfactant exerts its influences

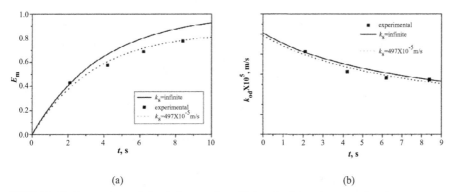

(a)   (b)

FIGURE 2.32 Comparison of simulation results with interfacial resistance taken into account or not with experimental data of: (a) extraction fraction E_m and (b) overall mass transfer coefficient k_{od} with mass transfer time ($d = 1.79$ mm, clean MIBK–acetic acid–water system) (Li et al., 2003).

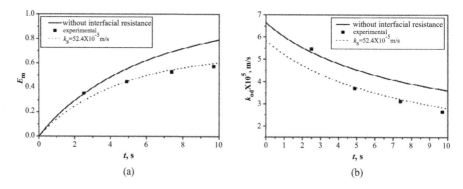

FIGURE 2.33 Comparison of simulation results with interfacial resistance taken into account or not with experimental data of: (a) extraction fraction E_m and (b) overall mass transfer coefficient k_{od} with mass transfer time (d = 1.56 mm, $c_{Tween\ 80}$ = 0.005 mol/m³, a heavily contaminated drop).

on mass transfer behavior of single drops via two mechanisms. The first is hydrodynamic: with surfactant adsorbed on the drop surface, the surface mobility is reduced, and the convective transport of the solute becomes weak, causing the driving force for interphase mass transfer (the normal gradient of solute concentration) to decrease. In addition to this, the surfactant molecules adsorbed on the drop surface decrease the effective mass transfer area. The second mechanism is the possible energy barrier created by the adsorbed surfactant, which depends largely on the physical and chemical interactions between the solute and the surfactant and differs from one solute–surfactant pair to another. Therefore, further CFD research on single drop mass transfer should be conducted in conjunction with the physicochemical study of solute–surfactant pairs.

2.5.3 Surfactant-induced Marangoni effect

Due to the complexity of the interfacial phenomena, the influence of surfactants on the Marangoni effect and mass transfer has not been well understood. However, much research has suggested that surfactants have a noticeable influence on interfacial instability and mass transfer efficiency (Liang and Slater, 1990; Li et al., 2003). The system with acetic acid transferred from an aqueous continuous phase to a rising MIBK drop is investigated in a droplet file column by the Schlieren technique, and the surfactants used are sodium dodecyl sulfate (SDS), Triton X-100 and Tween 80 (Wang, Z. H. et al., 2011). The effect of surfactants on terminal velocity and extraction fraction is obtained experimentally. It is found that surface-active agents also have great effects on the mass transfer behavior of MIBK drops.

For the system with surfactants, both ionic and nonionic surfactants dampen solute interfacial convection at low surfactant concentrations, which matches the trend of variation of extraction fraction shown in Figures 2.34 and 2.35a. However, SDS introduces new instability after t = 25 s as evidenced by Schlieren images, while no

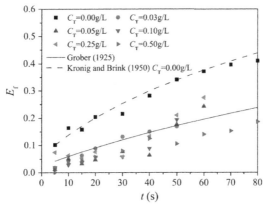

FIGURE 2.34 Influence of Triton X-100 on extraction fraction ($c_{A,d}$ = 3 g/L).

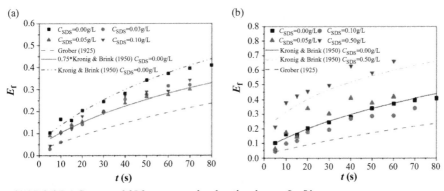

FIGURE 2.35 Influence of SDS on extraction fraction ($c_{A,d}$ = 3 g/L).

(a) Lower SDS concentration. (b) Higher SDS concentration.

interfacial instability is observed for Triton X-100 when $c_T < 0.5$ g/L. The presence of SDS at high concentrations induces intensified interfacial instability accompanied by drop oscillation, and the extraction fraction increases with surfactant concentration as shown in Figure 2.35b. The enhancement factor of extraction introduced by SDS at the concentration of $c_{SDS} = 0.5$ g/L is 3.37–6.3, larger than that provided by the solute at an initial concentration of 20 g/L in the pure system at the corresponding time. Meanwhile, increasing the concentration of Triton X-100 (c_T) to $c_T = 0.5$ g/L (higher than the critical micelle concentration of Triton X-100, 0.15 g/L), another mode of interfacial flow in the late period (around 45 s) of mass transfer is triggered, which looks like smoke rising up along the interface, but no drop oscillation is observed. Nevertheless, as shown in Figure 2.34, the extraction fraction after $t = 45$ s at $c_T = 0.25$ g/L is higher than the prediction from the stagnant mode while E_f at $c_T = 0.5$ g/L is lower than the stagnant model, perhaps due to the interfacial barrier at $c_T = 0.5$ g/L increasing as more surfactant molecules accumulate at the interface field.

2.6 BEHAVIOR OF PARTICLE SWARMS

2.6.1 Introduction

In practical multiphase processes, the dispersed phase is often present as small particles (including solid particles, bubbles, and drops) so that the large specific surface area generated leads to high interphase mass/heat transfer rates and high process efficiency. The interaction between particles must be included when estimating the interphase forces and transport rates between the dispersed and continuous phases. What are really needed are the correlations for the particles in a swarm or assemblage for macroscopic numerical simulation of process equipment instead of that for a single particle. It is thus necessary to establish the correlations for a swarm in parallel to those for single particles. Unfortunately, the work on particle swarms is not sufficient compared with its counterpart on single particles.

On one hand, we can conduct experiments on particle swarms and obtain data and empirical correlations. On the other hand, theoretical and numerical work may also help in achieving this goal. Based on numerical predictions the correlations for single particles may be modified by introducing the phase fraction as an additional parameter, if we can find a simple and accurate functional relationship between the properties of single particles and those in a swarm.

Theoretical and numerical efforts in this task include roughly three approaches, as exemplified in Figure 2.36, for a numerical study of gas–liquid flow using different numerical methods (Pan et al., 2002). The first is direct numerical simulation (DNS) of multi-particle systems, from a two-particle system (tandem or side-by-side

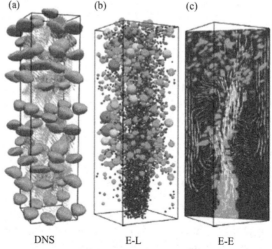

(a)   (b)   (c)

DNS   E-L   E-E

FIGURE 2.36 Gas–liquid multi-scale modeling hierarchy (Roghair et al., 2011).

(a) direct numerical simulations. (b) Eulerian–Lagrangian method. (c) Eulerian–Eulerian method.

configurations) up to thousands of particles. Due to the rapid progress in computer technology and numerical methods, the hydrodynamics of over a thousand solid particles (Pan et al., 2002) and 125 drops (boundary element method) (Zinchenko and Davis, 2000) have been simulated satisfactorily. Using the Eulerian–Lagrangian method (the second approach) the multiphase systems of even more particles have been simulated successfully (Portela and Oliemans, 2003; Chiesa et al., 2005). However, these numbers of particles are still smaller than a large-scale commercial chemical reactor (which can be millions of particles) to get reliable simulation results.

The third approach is the Eulerian–Eulerian method, which is simpler and computationally efficient. Interphase forces are used to describe the two-phase interaction in Eulerian–Eulerian two-fluid models. The dispersed phase can be either uniform or non-uniform in particle size. The uniform size method ignores coalescence and breakage behavior of the particles and economizes in terms of computational complexity. A multiphase system with only low phase holdup can be simulated based on the Euler–Larangrian method, due to the computational load increasing proportionately with the dispersed phase holdup. To promote computational efficiency, Zhang et al. (2008) developed an Eulerian–Eulerian two-fluid model using large eddy simulation (LES) for both gas and liquid phases in a Rushton impeller-driven stirred tank in a three-dimensional frame.

The cell model is perhaps the fourth approach to multi-particle systems. The so-called cell model takes the particle swarm as being composed of the same average cells, and deals only with a typical cell constituted by a central particle and the surrounding shell of the continuous phase in accordance with the average phase fraction (Happel, 1958; Happel and Brenner, 1973), as illustrated in Figure 2.37. Because of treating only an average cell, the cell model enjoys the advantages of simple formulations and low computational load. Its weakness is that it is difficult to express interparticle actions accurately because the means available for such manipulation are the outer cell boundary conditions only.

As the phase holdup increases, the particle size is no longer uniform and the coalescence and breakage behavior produce a significant effect on the hydrodynamics

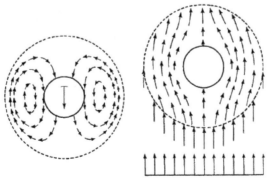

FIGURE 2.37 Spherical cell model in fixed and moving coordinate systems (Happel and Brenner, 1973).

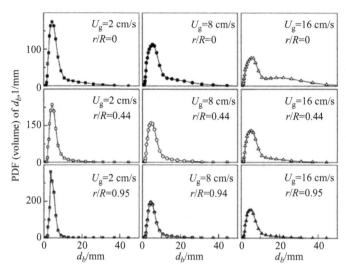

FIGURE 2.38 Bubble size distributions at different radial positions and superficial gas velocities (Wang, 2011).

of the system. Under these conditions, the population balance model (PBM) can be combined with CFD methods. The PBM is a common method to describe the dispersion size distribution in multiphase flow, first used by Hulburt and Katz (1964) in chemical engineering. The PBM can show the influence of particle behaviors on dispersion phase distribution and help in thoroughly understanding the mechanism of multiphase hydrodynamics. In recent years, with the rapid development of computing technology, the PBM has become a new research topic in multiphase flow fields, and is being applied more often in crystallization, polymerization and particle preparation systems. Figure 2.38 shows the bubble size distribution at different superficial gas velocities and radial positions based on the PBM. It can be confirmed that the superficial gas velocity has a great influence on the bubble size distribution in bubble columns, especially at larger superficial gas velocities. At higher superficial gas velocities, large bubbles in the center area increase in number, but in the wall regime the bubble distributions are little affected. Experimental measurements of bubble size distributions are difficult to achieve at high superficial gas velocities. As the results agree well with the qualitative experimental results, this numerical simulation could rectify this problem.

2.6.2 **Forces on single particles**

Many terminologies are used to describe the interactions between the particles and the surrounding continuous fluid phase, such as drag, lift force, virtual mass force, Basset force, turbulent dispersion force, lubrication force, pressure gradient force, etc. (Figure 2.39) (Mao and Yang, 2009). These interphase forces appear in the governing

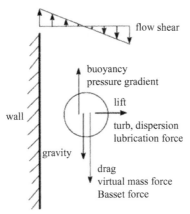

FIGURE 2.39 Forces arising from the interaction of a particle with a continuous fluid phase (Mao and Yang, 2009).

equations as body forces in Eulerian–Lagrangian models. Some of these forces have a sound physical basis, but some do not. Some forces have been correlated with simple formulae with sufficient accuracy, but some demand further attention.

2.6.2.1 Drag force

A steadily moving particle is subject to drag force from the surrounding fluid, which is usually calculated as

$$\mathbf{F}_D = -\frac{1}{2} C_D \rho_c \left(\frac{\pi d_p^2}{4} \right) |\mathbf{u}_p - \mathbf{u}_c| (\mathbf{u}_p - \mathbf{u}_c) \tag{2.95}$$

In fact, this equation is also the definition of the drag coefficient, C_D. Though it is well defined for a single particle, it is often used for particles with complex modes of motion. A single solid particle in steady motion is the most thoroughly studied case.

For the case of a spherical particle (including bubble and drop) in axisymmetric creeping flow (particle Reynolds number $Re_p \ll 1$) (Clift et al., 1978), the drag coefficient is analytically derived from the so-called Hadamard–Rybczynski solution of the flow field:

$$C_D = -\frac{8}{Re_p} \frac{2+3\lambda}{1+\lambda}, \quad \lambda = \frac{\mu_p}{\mu_c} \tag{2.96}$$

For a solid sphere, relative particle viscosity $\lambda \to \infty$, $C_D = 24/Re_p$; for a bubble, $\lambda \to 0$, $C_D = 16/Re_p$; the drag coefficient of a drop lies in between.

When Re_p is greater than 0.1, the assumption of creeping flow is not valid, and no analytical solution is available for the flow field around an axisymmetric particle. The only solution is experimental measurement and subsequent empirical correlation. For a solid sphere, a simple and popular method is the Schiller and Naumann correlation (Kwauk and Li, 2008):

$$C_D = \begin{cases} \dfrac{24}{Re_p}(1+0.15Re_p^{0.687}) & (Re_p < 1000) \\ 0.44 & (Re_p \geq 1000) \end{cases} \tag{2.97}$$

Much work has been devoted to improve this by piecewise correlation, but this makes little difference when used to simulate solid particle-laden engineering multiphase flows, particularly the time-averaged flow fields involved.

For gas bubbles, it is recognized that the drag coefficient is a function of Reynolds number and bubble shape, and the latter is in turn dependent on Re and other parameters of gas–liquid systems. Thus, a correlation valid under a wide range of regimes has been proposed by Tomiyama (1998):

$$C_D = \max \left\{ \min \left[\frac{24}{Re_p}(1+0.15Re_p^{0.687}), \frac{72}{Re_p} \right], \frac{8}{3}\frac{Eo}{Eo+4} \right\} \tag{2.98}$$

for slightly contaminated systems, and

$$C_D = \max \left\{ \frac{24}{Re_p}(1+0.15Re_p^{0.687}), \frac{8}{3}\frac{Eo}{Eo+4} \right\} \tag{2.99}$$

for contaminated systems ($10^{-2} < Eo < 10^3$, $10^{-14} < Mo < 10^7$). The situation is the same: the effects of many other parameters on C_D have not been examined to obtain generally accepted results.

Similarly, the drag exerted on single drops is also dependent on drop shape. When $Re_p < 600$ and surface tension is relatively large, a drop remains spherical or elliptical. When Re_p is in the range from 600 to 900, the drop velocity reaches a maximum and C_D a minimum. Further increase of Re_p induces greater change in shape and the drop pulsates, with C_D increasing sharply along with Re_p. When $Re_p > 1000$–3000, a drop would break up due to the hydrodynamic instability. Hu and Kintner (1955) proposed a correlation based on a large amount of experimental data:

$$C_D We E^{0.15} = \begin{cases} \dfrac{4}{3}\left(\dfrac{Re_p}{E^{0.15}}+0.75\right)^{1.275}, & 2 < C_D We E^{0.15} < 70 \\ 0.045\left(\dfrac{Re_p}{E^{0.15}}+0.75\right)^{2.37}, & C_D We E^{0.15} \geq 70 \end{cases} \tag{2.100}$$

with $E = \rho_c^2 \sigma^3 / \mu_c^4 g\Delta\rho$ and the Weber number $We = u^2 d_p \rho_c / \sigma$.

In the recent literature concerning the simulation of homogeneous multiphase flows, the effect of turbulence intensity is recognized to significantly reduce the drag coefficients of particles in a stagnant fluid, C_{D0}, to C_D in turbulent flow, and the extent of this reduction depends on the turbulence intensity. Brucato et al. (1998) made measurements of solid particle settling velocities in turbulent flow and proposed

$$\frac{C_D}{C_{D0}} = 1 - 8.76 \times 10^{-4}\left(\frac{d}{\lambda}\right)^3 \tag{2.101}$$

where d is the particle diameter and λ is the Kolmogorov microscale of turbulence. Lane et al. (2000) adopted this tentatively to simulate gas–liquid flow in a stirred tank with an empirical constant of 6.5×10^{-6}, which was used successfully by Zhang et al. (2008) in the Eulerian–Eulerian large eddy simulation of a gas–liquid stirred tank. Subsequently, Lane et al. (2005) proposed a more sophisticated correlation in terms of the integral time scale of turbulence and particle relaxation time on the basis of measurements on solid and gas particles and numerical simulation of gas bubbles in homogeneous turbulence. A common sense approach suggests that the effect of turbulence on particle behavior is a necessary ingredient in the numerical simulation of multiphase flow.

In summary, knowledge on single spheres in a quiescent fluid is relatively satisfactory, but other situations (fluid shear, particle rotation, oscillation and unsteady motion), particularly for bubbles and drops, have still not been sufficiently understood.

2.6.2.2 Unsteady forces

Added mass force (or virtual mass force) and Basset force appear when a particle is in rectilinear unsteady motion. These two forces are clarified as those with sound physical meaning. If a particle accelerates in an ideal fluid, extra force is needed to move the fluid around the particle, and such a force may be calculated as

$$\mathbf{F}_A = -C_A V_p \rho_c \frac{D\mathbf{u}_{\text{slip}}}{Dt} \tag{2.102}$$

where $\mathbf{u}_{\text{slip}}$ is the slip velocity of the particle and V_p is the particle volume. This equation is also the definition of the added mass force coefficient C_A. For bubbles with ρ_b far less than ρ_c, F_A is much greater than the force to accelerate the bubble mass itself.

History force (also called Basset force) is a force related to the history of acceleration of a particle. Compared to a steadily moving particle, an accelerating particle at the same slip velocity has a thinner momentum boundary layer, which is equivalent to larger friction at the particle surface. This force is also thought to be related to the inertia of the surrounding fluid mass. Its name derives from Basset in 1888 finding a force exerted on a solid particle in sinusoidal oscillation in a creeping flow expressed by the following integral (Zapryanov and Tabakova, 1999):

$$\mathbf{F}_H = -\frac{3d_p^2 \mu_c}{2} \sqrt{\frac{\pi}{\nu}} \int_0^t \frac{d\mathbf{u}_p/d\tau}{\sqrt{t-\tau}} d\tau \tag{2.103}$$

This is essentially part of the difference of the force to drive an accelerating sphere minus the drag at steady-state motion. For other unsteady flow situations, analytical expressions are not available for engineering applications. Further study on its evaluation and role in particle flow simulation is necessary.

2.6.2.3 Lift force

Lift force refers to the force exerted on a particle in the direction perpendicular to the particle motion path. Two mechanisms create the lateral force: the shear in the flow induces the Saffman force, and the rotation of a particle leads to the Magnus force.

Such a distinction seems to make theoretical sense, because a particle will rotate in shear flow so that two forces appear together for freely moving particles. The lift force is crucial in obtaining the correct spatial distribution of phase holdups, for example the radial profile of gas holdup in a bubble column.

Drew and Lahey (1987) gave a general expression for the Saffman force in an ideal fluid:

$$\mathbf{F}_L = C_L \rho_c (\mathbf{u}_p - \mathbf{u}_c) \times (\nabla \times \mathbf{u}_c) \tag{2.104}$$

Many factors influence the lift force coefficient C_L, and no general and reliable correlation for C_L is available. Moreover, the correlations reported in the literature give widely scattered values of C_L. So some authors ignored the lift force as compared with the drag, and sometimes the lift force coefficient was even used as an adjustable parameter to match gas holdup predictions with experimental data.

Besides the forces already mentioned, some other forces have been used to describe the interaction between the dispersed and continuous phases, although arguments exist on their physical soundness. For example, the turbulent dispersion force has been used to describe the spread of bubbles due to the effect of flow turbulence. This force is partly physical in the sense that turbulent vortices exert impact onto a particle and push it to move in the fluid. The wall (lubrication) force is conceived as a resistance to the motion of a particle when approaching a solid wall at a certain distance, and the flow of the continuous phase fluid in the narrow gap between the particle and solid wall creates a net force away from the wall. Different expressions for lubrication force were developed and tested, but no generally accepted relation is available. These two forces are sometimes used to compensate the lateral lift forces so that the gas holdup profiles with either wall peak or core peak can be derived from numerical simulation to match experimental data. Obviously, the physical mechanisms underlying these forces need further thorough investigation. In general, it is better to model these mechanisms based on their physical fundamentals instead of using analog and imitation, so that accurate constitutive equations can be found with fewer empirical constants and wider applicability.

2.6.3 **Cell model**

The accuracy of a cell model is critically reliant on the soundness of its outer cell boundary conditions. In the literature, (1) the external uniform flow, (2) the free-shear condition (model H) or (3) the zero-vorticity condition (model K) has been adopted for the outer cell boundary. However, these conditions fail to reflect the realities in dispersed multiphase flows. Some improvements of cell models have been suggested (Tal and Sirignano, 1982; Mao and Chen, 2002b), but further refinement seems necessary.

Cell models were applied to study solid sphere assemblages at the start (Happel, 1958; Happel and Brenner, 1973), including hydrodynamics (Mao and Chen, 2002b) and mass transfer (Mao and Wang, 2003). LeClair and Hamielec (1971) extended it to the swarms of spherical bubbles with $Re < 1000$ and $\varepsilon = 0.4$–1, and observed good agreement between prediction and experiment for small Re. Sankarana-rayanan and Sundaresan (2002) used the lattice Boltzmann method in combination with VOF (volume-of-fluid) for tracking the interface to examine the motion and lift force of single bubbles in water in a 3D periodical cubic domain under the condition of low voidage, and this is essentially a cell model with a cubic shape. It is also used to study mass transfer in a drop swarm (Mao and Chen, 2002a, 2005). In Figure 2.40 (Mao and Yang, 2009), the predicted drag coefficients (symbols) are in close agreement with the Kumar–Hartland correlation (Kumar and Hartland, 1985) for $Re < 100$ only.

Several physical parameters have a strong effect on the mass transfer in a particle swarm. The numerical predictions shown in Figure 2.41 (Mao and Wang, 2003) reveal that Pe and Re have significant influence on the mass transfer factor, particularly in the lower range of Pe. This indicates that the proposition of the mass transfer factor $j_D = Sh / ReSc^{1/3}$ (Pe not explicitly included) fails to account for the overall effect of

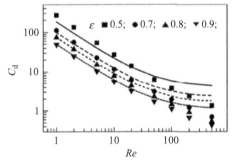

FIGURE 2.40 Drag coefficients of drops predicted by the cell model (Mao and Chen, 2005) with the Happel boundary condition (model H) versus the correlation (Kumar and Hartland, 1985).

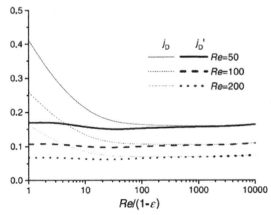

FIGURE 2.41 Dependence of mass transfer factor on Peclet number at constant Re (model H, $\varepsilon = 0.7$).

Pe. This phenomenon is easily understood. Since mass transfer with small *Pe* means a large diffusion coefficient and small concentration gradient as the driving force, the concentration boundary layer in this case would stretch far from the particle surface even into the outer cell boundary, where constant concentration is specified. Therefore, j_D will increase naturally due to the limited transfer distance to the center sphere. Small *Re* eases the concentration boundary layer stretching in all directions, particularly in the upstream direction, resulting in increased j_D, as expected. An effort has been made to find an approximate functional form to represent the effect of *Pe* that is not covered by the previous definition of j_D. It is found that the modified form of j_D can transform the varying j_D into a roughly constant j_D:

$$j_D' = j_D \Big/ \left(1 + \frac{1}{\varepsilon Pe^{2/3}}\right) \tag{2.105}$$

The role of voidage has not been included in the classical correlations. When j_D' is used to represent the effect of *Pe* and ε, the simulation results become easier to correlate (Figure 2.42) (Mao and Wang, 2003). From these results, it seems that a more accurate and reliable correlation may be obtained from straightforward numerical simulations on the basis of existing empirical correlations.

Despite the previous observation, it is still desirable to compare the present simulation results with some available nondimensional correlations recommended in the literature. In Figure 2.43a (Mao and Wang, 2003), comparison is made between the model H simulation for *Pe* in the range from 1 to 3000 and two correlations. The agreement is reasonable and it seems that the Sen Gupta and Thodos (1963) correlation represents the effect of voidage a little better. Since their data are on the water vapor–air system with small *Sc*, the correlation can reflect the behavior of j_D more accurately in the lower range of *Pe* in Figure 2.43. When the modified mass transfer

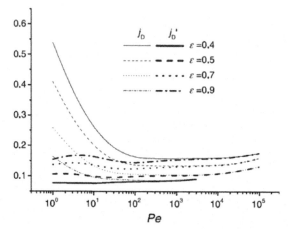

FIGURE 2.42 Dependence of mass transfer factor on Peclet number at different ε (model H, Re = 100).

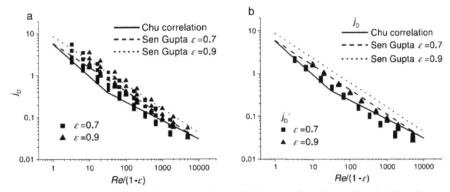

FIGURE 2.43 (a) Comparison of mass transfer coefficient predicted by cell model H and the empirical correlation ($\varepsilon = 0.7$ and 0.9). (b) Comparison of mass transfer coefficient predicted by cell model H and the empirical correlation in terms of j_D' ($\varepsilon = 0.7$ and 0.9).

factor in Eq. (2.105) is used to represent the simulation results, the scatter of the simulated j_D is significantly reduced.

Mao and Chen (2002b) commented that model H is closer to reality than model K, and suggested a modified version on the basis of model H. In their model M2, the free surface boundary condition $\tau_{\xi\eta} = 0$ is retained, while two modifications are proposed: (1) the central sphere is subject to a uniform flow of $\Psi = 0.4512y^2$; (2) the fore–aft symmetry of vorticity is enforced on the outer cell boundary. It has been demonstrated that model M2 predicts the drag coefficient of solid particles in a swarm with better accuracy than models H and K. More issues need to be tackled. Mao and Chen (2005) attempted to extend the cell models to an intermediate range of Reynolds numbers up to 500 by formulating the pseudo-turbulence in fluid flow in the inter-sphere interstices as a factor to increase the apparent viscosity of a flowing liquid. Formally considering the turbulence in a general particle swarm remains a topic for further investigation because many commercial units normally operate in a turbulent state. Mao and Yang (2006) also explored the position of the central particle in the cell and found that needed more stringent justification, so that the cell models can express the interaction between particles with reasonable accuracy.

2.7 SINGLE PARTICLES IN SHEAR FLOW AND EXTENSIONAL FLOW

Most previous studies on the mass/heat transfer from solid particles, drops, and bubbles are focused on particles in motion under buoyancy in a quiescent fluid. However, the fluid motion relative to a particle can be dominated by fluid shear and/or extension when the density difference between the particle and fluid is small. For the case of a neutrally buoyant spherical solid particle or a drop of radius R immersed in a planar creeping flow of large extent, the velocity field far from the drop particle is assumed to be the simple extensional flow, which is given in Cartesian coordinates by $\mathbf{u}^\infty = (-ex/2, -ey/2, ez)$, where e is the extension strength. For the case of simple shear flow, the velocity field far from the neutrally buoyant spherical particle is represented by $\mathbf{u}^\infty = (\dot{\gamma}y, 0, 0)$, where $\dot{\gamma}$ is the velocity gradient of the imposed flow.

2.7.1 Mass/heat transfer from a spherical particle in extensional flow

By making use of the known Stokes velocity field (Leal, 1992), the mass transfer process both outside and inside a sphere immersed in a simple extensional creeping flow can be numerically investigated in a spherical coordinate system (Zhang et al., 2012a). A finite difference method is adopted to solve the advection–diffusion transport equation, in which a fifth-order WENO scheme is applied in spatial discretization and a third-order TVD Runge–Kutta scheme is applied in time.

2.7.1.1 Steady transport

For the steady transport outside a solid sphere, the simulation results of Zhang et al. (2012a) show that the transport rate depends heavily on the Peclet number. Sh increases with increasing Pe, but its dependency on Pe is associated with the physical properties of the sphere. At high Peclet numbers, Sh is proportional to $Pe^{1/3}$ for a solid sphere, whereas it is proportional to $Pe^{1/2}$ for a liquid sphere. For the case of a drop, the value of the viscosity ratio affects the mass transfer rate because flow velocities are affected by λ. In terms of numerical results, two new correlations are derived to predict Sh:

$$Sh_{ex} = \frac{1}{\lambda+1}(0.207Pe_1^{1/2} - 0.201) + 0.467Pe_1^{1/2} + 1.053 \qquad (1 \leq Pe_1 \leq 10) \quad (2.106)$$

$$Sh_{ex} = \frac{1}{\lambda+1}[0.6 + (0.16 + 0.48Pe_1)^{1/2}] + \frac{\lambda}{\lambda+1}[0.5 + (0.125 + 0.745Pe_1)^{1/3}]$$
$$+ f_1 + f_2 \exp(-Pe^{1/6}/f_3) \qquad\qquad\qquad (Pe_1 > 1000) \qquad (2.107)$$

The fitting parameters in Eq. (2.107) are related to the viscosity ratio and are given by

$$f_1 = -19.844 + 17.846\frac{1}{\lambda+1} + 19.491\exp\left(\frac{-2.174}{\lambda+1}\right) \qquad (2.108)$$

$$f_2 = -1.781 + 2.746\exp\left[-\left(\frac{1.336}{\lambda+1} - 0.664\right)^2\right] \qquad (2.109)$$

$$f_3 = -1.478 - 0.371\exp(-0.274\lambda) - 0.251\exp(-0.072\lambda) \qquad (2.110)$$

2.7.1.2 Unsteady transport

For unsteady transport inside a spherical drop, the simulation results of Zhang et al. (2012a) show that Sh also increases with increasing modified Peclet number Pe' – here $Pe' = Pe/(1 + \lambda)$ – whereas Sh would approach an asymptotic value ($Sh_\infty \rightarrow 15$)

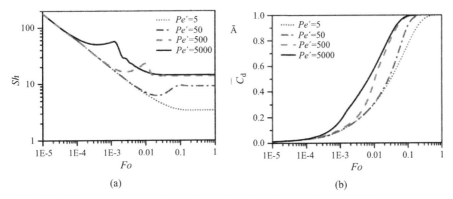

FIGURE 2.44 Transient transport behavior inside a drop for $\lambda = 1$ in a uniaxial extensional flow.

(a) Variation of Sh vs. Fo. (b) Variation of $\bar{C}_d$ vs. Fo.

as $Pe' \rightarrow \infty$. As shown in Figure 2.44, the transport process can be divided into three stages: diffusion-dominated, transitional, and quasi-steady transport stages.

According to the numerical results (Zhang et al., 2012a), one correlation applicable with an error of 3% is derived as

$$Sh_{in,\infty} = 14.09 - \frac{10.87}{1 + \left[\dfrac{Pe/(\lambda+1)}{45.81}\right]^{1.83}} \tag{2.111}$$

For transient conjugate mass transfer from a liquid drop, the numerical results of Zhang et al. (2012b) reveal that the dependence of Sh_i with Pe_i is weaker at higher Pe_i and the increasing rate of Sh_i with Pe_i increases with increasing K and/or decreasing m, where K is the interior-to-exterior diffusivity ratio and m is the distribution coefficient. The following formula is proposed to compute the conjugate Sherwood number in the range of $10 \leq Pe_i \leq 1000$, $\lambda > 1$, and $K > 1$:

$$\frac{1}{Sh_{i,\infty}} = \frac{D_i}{D_1}\left(\frac{m}{Sh_{ex}} + \frac{1}{KSh_{in,\infty}}\right) \tag{2.112}$$

where Sh_{ex} is the Sherwood number for the steady-state external problem and $Sh_{in,\infty}$ is the asymptotic Sherwood number for the internal problem.

2.7.2 Flow and transport from a sphere in simple shear flow

2.7.2.1 Flow field

For the particular geometry of a single sphere of radius R immersed in a simple shear flow, the numerical results of Yang et al. (2011) reveal that the rotation rate of a sphere and the hydrodynamic stresslet agree with the simulated results of Mikulencak and Morris (2004) at finite Re and the analytical predictions of Subramanian and Koch

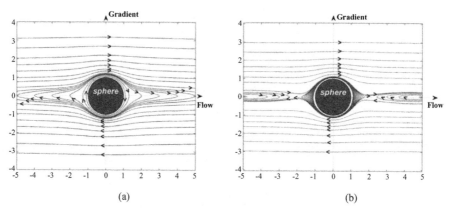

FIGURE 2.45 Streamlines in the *x–y* plane for a single solid sphere in simple shear flow.

(a) *Re* = 0. (b) *Re* = 1.

(2006) at small *Re*. As shown in Figure 2.45 (Yang et al., 2011), a set of closed streamlines envelops the sphere at zero Reynolds number. However, at small, finite Reynolds numbers, the numerical results show that the streamlines in the flow-gradient plane spiral away from the particle.

2.7.2.2 Mass/heat transfer

Yang et al. (2011) computed the mass transfer from a sphere using a boundary layer analysis coupled with numerical simulation at sufficiently large *Pe* and *Re* = $O(1)$. For *Re* ≤ 1, numerical results agree well with *Nu* = $(0.325 − 0.126Re^{1/2})(RePe)^{1/3} + O(1)$ derived by Subramanian and Koch (2006) at *Re* << 1 and *PeRe* >> 1.

As shown in Figure 2.46, the predicted *Nu* at *Re* = 0 increases with increasing Peclet number and eventually approaches an asymptote of about 4.5 in agreement with Acrivos and Goddard's (1965) asymptotic analysis (Subramanian and Koch, 2006),

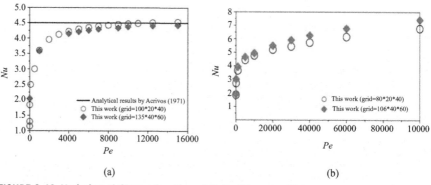

FIGURE 2.46 Variation of *Sh* as a function of *Pe* in different grids.

(a) *Re* = 0. (b) *Re* = 0.3.

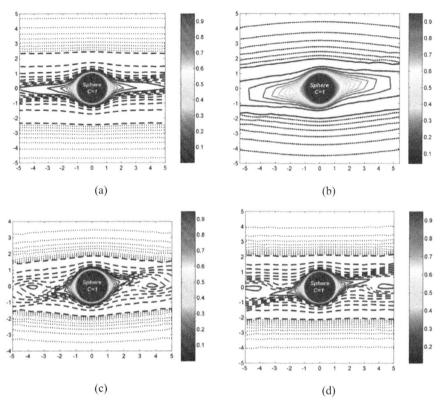

FIGURE 2.47 Contour of solute concentration around a neutrally buoyant sphere embedded in simple shear flow.

(a) $Re = 0$, $Pe = 9000$. (b) $Re = 0.01$, $Pe = 100$, $Sc = 10,000$. (c) $Re = 1$, $Pe = 10,000$, $Sc = 10,000$. (d) $Re = 1$, $Pe = 3000$, $Sc = 3000$.

which validates the numerical method. At finite Re, the Nusselt number for the mass/heat transfer continues to increase with increasing Peclet number without approaching a constant value, as is expected due to the broken symmetry of streamlines.

Figure 2.47 shows the concentration profiles around a neutrally buoyant sphere in simple shear flow at different Schmidt numbers and Re (or Pe) (Yang et al., 2011). From these figures, one can observe that the distortion of the concentration contour lines develops gradually with increasing Re to form a series of closed contours.

2.8 SUMMARY AND PERSPECTIVE

2.8.1 Summary

The classical theory of fluid mechanics and transport phenomena has provided fundamental formulations that apply to many problems involved with motion and mass/heat transfer of solid and fluid particles. The mathematical models usually consist of partial differential equations with appropriate boundary and initial conditions. With the present rapid development of computer hardware and progress in numerical methods, the motion and transport problems of individual particles can be simulated reliably by numerical techniques with reasonable accuracy. Among the numerical methods developed in recent decades, the level set method, the mirror fluid method, the orthogonal boundary-fitted coordinates method, and the cell model are only a few from the available arsenal for engineering science. Choosing a proper numerical method for a specific problem of particle motion and transport is as important as building a solid bridge to the other bank of success.

Consider the level set method (LSM) as an example. Both the LSM and the volume-of-fluid (VOF) method are popularly used for tracking a moving and deforming interface. The LSM has the advantage of easily capturing the moving and deformable boundary (e.g., more accurate calculation of the curvature of the interface) and is easily extended to three-dimensional cases. Of course, it has difficulty retaining mass conservation as compared with the VOF method. Therefore, we proposed several improvements to the LSM in refining the reinitialization techniques for better phase conservation (Yang and Mao, 2002; Li et al., 2008) and the evaluation of the surface force (Wang et al., 2006), etc. The concentration transformation method (Yang and Mao, 2005b) was proposed to improve the accuracy of interphase mass transfer simulation by mitigating the difficulty caused by the double discontinuity of solute concentration and its gradient across the interface. The surface tension gradient-induced Marangoni convection at the interface was then simulated by the improved LSM coupled with the concentration transformation method (Wang, J. F. et al., 2011). In general, other interface capturing algorithms can do the job equally well, if the target is right in its shooting range.

The orthogonal boundary-fitted coordinates method has so far been used successfully for two-dimensional or axisymmetric problems of particle motion and mass transfer. It is particularly efficient when combined with the stream function–vorticity formulation in which the pressure is not explicitly present so that the numerical solution is simple. It is more troublesome in three-dimensional flow because the stream function and vorticity have to be expressed as a vector so that the partial differential equations increase in number and are coupled more strongly. Three-dimensional orthogonal meshing remains possible with moderate computational loads. Perhaps a feasible way to solve 3D problems is to solve the Navier–Stokes equation in terms of primary parameters ($\mathbf{u}$ and p instead of ψ and ω) on an orthogonal boundary-fitted coordinate system. In general, the transformed PDE on an orthogonal grid would give less error in discretization than that on a non-orthogonal grid. Thus, numerical simulation on an orthogonal grid is preferred.

The mirror fluid method is beneficial in solving fluid flow around an irregular solid particle, with the no-slip boundary condition enforced in an implicit way. This method was first proposed for solid particles in laminar Newtonian flows (Yang and Mao, 2005a), but it has been adopted successfully for cases of non-Newtonian fluid flows in ducts with obstacles (Yang and Mao, 2003) and turbulent flows in a stirred tank driven by an axial impeller (Wang, T. et al., 2013). The mirror relation needs careful scrutiny so that the low turbulence wall rules are correctly encompassed into the mirror relation. It is expected that this technique will play a more important role in simulating multiphase turbulent flows in stirred tanks with irregular internals and impellers.

2.8.2 Perspective

Engineers are faced with diverse problems of hydrodynamics or transport involved with single particles or particle swarms in quiescent, uniform, shear or extensional flows under laminar or turbulent conditions. Obviously, a single numerical method cannot handle all those tasks with satisfactory efficiency and accuracy. The best strategy is to select the right numerical method for a specific problem. Also, new numerical techniques should be proposed and validated, while existing methods, including the numerical methods that are the focus in this chapter, need to be redeveloped to give higher accuracy and faster computational speed.

Constitutive equations of interphase forces. No matter what numerical approach is used to simulate dispersed two-phase flows, there is an indispensable need for accurate constitutive equations of interphase forces that are valid over a wide range covering turbulent and/or accelerating motion subject to breakage and coalescence of fluid particles. Therefore, more systematic efforts in experimental, theoretical, and simulation studies on the particle scale are necessary.

The present philosophy of formulating the interphase force is by linear addition of force components corresponding to different physical mechanisms. Thus, in research work a formula often encountered is the following:

$$\mathbf{F} = \mathbf{F}_{gravity} + \mathbf{F}_{drag} + \mathbf{F}_{virtual\ mass} + \mathbf{F}_{Basset} + \mathbf{F}_{Magnus} + \mathbf{F}_{Saffman} + \mathbf{F}_{lubrication} + \cdots \qquad (2.113)$$

This assumption of linearity causes large errors of overestimation or underestimation. In real flow, when the mechanisms of Magnus and Saffman forces act simultaneously, a linear sum of their values when they occur separately is not reliable from a theoretical point of view. In view of this, numerical investigation of the cases of separate force and/or co-occurring forces is indeed necessary. The numerical simulation of single particles has been validated in many tests, and the numerical evaluation of these two forces is perceived as reliable. At the same time, experiments on controlling the rotation of free particles and creation of simple shear flow of good quality are technically difficult. Thus, a reliable correlation of the lift force may be derived after skillful analysis of numerical simulation results.

The same strategy may apply to obtaining the constitutive equations for solid and fluid particles in different flow environments. In this case, homogeneous dispersion is easier. Both cell models and direct numerical simulation of multi-particle systems can contribute to constitutive equations with better accuracy. However, the effects of non-uniformity of phase holdup distribution on the constitutive equations and the numerical procedures have not received enough attention so far.

Non-Newtonian fluids are popularly used in chemical engineering as a stream of reactants or as the media for enhancing mass and heat transfer. Numerical study of particle motion and related transport phenomena are essential to scientific scale-up and operation of industrial processing equipment. Due to the complex rheological behavior of non-Newtonian fluids, such simulations can be made using a similar strategy as for Newtonian fluids, with added difficulty in programming and computational loads.

Particles in shear flow near wall. This situation is important for two fluid reactions over a solid catalyst. Numerical study of this case is subject to two difficulties in conducting a 3D particle situation in the near-wall turbulent shear flow. Considering the significance of this type of simulation in successful design of catalytic reactors with an optimized catalyst arrangement, this effort is worthwhile.

Properties of the interface, more than an ideal surface. The behavior of fluid particles is more complicated than its counterpart for solid particles because of the presence of a free interface. In the traditional way, the particle surface is ideal, without mass and thickness, and it has only a few physical properties such as surface tension (equivalently surface energy). However, this property is not enough to characterize the zig-zag motion and shape oscillation of a fluid particle. Other properties like surface viscosity and elasticity may also be responsible for this behavior. Therefore, surface properties should be approached from a chemical engineering perspective and incorporated into the pertinent numerical scheme.

3D simulation. More simulation efforts should be devoted to three-dimensional scenarios to reveal the true quantitative nature of the complexity of particle motion and transport phenomena. From the comparative simulation of solute extraction processes in the same two-liquid layer system, it is observed that the Marangoni effect is more difficult to trigger in a 3D framework (Mao et al., 2008a) than in a 2D situation (Mao et al., 2008b), which demands less computational resources. Also, more spatial and temporal patterns of instability are present in 3D cases. By the a priori setup of a 2D or axisymmetry, some natural patterns of particle motion and hydrodynamic instability are excluded at the beginning of a study to explore the unknowns. For example, the zig-zag motion of single particles and accompanying asymmetric shape undulation can be simulated only in a three-dimensional scenario. The quantitative knowledge desired in engineering circles is expected to be gleaned from realistic 3D simulations based on up-to-date computational techniques and skills.

The above observations also apply to numerical simulation of heat and mass transfer processes in process equipment. It is expected that fundamental studies on the particle (and particle swarm) scale will eventually contribute to accurate numerical simulation of macroscopic multiphase process equipment, so that operation optimization, renovation, and scale-up of process equipment can be done primarily on the basis of accurate and efficient numerical simulation with reduced cost and computation time.

NOMENCLATURE

A	area	m^2
Bi	Biot number	–
c	concentration of solute	wt%
C	dimensionless concentration	–
C_A	added mass coefficient	–
C_D	drag coefficient	–
C_L	lift coefficient	–
d	diameter	m
D	molecular diffusivity of solute	m^2/s
Fr	Froude number	–
F	body force	N/m^3
h_ξ, h_η	scaling factor	m
H_ξ, H_η	dimensionless scaling factor	m
k	mass transfer coefficient	m/s
m	distribution coefficient	
N	molar flux	$mol/(m^2.s)$
n	unit vector normal to a surface	–
p	pressure	Pa
Pe	Peclet number	–
r	radial coordinate	m
Re	Reynolds number	–
Sh	Sherwood number	–
S	source term in the discretized equation	
t	time	s
u	axial velocity component	m/s
U	terminal velocity	m/s
u	velocity vector	m/s
v	radial or transverse velocity component	m/s
V_d	volume of drop	m^3
We	Weber number	–
x	axial coordinate	m
X	dimensionless axial coordinate	–
y	radial or transverse coordinate	m
Y	dimensionless radial or transverse coordinate	–

Greek letters

$\dot{\gamma}$	shear rate	s^{-1}
δ	Dirac delta function	–
ε	"thickness" of interface, domain adjusting factor	–
ζ	relative density	–
η	coordinate in computational plane	–
θ	dimensionless time	–
κ	Gaussian curvature	–

λ	relative viscosity, parameter of Carreau model	–, s
μ	viscosity	Pa·s
ξ	coordinate in computational plane	–
ρ	density	kg/m^3
σ	surface tension	N/m
τ	stress tensor	–
ψ	stream function	m^3/s
ω	vorticity	s^{-1}
ϕ	level set function	m
Γ	dimensionless surface adsorption	–
Ψ	dimensionless stream function	–

Subscripts

1	continuous phase
2	dispersed phase
A	added force
D	drag force
L	lift force
in	first measurement location
s	surface of bubble or drop
od	overall
out	second measurement location
x	axial direction
y	radial or transverse direction

Superscripts

*	in equilibrium with other phase
T	matrix transposition
∞	remote boundary

REFERENCES

Acrivos, A., & Goddard, J. D. (1965). Asymptotic expansions for laminar forced-convection heat and mass transfer. *J. Fluid Mech.*, *23*, 273–291.

Bhaga, D., & Webber, M. E. (1981). Bubbles in viscous liquids: Shapes, wakes and velocities. *J. Fluid Mech.*, *105*(1), 61–85.

Brucato, A., Grisafi, F., & Montante, G. (1998). Particle drag coefficients in turbulent fluids. *Chem. Eng. Sci.*, *53*, 3295–3314.

Chang, Y. C., Hou, T. Y., Merriman, B., & Osher, S. (1996). A level set formulation of Eulerian interface capturing methods for incompressible fluid flows. *J. Comput. Phys.*, *124*(2), 449–464.

Chen, Y., Mertz, R., & Kulenovic, R. (2009). Numerical simulation of bubble formation on orifice plates with a moving contact line. *Int. J. Multiphase Flow*, *35*(1), 66–77.

Chhabra, R. P. (1993). *Bubbles, drops and particles in non-Newtonian fluids*. Boca Raton, FL: CRC Press.

Chiesa, M., Mathiesen, V., Melheim, J. A., & Halvorsen, B. (2005). Numerical simulation of particulate flow by the Eulerian–Lagrangian and the Eulerian–Eulerian approach with application to a fluidized bed. *Comput. Chem. Eng.*, *29*, 291–304.

Clift, R., Grace, J. R., & Weber, M. E. (1978). *Bubbles, drops, and particles* (pp. 30–33). New York: Academic Press.

Dandy, D. S., & Leal, L. G. (1989). Buoyancy-driven motion of a deformable drop through a quiescent liquid at intermediate Reynolds number. *J. Fluid Mech.*, *208*, 161–192.

Drew, D. A., & Lahey, R. T, Jr. (1987). The virtual mass and lift force on a sphere in rotating and straining flow. *Int. J. Multiphase Flow*, *13*(1), 113–121.

Glowinski, R., Pan, T. W., Hesla, T. I., & Joseph, D. D. (1999). Experimental investigation of transverse flow through aligned cylinders. *Int. J. Multiphase Flow*, *25*(5), 755–794.

Glowinski, R., Pan, T. W., Hesla, T. I., Joseph, D. D., & Périaux, J. (2001). A fictitious domain approach to the direct numerical simulation of incompressible viscous flow past moving rigid bodies: Application to particulate flow. *J. Comput. Phys.*, *169*(2), 363–426.

Happel, J. (1958). Viscous flow in multiparticle systems: Slow motion of fluids relative to beds of spherical particles. *AIChE J.*, *4*(2), 197–201.

Happel, J., & Brenner, H. (1973). *Low Reynolds number hydrodynamics.* (2nd ed.). Leyden, the Netherlands: Noordhoff.

He, Z., Maldarelli, C., & Dagan, Z. (1991). The size of stagnant caps of bulk soluble surfactant on the interfaces of translating fluid droplets. *J. Colloid Interface Sci.*, *146*(2), 442–451.

Hu, S., & Kintner, R. C. (1955). The fall of single liquid drops through water. *AIChE J.*, *1*(1), 42–48.

Huang, P. Y., Hu, H. H., & Joseph, D. D. (1998). Direct simulation of the sedimentation of elliptic particles in Oldroyd-B fluids. *J. Fluid Mech.*, *362*, 297–326.

Hulburt, H. M., & Katz, S. (1964). Some problems in particle technology: A statistical mechanical formulation. *Chem. Eng. Sci.*, *19*(8), 555–574.

Kim, J., Kim, D., & Choi, H. (2001). An immersed-boundary finite-volume method for simulations of flow in complex geometries. *J. Comput. Phys.*, *171*(1), 132–150.

Kulkarni, A. A., & Joshi, J. B. (2005). Bubble formation and bubble rise velocity in gas–liquid system: A review. *Ind. Eng. Chem. Res.*, *44*(16), 5873–5931.

Kumar, A., & Hartland, S. (1985). Gravity settling in liquid/liquid dispersions. *Can. J. Chem. Eng.*, *63*(3), 368–376.

Kwauk, M. S., & Li, H. Z. (2008). *Handbook of fluidization* (p. 106). Beijing: Chemical Industry Press (in Chinese).

Lane, G. L., Schwarz, M. P., & Evans, G. M. (2000). Modeling of the interaction between gas and liquid in stirred vessels. *Proceedings of the 10th European conference on mixing* (pp. 197–204), The Netherlands.

Lane, G. L., Schwarz, M. P., & Evans, G. M. (2005). Numerical modeling of gas–liquid flow in stirred tanks. *Chem. Eng. Sci.*, *60*, 2203–2214.

Leal, L. G. (1989). The stability of drop shapes for translation at zero Reynolds number through a quiescent fluid. *Phys. Fluids A*, *1*(8), 1309–1313.

Leal, L. G. (1992). *Laminar flow and convective transport processes.* London: Butterworth-Heinemann.

LeClair, B. P., & Hamielec, A. E. (1971). Viscous flow through particle assemblages at intermediate Reynolds numbers: A cell model for transport in bubbles swarms. *Can. J. Chem. Eng.*, *49*(6), 713–720.

Li, T. W., Mao, Z.-S., Chen, J. Y., & Fei, W. Y. (2001). Terminal effect of drop coalescence on single drop mass transfer measurement and its minimization. *Chinese J. Chem. Eng.*, *9*(3), 204–207.

Li, T. W., Sun, C. G., Mao, Z. -S., & Chen, J. Y. (2000). Influence of distortion function on the accuracy of numerical simulation of the motion of a single buoyancy-driven deformable drop. *Sel. pap. eng. chem. metall.* (China), 1999, (pp. 92–102). Beijing: Science Press.

Li, X. J., & Mao, Z. -S. (2001). The effect of surfactant on the motion of a buoyancy-driven drop at intermediate Reynolds numbers: A numerical approach. *J. Colloid Interface Sci., 240*(1), 307–322.

Li, X. J., Mao, Z. -S., & Fei, W. Y. (2003). Effect of surface-active agents on mass transfer of a solute into single buoyancy driven drops in solvent extraction systems. *Chem. Eng. Sci., 58*(19), 3793–3806.

Li, X. Y., Wang, Y. F., Yu, G. Z., Yang, C., & Mao, Z. -S. (2008). A volume-amending method to improve mass conservation of level approach for incompressible two-phase flows. *Sci. China Ser. B-Chem., 51*(11), 1132–1140.

Liang, T. B., & Slater, M. J. (1990). Liquid–liquid extraction drop formation: Mass transfer and the influence of surfactant. *Chem. Eng. Sci., 45*(1), 97–105.

Lin, T. J., Reese, J., Hong, T., & Fan, L. S. (1996). Quantitative analysis and computation of two-dimensional bubble columns. *AIChE J., 42*(2), 301–318.

Lu, P., Wang, Z. H., Yang, C., & Mao, Z. -S. (2010). Experimental investigation and numerical simulation of mass transfer during drop formation. *Chem. Eng. Sci., 65*(20), 5517–5526.

Manga, M., & Stone, H. A. (1993). Buoyancy-driven interactions between two deformable viscous drops. *J. Fluid Mech., 256*(3), 647–683.

Mao, Z. -S., & Chen, J. Y. (1997). Numerical solution of viscous flow past a solid sphere with the control volume formulation. *Chinese J. Chem. Eng., 5*(2), 105–116.

Mao, Z. -S., & Chen, J. Y. (2002 a). Numerical approach to the motion and external mass transfer of a drop swarm by the cell model. *International solvent extraction conference (ISEC'2002)* (pp 227–232), Johannesburg, South Africa.

Mao, Z. -S., & Chen, J. Y. (2002 b). Numerical simulation of viscous flow through spherical particle assemblage with the modified cell model. *Chinese J. Chem. Eng., 10*(2), 149–162.

Mao, Z. -S., & Chen, J. Y. (2004). Numerical simulation of the Marangoni effect on mass transfer to single slowly moving drops in the liquid–liquid system. *Chem. Eng. Sci., 59*(8–9), 1815–1828.

Mao, Z. -S., & Chen, J. Y. (2005). An attempt to improve the cell model for motion and external mass transfer of a drop in swarms at intermediate Reynolds numbers. *International solvent extraction conference (ISEC'2005).* Beijing: China, A417.

Mao, Z. -S., & Wang, Y. F. (2003). Numerical simulation of mass transfer in a spherical particle assemblage with the cell model. *Powder Technol., 134*(1–2), 145–155.

Mao, Z. -S., & Yang, C. (2006). The cell model approach to solid particle assemblages: further justifications. *International conference of computational methods in sciences and engineering 2006* (ICCMSE 2006), Chania, Greece.

Mao, Z. -S., & Yang, C. (2009). Challenges in study of single particles and particle swarms. *Chinese J. Chem. Eng., 17*(4), 535–545.

Mao, Z. -S., Li, T. W., & Chen, J. Y. (2001). Numerical simulation of steady and transient mass transfer to a single drop dominated by external resistance. *Int. J. Heat Mass Transfer, 44*(6), 1235–1247.

Mao, Z. -S., Lu, P., Zhang, G. J., & Yang, C. (2008 a). Numerical approach to the three-dimensional Marangoni effect in a two liquid layer system. *International solvent extraction conference (ISEC'2008)*, Tucson, USA.

Mao, Z. -S., Lu, P., Zhang, G. J., & Yang, C. (2008 b). Numerical simulation of the Marangoni effect with interphase mass transfer between two planar liquid layers. *Chinese J. Chem. Eng., 16*(2), 161–170.

Mikulencak, D. R., & Morris, J. F. (2004). Stationary shear flow around fixed and free bodies at finite Reynolds number. *J. Fluid Mech.*, *520*, 215–242.

Ohta, M., Iwasaki, E., Obata, E., & Yoshida, Y. (2005). Dynamic processes in a deformed drop rising through shear-thinning fluids. *J. Non-Newtonian Fluid Mech.*, *132*(1–3), 100–107.

Oka, H., & Ishii, K. (1999). Numerical analysis on the motion of gas bubbles using level set method. *J. Phys. Soc. (Japan)*, *68*(3), 823–832.

Osher, S., & Sethian, J. A. (1988). Fronts propagating with curvature-dependent speed: Algorithms based on Hamilton–Jacobi formulations. *J. Comput. Phys.*, *143*(2), 495–518.

Pan, T. W., Joseph, D. D., Bai, R., Glowinski, R., & Sarin, V. (2002). Fluidization of 1204 spheres: Simulation and experiment. *J. Fluid Mech.*, *451*, 169–191.

Peskin, C. S. (2002). The immersed boundary method. *Acta Numerica*, *11*, 479–517.

Petera, J., & Weatherley, L. R. (2001). Modeling of mass transfer from falling drops. *Chem. Eng. Sci.*, *56*(19), 4929–4947.

Portela, L. M., & Oliemans, R. V. A. (2003). Eulerian–Lagrangian DNS/LES of particle–turbulence interactions in wall-bounded flows. *Int. J. Numer. Meth. Fluids*, *43*, 1045–1065.

Qi, D. (1999). Lattice-Boltzmann simulations of particles in non-zero-Reynolds-number flows. *J. Fluid Mech.*, *385*, 41–62.

Roghair, I., Lau, Y. M., Deen, N. G., Slagter, H. M., Baltussen, M. W., Van Sint Annaland, M., & Kuipers, J. A. M. (2011). On the drag force of bubbles in bubble swarms at intermediate and high Reynolds numbers. *Chem. Eng. Sci.*, *66*(14), 3204–3211.

Ryskin, G., & Leal, L. G. (1983). Orthogonal mapping. *J. Comput. Phys.*, *50*(1), 71–100.

Ryskin, G., & Leal, L. G. (1984 a). Numerical solution of free boundary problems in fluid mechanics: Part 1. The finite-difference technique. *J. Fluid Mech.*, *148*, 1–17.

Ryskin, G., & Leal, L. G. (1984 b). Numerical solution of free boundary problems in fluid mechanics: Part 2. Buoyancy-driven motion of a gas bubble, through a quiescent liquid. *J. Fluid Mech.*, *148*(1), 19–35.

Ryskin, G., & Leal, L. G. (1984 c). Numerical solution of free boundary problems in fluid mechanics: Part 3. Bubble deformation in an axisymmetric straining flow. *J. Fluid Mech.*, *148*(1), 37–43.

Sankaranarayanan, K., Shan, X., Kevrekidis, I. G., & Sundaresan, S. (1999). Bubble flow simulation with the lattice Boltzmann method. *Chem. Eng. Sci.*, *54*(21), 4817–4823.

Sankaranarayanan, K., & Sundaresan, S. (2002). Lift force in bubbly suspensions. *Chem. Eng. Sci.*, *57*(17), 3521–3542.

Sen Gupta, A., & Thodos, G. (1963). Direct analogy between mass and heat transfer to beds of spheres. *AIChE J.*, *9*, 751–754.

Shyy, W., Tong, S. S., & Correa, S. M. (1985). Numerical recirculating flow calculation using a body-fitted coordinate system. *Numer. Heat Transfer*, *8*(1), 99–113.

Sternling, C. V., & Scriven, L. E. (1959). Interfacial turbulence: Hydrodynamic instability and the Marangoni effect. *AIChE J.*, *5*(4), 514–523.

Subramanian, G., & Koch, D. L. (2006). Inertial effects on the transfer of heat or mass from neutrally buoyant spheres in a steady linear velocity field. *Phys. Fluids*, *18*, 073302.

Sussman, M., Smereka, P., & Osher, S. (1994). A level set approach for computing solutions to incompressible two-phase flow. *J. Comput. Phys.*, *114*(1), 146–159.

Tal (Thau), R., & Sirignano, W. A. (1982). Cylindrical cell model for the hydrodynamics of particle assemblages at intermediate Reynolds numbers. *AIChE J.*, *28*(2), 233–237.

Thames, F. C., Thompson, J. F., Mastin, C. W., & Walker, R. L. (1977). Numerical solutions for viscous and potential flow about arbitrary two-dimensional bodies using body-fitted coordinate systems. *J. Comput. Phys.*, *24*(3), 245–273.

Tomiyama, A. (1998). Struggle with computational bubble dynamics. *Multiphase Sci. Technol.*, *10*(4), 369–405.

Tomiyama, A., Zun, I., & Sou, A. (1993). Numerical analysis of bubble motion with the VOF method. *Nucl. Eng. Des.*, *141*(1–2), 69–82.

Unverdi, S. O., & Tryggvason, G. (1992). A front-tracking method for viscous, incompressible, multi-fluid flows. *J. Comput. Phys.*, *100*(1), 25–37.

Wang, J. F., Yang, C., & Mao, Z. -S. (2006). A simple weighted integration method for calculating surface tension force to suppress parasitic flow in the level set approach. *Chinese J. Chem. Eng.*, *14*(6), 740–746.

Wang, J. F., Lu, P., Wang, Z. H., Yang, C., & Mao, Z. -S. (2008 a). Numerical simulation of unsteady mass transfer by the level set method. *Chem. Eng. Sci.*, *63*, 3141–3151.

Wang, J. F., Yang, C., & Mao, Z. -S. (2008 b). Numerical simulation of Marangoni effects of single drops induced by interphase mass transfer in liquid–liquid extraction systems by the level set method. *Sci. China Ser. B-Chem.*, *51*(7), 684–694.

Wang, J. F., Wang, Z. H., Lu, P., Yang, C., & Mao, Z. -S. (2011). Numerical simulation of the Marangoni effect on transient mass transfer from single moving deformable drops. *AIChE J.*, *57*(10), 2670–2683.

Wang, T. (2011). Simulation of bubble column reactors using CFD coupled with a population balance model. *Frontiers Chem. Sci. Eng.*, *5*(2), 162–172.

Wang, T., Cheng, J. C., Li, X. Y., Yang, C., & Mao, Z. -S. (2013). Numerical simulation of a pitched-blade turbine stirred tank with mirror fluid method. *Can. J. Chem. Eng.*, *91*(5), 902–914.

Wang, Y. F., Yang, C., Mao, Z. -S., & Chen, J. Y. (2004). Application of the level set approach for numerical simulations of two-phase flow (in Chinese). *Prog. Nat. Sci.*, *14*(2), 220–222.

Wang, Z. H., Lu, P., Zhang, G. J., Yong, Y. M., Yang, C., & Mao, Z. -S. (2011). Experimental investigation of Marangoni effect in 1-hexanol/water system. *Chem. Eng. Sci.*, *66*, 2883–2887.

Wang, Z. H., Lu, P., Wang, Y., Yang, C., & Mao, Z. -S. (2013). Experimental investigation and numerical simulation of Marangoni effect induced by mass transfer during drop formation. *AIChE J.*, (doi: 10.1002/aic.14161).

Wegener, M., Eppinger, T., Bäumler, K., Kraume, M., Paschedag, A. R., & Bänsch, E. (2009). Transient rise velocity and mass transfer of a single drop with interfacial instabilities: Numerical investigations. *Chem. Eng. Sci.*, *64*(23), 4835–4845.

Yang, C., & Mao, Z. -S. (2002). An improved level set approach to the simulation of drop and bubble motion. *Chinese J. Chem. Eng.*, *10*(3), 263–272.

Yang, C. & Mao, Z. -S. (2003). Numerical simulation of viscous flow of a non-Newtonian fluid past an irregular solid obstacle by the mirror fluid method. In P.J. Witt & M.P. Schwarz, (Eds.), *Proceedings of the third international conference on CFD in the minerals and process industries* (pp. 391–396), Australia.

Yang, C., & Mao, Z. -S. (2005 a). Mirror fluid method for numerical simulation of sedimentation of a solid particle in a Newtonian fluid. *Phys. Rev. E.*, *71*, 036704.

Yang, C., & Mao, Z. -S. (2005 b). Numerical simulation of interphase mass transfer with the level set approach. *Chem. Eng. Sci.*, *60*(10), 2643–2660.

Yang, C., Lu, P., Mao, Z. -S., Yu, G. Z., & Zhang, G. J. (2005). Numerical simulation of two-phase flow during drop formation stages. *International solvent extraction conference (ISEC'2005)*, Beijing, China.

Yang, C., Zhang, J. S., Koch, D. L., & Yin, X. L. (2011). Mass/heat transfer from a neutrally buoyant sphere in simple shear flow at finite Reynolds and Peclet numbers. *AIChE J.*, *57*(6), 1419–1433.

Zapryanov, Z., & Tabakova, S. (1999). *Dynamics of bubbles, drops and rigid particles.* Dordrecht: Kluwer Academic (pp. 338–343).

Zhang, J. S., Yang, C., & Mao, Z. -S. (2012 a). Mass and heat transfer from or to a single sphere in simple extensional creeping flow. *AIChE J., 58*(10), 3214–3223.

Zhang, J. S., Yang, C., & Mao, Z. -S. (2012 b). Unsteady conjugate mass transfer from a spherical drop in simple extensional creeping flow. *Chem. Eng. Sci., 79*, 29–40.

Zhang, L., Yang, C., & Mao, Z. -S. (2010). Numerical simulation of a bubble rising in shear-thinning fluids. *J. Non-Newtonian Fluid Mech., 165*(11–12), 555–567.

Zhang, Y. H., Yang, C., & Mao, Z. -S. (2008). Large eddy simulation of the gas–liquid flow in a stirred tank. *AIChE J., 54*(8), 1963–1974.

Zinchenko, A. Z., & Davis, R. H. (2000). An efficient algorithm for hydrodynamical interaction of many deformable drops. *J. Comput. Phys., 157*(2), 539–587.

Multiphase stirred reactors

3.1 INTRODUCTION

Stirred tanks are the most widely used type of reactor in the process industry. A stirred tank typically contains one or more impellers mounted on a shaft, sometimes baffles, and other internals such as spargers, coils and draft tubes. Numerous parameters like tank and impeller shapes, the aspect ratio of the tank, the number, type, location and size of impellers, the degree of baffling, etc. provide unmatched flexibility and control over the performance of stirred reactors and also present great challenges to the design and scale-up (or scale-down) of such reactors. Thus, it is essential to understand the underlying physics and the quantitative relationships between the reactor parameters and the design objectives.

Although various measurement techniques for multiphase reactors have been developed, including invasive and non-invasive techniques, they all have some limitations (Li et al., 2012). For example, invasive measuring techniques always disturb the flow field and non-invasive techniques become ineffective in the case of nearly industrial operating conditions (particular physicochemical characteristics, opaque walls, high gas holdups or solid concentrations, etc.) and are often too expensive. CFD (computational fluid dynamics) is a body of knowledge and techniques to solve mathematical models of fluid dynamics on digital computers. CFD has been identified as one of the critical enabling technologies for the chemical industry. There are some potential applications of CFD modeling of stirred reactors: (1) resolving conflicting processing requirements; (2) translating batch data for continuous reactors; (3) scale-down/scale-up analysis; (4) testing of new reactor concepts; (5) development of theories for heat and mass transfer (Joshi et al., 2011). Therefore, much effort has been devoted to numerically resolving the hydrodynamics in stirred tanks.

Up to now, much progress has been achieved with respect to single-phase liquid flow. However, multiphase flow and transport in stirred tanks should be paid more attention because these conditions are more frequent and important in industry, and also more complex, leading to more difficulties in design. In most work, the Eulerian multi-fluid approach and the $k-\varepsilon$ turbulence model (standard, RNG, or realizable $k-\varepsilon$ model) are usually employed to simulate multiphase flow (Paul et al., 2004). The $k-\varepsilon$ turbulence model based on isotropy (Boussinesq hypothesis) is sufficiently accurate in predicting flow characteristics around the impeller region. Assuming that the turbulent viscosity is anisotropic, the Reynolds stress model computes the stresses individually. For 2D models this amounts to four additional transport equations, and for 3D models six additional transport equations are required. As computer capacity and speed have increased rapidly during recent years, use of the Reynolds stress turbulence model has become more widespread, giving rise to improved accuracy over other RANS (Reynolds average Navier–Stokes)-based turbulence models, as has been proved with closer matching to experimental results for a number of applications. Large eddy simulation (LES) recognizes that turbulent eddies occur on many scales in a flow field. With the LES model, the continuity and momentum equations are filtered prior to being solved in a transient fashion. As a consequence of the use of finer resolution in the space and time domains, the accuracy of numerical simulation of turbulent flow in stirred tanks is significantly improved.

In the remainder of this chapter, mathematical models and numerical methods including governing equations, interphase momentum exchange, RANS methods such as the k–ε model and EASM (explicit algebraic stress model), LES model, impeller treatment, methods to deal with axial flow impellers and some other numerical details are first considered. Then CFD simulations of multiphase flow in stirred tanks are considered in a sequence from two-phase systems (solid–liquid, gas–liquid, and liquid–liquid) to three-phase systems (liquid–liquid–solid, gas–liquid–liquid, liquid–liquid–liquid, and gas–liquid–solid). Finally, the present status of CFD simulations of multiphase flow in stirred tanks is summarized and possible suggestions are given for future work in the area of CFD modeling of stirred reactors.

3.2 MATHEMATICAL MODELS AND NUMERICAL METHODS

Two-phase flow is a simple but typical case in multiphase systems. Compared with single-phase systems, the flow becomes more complex if a second phase is introduced. Therefore, the mathematical treatment of multiphase flow becomes more complex, involving the crucial problem of describing dispersed phase dynamics, especially in high holdup situations. In general, there are two approaches to simulate multiphase flow: Eulerian–Eulerian and Eulerian–Lagrangian approaches. The Eulerian–Eulerian approach, which is also called the two-fluid model, considers both the continuous phase and the dispersed phase as interacting interpenetrating continua. The resulting governing equations of the dispersed phase are solved in an Eulerian frame analogously to the continuous phase. In contrast, the Eulerian–Lagrangian method describes the continuous phase by Eurlerian equations but treats the dispersed phase as a large number of individual particles. Lagrangian tracking of a particle using Newton's second law of motion allows the interactions between two phases or particle–particle interactions to be easily and appropriately dealt with. However, the computational requirements of particle tracking are greatly dependent on the number of particles, which limits the use of this approach to low dispersed phase holdups only. Therefore, the Eulerian–Eulerian approach is often preferred for its lower computational need, faster numerical resolution, and especially its capability to handle high dispersed phase loading conditions.

As for the modeling of turbulence, direct numerical simulation (DNS), large eddy simulation (LES), and Reynolds average Navier–Stokes (RANS) models are commonly encountered and these various approaches (relative to scale resolved) can be best represented on an energy spectrum, as shown in Figure 3.1 (Joshi et al., 2011). The DNS or LES is a straightforward way to understand turbulence. The DNS directly solves the time-resolved and full-length scale Navier–Stokes equation without any models or assumptions. Accurate flow field and time-dependent flow information can be obtained at the expense of huge computational cost. The computational requirements increase so rapidly with Reynolds number that the applicability of this approach is limited to flow with low or moderate Reynolds numbers. In the LES, the turbulent motion is decomposed into two scales, large-scale and small-scale eddies, by filtering the Navier–Stokes equation. The large-scale motions are directly solved, whereas the effects of the smaller-scale motions are modeled using a subgrid scale stress. Compared with the DNS, the LES can be used to simulate flow with high Reynolds numbers, because the vast computational cost of the small-scale motions in the DNS is avoided in the LES. However, in contrast to the RANS model, the great computational cost of the LES is still a major limitation in industrial applications. The RANS model averages the unsteady equation by time scale, producing some unknown fluctuation terms. In order to close the governing equations, some assumptions or models should be adopted. In practical engineering applications, there is a greater focus on mean flow characteristics rather than the detailed generation and evolution of turbulence. Therefore, the RANS model combined with the Eulerian–Eulerian approach is commonly used to simulate multiphase flow in industrial reactors.

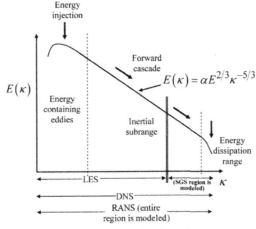

FIGURE 3.1 CFD simulation using RANS, LES, and DNS models: relative scale resolutions in a typical energy spectrum (Joshi et al., 2011).

3.2.1 Governing equations

Based on the Eulerian–Eulerian two-fluid model, the Reynolds-averaged continuity and momentum equations for phase k are written as

$$\frac{\partial(\rho_k \alpha_k)}{\partial t} + \frac{\partial(\rho_k \alpha_k u_{kj} + \rho_k \overline{\alpha_k' u_{kj}'})}{\partial x_j} = 0 \tag{3.1}$$

$$\frac{\partial(\rho_k \alpha_k u_{ki})}{\partial t} + \frac{\partial(\rho_k \alpha_k u_{ki} u_{kj})}{\partial x_j} = -\alpha_k \frac{\partial P}{\partial x_i} + \frac{\partial(\alpha_k \overline{\tau_{kij}})}{\partial x_j} + F_{ki} + \rho_k \alpha_k g_i -$$

$$\rho_k \frac{\partial}{\partial x_j} (\alpha_k \overline{u_{kj}' u_{ki}'} + u_{ki} \overline{\alpha_k' u_{kj}'} + u_{kj} \overline{\alpha_k' u_{ki}'} + \overline{\alpha_k' u_{kj}' u_{ki}'}) \tag{3.2}$$

$$\sum \alpha_k = 1.0 \tag{3.3}$$

where α_k is the phase volume fraction.

The correlation term of phase holdup and velocity fluctuations $\overline{\alpha_k' u_{ki}'}$ in both continuity and momentum equations represents the transport of both mass and momentum by dispersion. Since the influence of the dispersed phase on turbulence structure is not well understood, a simple gradient assumption can be adopted to model $\overline{\alpha_k' u_{ki}'}$, which is given by

$$\overline{u_{ki}' \alpha_k'} = -\frac{v_{k,t}}{\sigma_t} \frac{\partial \alpha_k}{\partial x_i} \tag{3.4}$$

where σ_t is the turbulent Schmidt number for the phase dispersion. The value of this number depends on the size of the dispersed phase and the scale of turbulence. It was found that the simulation results were sensitive to σ_t in solid–liquid flow simulation and a value between 1.0 and 2.0 was suggested (Shan et al., 2008). In gas–liquid systems, the value of 1.0 was recommended (Ranade and Van den Akker, 1994), but Wang and Mao (2002) suggested a value of 1.6 was suitable.

3.2.2 Interphase momentum exchange

The interphase coupling terms make multiphase flows fundamentally different from single-phase flows. For a dispersed two-phase flow, there are mainly four kinds of interphase forces that should be considered, namely drag force, virtual mass force, Basset force, and lift force. Basset force arises as a result of the development of a boundary layer around bubbles. It is only relevant to unsteady flows involving a history integral. In most cases, the magnitude of the Basset force is much smaller than that of the interphase drag force. Lift force results from the vorticity and shear in the continuous flow field. It is proportional to the vector product of the slip velocity and the curl of the continuous liquid velocity. If the velocity gradients are large, lift force will be significant. An order of magnitude analysis indicates that the magnitude of lift force is also much smaller in contrast to drag force. Thus, Basset force and lift force are often ignored in two-phase flow simulations.

When a secondary phase accelerates relative to the primary phase, virtual mass force needs to be considered. In gas–liquid stirred tanks, where the gas-phase density is much smaller than the liquid-phase density, this force becomes much more significant. Many reports indicate that the virtual mass force has little effect on the flow field in the bulk region (Ljungqvist and Rasmuson, 2001), but cannot be neglected in the region near the impeller fringes (Zhang et al., 2013). The virtual mass force is modeled as follows:

$$F_{VM} = -C_V \rho_c \alpha_d \left(\frac{Du_d}{Dt} - \frac{Du_c}{Dt} \right)$$ (3.5)

where C_V is the virtual mass coefficient and is commonly set to 0.5.

The drag force results from the velocity slip between the continuous phase and the dispersed phase. A simple expression is often given in some early works (Ranade et al., 2001) for a gas–liquid system:

$$F_{di,drag} = -F_{ci,drag} = C_f \alpha_c \alpha_d (u_{di} - u_{ci})$$ (3.6)

where C_f is equal to 5×10^4 obtained from an experimental method (Schwarz and Turner, 1989). In general cases, the drag force is commonly expressed as

$$F_{ci,drag} = \frac{3 \rho_d \alpha_c \alpha_d C_D |\mathbf{u}_d - \mathbf{u}_c| (u_{di} - u_{ci})}{4 d_d}$$ (3.7)

where C_D is the drag force coefficient and d_d is the equivalent diameter of particles (bubbles or droplets). There are various empirical correlations for the drag force coefficient due to differences in experimental systems, methods and ranges of parameters, etc. Thus, an appropriate drag correlation is crucial to simulate multiphase flows. A classical model for C_D in a stagnant fluid is given by Clift et al. (1978):

$$\begin{aligned} C_D &= \frac{24(1 + 0.15 Re_d^{0.687})}{Re_d} \quad \text{(for } Re_d < 1000) \\ C_D &= 0.44 \quad \text{(for } Re_d \geq 1000) \end{aligned}$$ (3.8)

This model is widely employed in simulation of solid–liquid (Wang, F. et al., 2004a; Micale et al., 2004; Guha et al., 2008), gas–liquid (Deen et al., 2002), and liquid–liquid (Laurenzi et al., 2009) flows in stirred tanks.

In turbulent two-phase stirred tanks, the prevailing bulk turbulence has a great influence on the drag force coefficient. Bakker and van den Akker (1994) employed a modified Reynolds number to relate the influence of turbulence on drag coefficient in a usual correlation developed for stagnant liquid. The correlation is expressed as

$$C_D = \frac{24(1 + 0.15(Re^*)^{0.687})}{Re^*}$$ (3.9)

with

$$Re^* = \frac{\rho_c U_{slip} d_d}{\mu_c + \frac{2}{9} \mu_t}$$

The effective viscosity is calculated by adding some fraction of turbulent viscosity, where the fraction is an adjustable parameter and the value of 2/9 is recommended.

In another approach to the effect of turbulence, Brucato et al. (1998a) found that the magnitude of the effect of the prevailing bulk turbulence increased with both particle size and mean turbulent energy dissipation rate. Taking this effect into account, a drag coefficient correction factor is given by

$$\frac{C_D - C_{D0}}{C_{D0}} = K \left(\frac{d_d}{\lambda} \right)^3, \qquad \lambda = \left(\frac{v_c^3}{\varepsilon_c} \right)^{0.25} \tag{3.10}$$

in which C_D is the drag coefficient in a turbulent liquid, C_{D0} is the drag coefficient in a stagnant fluid, d_d is the bubble/particle diameter, and λ is the Kolmogorov length scale. Experimental data of Brucato et al. (1998a) indicate that only microscale turbulence affected the particle drag and a correction parameter K of 8.76×10^{-4} was recommended. Khopkar et al. (2006a) considered that the correlation proposed by Brucato et al. (1998a) was not universal because it was obtained in a Taylar–Couette apparatus, in which the distribution of energy dissipation rates is quite different from that in stirred vessels. They developed a unit cell approach to model single-phase flow through regularly arranged cylindrical objects to investigate the effect of turbulence on drag coefficient. A much lower proportionality constant of 8.76×10^{-5} was found for solid–liquid systems (Khopkar et al., 2006a) and 6.5×10^{-6} for gas–liquid flow (Khopkar and Ranade, 2006). Another method used by Lane et al. (2005) is to correlate the available data of setting velocities of particles and rise velocities of bubbles in a turbulent flow with particle relaxation time (τ_p) and integral time scale of turbulence (T_L) as

$$\frac{C_D}{C_{D0}} = \left[1 - 1.4 \left(\frac{\tau_p}{T_L} \right)^{0.7} \exp \left(-0.6 \frac{\tau_p}{T_L} \right) \right]^{-2} \tag{3.11}$$

3.2.3 **RANS method**

When the RANS method is employed, the velocity fluctuation correlation term $\overline{u'_{ki} u'_{kj}}$, namely the Reynolds stress, will appear. For the closure of momentum equations, this term should be treated by involving known or calculable quantities. This is done through various turbulence models.

3.2.3.1 *k–ε model*

The k–ε model is widely known and popularly used for single-phase flows. Referring to the two-phase k–ε model, the Reynolds stresses are also treated based on the Boussinesq gradient hypothesis as in the single-phase k–ε model:

$$\overline{u'_{ki} u'_{kj}} = \tfrac{2}{3} k \delta_{ij} - v_{k,t} \left(\frac{\partial u_{kj}}{\partial x_i} + \frac{\partial u_{ki}}{\partial x_j} \right) \tag{3.12}$$

Due to the different methods of calculating turbulence viscosity, turbulence kinetic energy, and dissipation rate, three expansions of the standard k–ε model to multiphase flows, namely each phase model, mixture model and dispersed model, are often encountered.

The each phase model is rigorous and complex when treating multiphase turbulence. Based on the Eulerian–Eulerian method, the dispersed phase is deemed to be a continuum with pseudo-fluid properties. It is hence reasonable to derive the turbulence equations of k and ε in the light of the single-phase theory. The turbulence viscosities are calculated by solving the k and ε transport equations of both phases. This model is more suitable for dense multiphase systems, as the interface between the continuous phase and the dispersed phase is not obvious. For example, the k–ε–k_p model is an example of an each phase model (Zhou et al., 1994). The turbulence viscosity of the dispersed phase is written as

$$V_{pt} = C_{\mu p} \alpha_p \frac{k_p^2}{\varepsilon_p} \tag{3.13}$$

The k and ε equations of the dispersed phase are given by

$$\frac{\partial}{\partial x_j}(\rho_p \alpha_p u_{pj} k_p) = \frac{\partial}{\partial x_j}\left(\frac{V_{pt}\rho_p}{\sigma_{kp}}\frac{\partial k_p}{\partial x_j}\right) + G_{kp} - \rho_p \alpha_p \varepsilon_p \tag{3.14}$$

$$\varepsilon_p = -\frac{2}{\tau_{rg}}\left[\left(C_g^k \sqrt{kk_p} - k_p\right) - (V_{1i} - V_{pi})\frac{V_{pt}}{\sigma_{kp}}\frac{\partial \alpha_p}{\partial x_i}\right] \tag{3.15}$$

The mixture model is also called the homogeneous model, in which the continuous phase and the dispersed phase share the same values of turbulence quantities. The velocity and properties of the mixture are adopted for calculation. The mixture k and ε equations can be written as

$$\frac{\partial}{\partial t}(\rho_m k) + \nabla \cdot (\rho_m \mathbf{u}_m k) = \nabla \cdot \left(\frac{\mu_{t,m}}{\sigma_k}\nabla k\right) + G_m - \rho_m \varepsilon \tag{3.16}$$

$$\frac{\partial}{\partial t}(\rho_m \varepsilon) + \nabla \cdot (\rho_m \mathbf{u}_m \varepsilon) = \nabla \cdot \left(\frac{\mu_{t,m}}{\sigma_\varepsilon}\nabla \varepsilon\right) + (C_1 G_m - C_2 \rho_m \varepsilon)\frac{\varepsilon}{k} \tag{3.17}$$

The velocity and density of the mixture are given by

$$\mathbf{u}_m = \frac{\displaystyle\sum_{k=1}^{N}\alpha_k \rho_k \mathbf{u}_k}{\displaystyle\sum_{k=1}^{N}\alpha_k \rho_k} \tag{3.18}$$

$$\rho_m = \sum_{k=1}^{N}\alpha_k \rho_k \tag{3.19}$$

The viscosity and turbulence generation of the mixture are written as

$$\mu_{t,m} = \rho_m C_\mu \frac{k^2}{\varepsilon} \tag{3.20}$$

$$G_m = \mu_{t,m} (\nabla \mathbf{u}_m + \nabla \mathbf{u}_m^T) : \nabla \mathbf{u}_m \tag{3.21}$$

The dispersed model is intermediate between the each phase model and the mixture model, in which the k and ε equations of the continuous phase are solved. The turbulence viscosity of the dispersed phase is given in terms of that of the continuous phase. Also, the influence of the dispersed phase on the continuous phase turbulence is considered by including a source term in the turbulence equations. Based on the Hinze–Tchen theory, a simple A_p model (or k–ε–A_p model) is given by

$$\frac{v_{dt}}{v_{ct}} = \left(\frac{k_d}{k_c} \right)^2 = \left(1 + \frac{\tau_p}{\tau_1} \right)^{-1} \tag{3.22}$$

$$\tau_1 = \frac{k}{\varepsilon} \tag{3.23}$$

$$\tau_p = \frac{\rho_d d_d^2}{18\mu_{c,lam}} \tag{3.24}$$

where τ_1 is the mean eddy lifetime and τ_p is the particle response time. Another way to express the turbulence viscosity of the dispersed phase involves the velocity fluctuation of each phase as follows (Grienberger and Hofmann, 1992):

$$\mu_{d,t} = \mu_{c,t} \frac{\rho_d \overline{u'_{d,i} u'_{d,i}}}{\rho_c \overline{u'_{c,i} u'_{c,i}}} \tag{3.25}$$

Gosman et al. (1992) proposed a correlation of u'_d to u'_c derived from a Lagrangian analysis of particle response to eddies that are much larger than the particle diameter:

$$u'_{d,i} = u'_{c,i} \left[1 - \exp\left(-\frac{\tau_1}{\tau_p} \right) \right] \tag{3.26}$$

where $\tau_1 = 0.41 k/\varepsilon$ is the mean eddy lifetime, and τ_p is the particle response time obtained by Lagrangian integration of the motion equation of a swarm of particles moving through a fluid eddy of given velocity distribution with the expression:

$$\tau_p = \frac{4\rho_d d_d}{3\rho_c C_D \alpha_d |\mathbf{u}_d - \mathbf{u}_c|} \tag{3.27}$$

There are also some simplified methods to describe the turbulence viscosity of the dispersed phase. For example, Schwarz and Turner (1989) made the gas-phase turbulence dynamic viscosity equal to the liquid-phase viscosity in their simulation of gas–liquid bubble flow:

$$v_{gt} = v_{lt} \tag{3.28}$$

or

$$\mu_{gt} = \mu_{lt} \frac{\rho_g}{\rho_l} \tag{3.29}$$

To treat the turbulent two-phase flow rigorously, the turbulent model adopted should include interphase turbulence transfer terms accounting for turbulence promotion or damping due to the presence of the dispersed phase. However, there is no reliable information on such terms, and a proper turbulence model for turbulent multiphase systems has not been found. In multiphase stirred tanks, the turbulence is mainly attributed to velocity fluctuation of the liquid phase because the holdup of the dispersed phase is often quite low in most parts of the tank. The dispersed phase can affect the turbulence of the system via interphase momentum exchange. For the dispersed k–ε model, the k and ε equations can be written in a general form as

$$\frac{\partial}{\partial t}(\rho_c \alpha_c k) + \frac{\partial}{\partial x_i}(\rho_c \alpha_c u_{ci} k) = \frac{\partial}{\partial x_i}\left(\alpha_c \frac{\mu_{ct}}{\sigma_k}\frac{\partial k}{\partial x_i}\right) + \frac{\partial}{\partial x_i}\left(k \frac{\mu_{ct}}{\sigma_k}\frac{\partial \alpha_c}{\partial x_i}\right) + S_k \tag{3.30}$$

$$\frac{\partial}{\partial t}(\rho_c \alpha_c \varepsilon) + \frac{\partial}{\partial x_i}(\rho_c \alpha_c u_{ci} \varepsilon) = \frac{\partial}{\partial x_i}\left(\alpha_c \frac{\mu_{ct}}{\sigma_\varepsilon}\frac{\partial \varepsilon}{\partial x_i}\right) + \frac{\partial}{\partial x_i}\left(\varepsilon \frac{\mu_{ct}}{\sigma_\varepsilon}\frac{\partial \alpha_c}{\partial x_i}\right) + S_\varepsilon \tag{3.31}$$

where $\sigma_k = 1.3$ and $\sigma_\varepsilon = 1.0$. The source terms in the above equations are

$$S_k = \alpha_c[(G + G_e) - \rho_c \varepsilon] \tag{3.32}$$

$$S_\varepsilon = \alpha_c \frac{\varepsilon}{k}[C_1(G + G_e) - C_2 \rho_c \varepsilon] \tag{3.33}$$

where G is the turbulent generation and G_e is the extra production term due to the dispersion phase. Based on the analysis of Kataoka et al. (1992), G_e is mainly dependent on the drag force between the continuous phase and the dispersed phase:

$$G = -\rho_c \alpha_c \overline{u'_{ci} u'_{cj}} \frac{\partial u_{ci}}{\partial x_j} \tag{3.34}$$

$$G_e = \sum_d C_b |\mathbf{F}| \left(\sum (u_{di} - u_{ci})^2\right)^{0.5} \tag{3.35}$$

where C_b is an empirical coefficient. When $C_b = 0$, the energy induced by the dispersed phase dissipates at the interface and has no influence on the turbulent kinetic energy of the continuous phase. According to the analysis in the literature, the value of C_b has always been set as 0.02. The other model constants are well accepted: $C_\mu = 0.09$, $C_1 = 1.44$, and $C_2 = 1.92$.

3.2.3.2 EASM

Although the k–ε model is robust, economical, and has served the engineering community widely for many years, it is not recommended for highly swirling flows such as rotational flow in stirred tanks. In order to account for the anisotropy of turbulence, the Reynolds stress model (RSM) and algebraic stress model (ASM) are often encountered and perform well in single-phase flows. Since Reynolds stress components are directly solved from a differential equation or an algebraic equation rather than modeling using an isotropic hypothesis like the k–ε model, anisotropic turbulence can be successfully predicted. Unfortunately, neither the RSM nor ASM are computationally robust and it is difficult to get convergent solutions. Pope (1975) proposed an explicit algebraic stress model (EASM) for two-dimensional flows, which was derived from the RSM by using a tensor polynomial expansion theory. The Reynolds stress components are expressed as an explicit algebraic correlation of mean strain rate tensor, rotation rate tensor, and turbulence characteristic quantities. As a result, the computational stability is improved and the computational cost is greatly reduced. Following Pope's theory, the EASMs for three-dimensional flows have been developed by Gatski and Speziale (1993) and Wallin and Johansson (2000). Recently, Feng et al. (2012a) employed Wallin and Johansson's EASM to successfully simulate single-phase turbulent flow in stirred tanks. Extensively quantitative comparisons of EASM predictions with experimental data and other results using different turbulence models like the k–ε model, ASM, RSM, and LES were conducted in order to assess the comprehensive performance of the EASM. Satisfactory agreement with experimental data was found and the superiority of the EASM among turbulence models was demonstrated. It is concluded that the EASM can become an alternative tool for turbulence modeling of industrial stirred tanks.

As for the anisotropic turbulence models for multiphase flow, a two-phase Reynolds stress model, also known as a second-order moment model, is commonly used. In this model, the turbulence viscosity of each phase is a tensor rather than a constant. The Reynolds stresses of each phase are solved directly by establishing Reynolds stress differential equations. In addition, the interphase interaction terms appearing in the differential equations also need to be closed, which is a specific problem different from the single-phase model. However, the two-phase RSM is quite complex and requires huge computational effort, especially in 3D simulations. Based on the two-phase RSM, Feng et al. (2012b) developed a two-phase EASM to simulate solid–liquid flow in stirred tanks.

The Reynolds stress tensor for two-phase systems can first be simplified by defining the Reynolds stress anisotropy tensor a_{qij} as

$$a_{qij} = \frac{\overline{u'_{qi}u'_{qj}}}{k} - \frac{2}{3}\delta_{ij} \tag{3.36}$$

Then the Reynolds stress anisotropy tensor can be expressed as a sum of expansion bases multiplying expansion coefficients. Following Wallin and Johansson's theory, a_{qij} in the inertial reference frame is given as

$$a_{qij} = \beta_1 S_{ij} + \beta_3 (\omega_{il}\omega_{lj} - \tfrac{1}{3}\eta_2 I) + \beta_4 (S_{il}\omega_{lj} - \omega_{il}S_{lj}) +$$
$$\beta_6 (S_{il}\omega_{lm}\omega_{mj} + \omega_{il}\omega_{lm}S_{mj} - \tfrac{2}{3}\eta_4 I) + \beta_9 (\omega_{il}S_{lm}\omega_{mn}\omega_{nj} - \omega_{il}\omega_{lm}S_{mn}\omega_{nj}) \quad (3.37)$$

S_{ij} and ω_{ij} are the normalized mean strain rate tensor and the mean rotation rate tensor respectively, which are also defined as two-phase forms:

$$S_{ij} = \frac{1}{2}\frac{k}{\varepsilon}\left(\frac{\partial u_{qi}}{\partial x_j} + \frac{\partial u_{qj}}{\partial x_i}\right), \qquad \omega_{ij} = \frac{1}{2}\frac{k}{\varepsilon}\left(\frac{\partial u_{qi}}{\partial x_j} - \frac{\partial u_{qj}}{\partial x_i}\right) \quad (3.38)$$

where k and ε are the shared values of both phases.

For the non-inertial frame, the Reynolds stress anisotropy tensor can be written as

$$a_{qij} = \beta_1 S_{ij} + \beta_3 (W_{il}W_{lj} - \tfrac{1}{3}\eta_2 I) + \beta_4 (S_{il}W_{lj} - W_{il}S_{lj}) +$$
$$\beta_6 (S_{il}W_{lm}W_{mj} + W_{il}W_{lm}S_{mj} - \tfrac{2}{3}\eta_4 I) + \beta_9 (W_{il}S_{lm}W_{mn}W_{nj} - W_{il}W_{lm}S_{mn}W_{nj}) \quad (3.39)$$

where W_{ij} is the absolute rotation rate tensor:

$$W_{ij} = \omega_{ij} + \frac{k}{\varepsilon}c_w \varepsilon_{jil}\Omega_l \quad (3.40)$$

where c_w is a parameter related directly to the pressure–strain rate model and equals 3.25 when the model of Launder, Reece, and Rodi (LRR) (Launder et al., 1975) is employed (Wallin and Johansson, 2000). Ω_l is the constant angular rotation rate vector of the non-inertial frame. The expansion β coefficients are functions of five independent invariants of S_{ij} and W_{ij}, which are given by

$$\eta_1 = S_{ij}S_{ji}, \quad \eta_2 = W_{ij}W_{ji}, \quad \eta_3 = S_{ij}S_{jl}S_{li}, \quad \eta_4 = S_{ij}W_{jl}W_{li}, \quad \eta_5 = S_{ij}S_{jl}W_{lm}W_{mi} \quad (3.41)$$

For three-dimensional flows, the β coefficients are

$$\beta_1 = -\frac{N_c(2N_c^2 - 7\eta_2)}{Q}, \quad \beta_3 = -\frac{12N_c^{-1}\eta_4}{Q}, \quad \beta_4 = -\frac{2(N_c^2 - 2\eta_2)}{Q}$$
$$\beta_6 = -\frac{6N_c}{Q}, \quad \beta_9 = \frac{6}{Q}, \quad Q = \tfrac{5}{6}(N_c^2 - 2\eta_2)(2N_c^2 - \eta_2) \quad (3.42)$$

where N_c is a key parameter related closely to the production-to-dissipation ratio. The solution of N_c is quite complex, involving a six-order equation for three-dimensional flows. For simplification purposes, N_c is first calculated in a two-dimensional frame by solving a cubic nonlinear equation. Then a correction of N_c is conducted by making a perturbation solution. The procedure for solving N_c can be found in Wallin and Johansson (2000).

For the two-phase EASM, the transport equations of k and ε are also solved only for the continuous phase, which is similar to the "dispersed" k–ε model, and is written as

$$\frac{\partial}{\partial t}(\rho_c\alpha_c\phi) + \frac{\partial}{\partial x_i}(\rho_c\alpha_c u_{ci}\phi) = \frac{\partial}{\partial x_i}\left(\rho_c\alpha_c c_\phi \frac{k^2}{\varepsilon}\frac{\partial\phi}{\partial x_i}\right) + S_\phi \quad (3.43)$$

in which $c_k = 0.25$ and $c_\varepsilon = 0.15$ were suggested by Launder et al. (1975) and the source terms are the same as those given in the k–ε model.

3.2.4 LES model

Initially, the objectives of computational studies were focused on macroscopic hydrodynamics such as the gross flow parameters and time-averaged fields. However, during the past 10 years, deeper understanding has been pursued, such as the characterization of turbulent flow, flow instabilities, and tailoring the impeller design so as to get the desired flow field. It is well known that the $k-\varepsilon$ model assumes isotropic turbulence, and therefore anisotropic models such as the RSM and LES model are recommended for simulation of complex three-dimensional flows. However, the RSM has shortcomings such as non-universal model parameters and numerical difficulties, and is computationally expensive by an order of magnitude as compared to the $k-\varepsilon$ model. Furthermore, the RSM does not capture the transient nature of the flow. This limitation is overcome using the LES approach. During the last few years, the ability to resolve all but the smallest turbulence scales using LES has become more viable. LES can potentially produce more accurate results by modeling only the smallest scales, which is more isotropic, while fully resolving the turbulence at other larger scales.

Murthy and Joshi (2008) have assessed the standard $k-\varepsilon$ model, RSM, and LES turbulence models in a baffled stirred vessel agitated by various impeller designs. They found that both the standard $k-\varepsilon$ model and the anisotropic RSM failed to predict the turbulent kinetic energy profiles in the impeller region when the flow was dominated by unsteady coherent flow structures. However, the LES model provided predictions that agreed well with the measurements, and captured many flow features. The standard Smagorinsky model (Smagorinsky, 1963), though relatively preliminary compared with more advanced subgrid turbulence models such as the dynamic model and the scale similarity SGS model, was proven to be able to accurately simulate the fluid phenomena both qualitatively and quantitatively. So up to now, the standard Smagorinsky model is still prevalent in the large eddy simulation of stirred vessels.

In the LES, large-scale eddies are resolved and the smaller scale ones, which are isotropic in nature, are modeled using subgrid scale models. The major role of subgrid scale models is to provide proper dissipation for the energy transferred from the large scales to the small scales. The space-filtered equations for conservation of mass and momentum of an incompressible Newtonian fluid can be written as

$$\rho_m \frac{\partial(\bar{\alpha}_m)}{\partial t} + \rho_m \frac{\partial}{\partial x_j}(\overline{\alpha_m u_{mi}}) = 0 \tag{3.44}$$

$$\frac{\partial}{\partial t}(\rho_m \overline{\alpha_m u_{mi}}) + \left(\frac{\partial}{\partial x_j}(\rho_m \overline{\alpha_m u_{mi} u_{mj}})\right) = -\frac{\partial(\bar{p})}{\partial x_i} + \rho_m \bar{\alpha}_m g_i + \bar{F}_{mi}$$
$$+ \frac{\partial}{\partial x_j}\left(\mu_m \frac{\partial(\overline{\alpha_m u_{mj}})}{\partial x_i} + \frac{\partial(\overline{\alpha_m u_{mi}})}{\partial x_j}\right) \tag{3.45}$$

The flow field is decomposed into a large-scale or resolved component and a small-scale or subgrid-scale component:

$$\phi = \bar{\phi} + \phi' \tag{3.46}$$

where $\bar{\phi}$ ($\phi = u_{mj}$, α_m, p) represents the part that will be resolved in the simulation and ϕ' the unresolved part on a scale smaller than the mesh. So, Eqs. (3.44) and (3.45) can be rewritten as

$$\frac{\partial(\rho_m\bar{\alpha}_m)}{\partial t} + \frac{\partial}{\partial x_j}(\rho_m\bar{\alpha}_m\bar{u}_{mj}) + \frac{\partial}{\partial x_j}(\rho_m\overline{\alpha'_m u'_{mj}}) + \frac{\partial}{\partial x_j}(\rho_m\overline{\alpha'_m}\bar{u}_{mj}) + \frac{\partial}{\partial x_j}(\rho_m\bar{\alpha}_m\overline{u'_{mj}}) = 0 \tag{3.47}$$

$$\frac{\partial}{\partial t}(\rho_m\bar{\alpha}_m\bar{u}_{mi}) + \frac{\partial}{\partial t}(\rho_m\overline{\alpha'_m u'_{mi}}) + \frac{\partial}{\partial t}(\rho_m\overline{\alpha'_m}\bar{u}_{mi}) + \frac{\partial}{\partial t}(\rho_m\bar{\alpha}_m\overline{u'_{mi}}) + \frac{\partial}{\partial x_j}(\rho_m\bar{\alpha}_m\bar{u}_{mi}\bar{u}_{mj} + \rho_m\overline{\alpha'_m u_{mi} u_{mj}})$$

$$= -\frac{\partial(\bar{p})}{\partial x_i} + \rho_m\bar{\alpha}_m g_i + \bar{F}_{mi} + \mu_m\frac{\partial}{\partial x_j}\left(\frac{\partial(\bar{\alpha}_m\bar{u}_{mj})}{\partial x_i} + \frac{\partial(\bar{\alpha}_m\bar{u}_{mi})}{\partial x_j}\right) + \tag{3.48}$$

$$\mu_m\frac{\partial}{\partial x_j}\left(\frac{\partial(\overline{\alpha'_m}\bar{u}_{mj} + \bar{\alpha}_m\overline{u'_{mj}})}{\partial x_i} + \frac{\partial(\overline{\alpha'_m}\bar{u}_{mi} + \bar{\alpha}_m\overline{u'_{mi}})}{\partial x_j}\right)$$

The unresolved parts of numerical simulations are the fluctuating velocity and holdup, which have to be omitted because there is no proper closing method. The mean and the fluctuating velocities are the grid scale and the subgrid scale terms through a filtering operation. The terms with $\overline{u_{mi} u_{mj}}$ are formulated using an LES subgrid model. Omitting the terms with α'_m and u'_{mi}, the above equations become

$$\frac{\partial(\rho_m\bar{\alpha}_m)}{\partial t} + \frac{\partial}{\partial x_j}(\rho_m\bar{\alpha}_m\bar{u}_{mj}) = 0 \tag{3.49}$$

$$\frac{\partial}{\partial t}(\rho_m\bar{\alpha}_m\bar{u}_{mi}) + \frac{\partial}{\partial x_j}(\rho_m\bar{\alpha}_m\bar{u}_{mi}\bar{u}_{mj}) = -\frac{\partial(\bar{p})}{\partial x_i} + \rho_m\bar{\alpha}_m g_i + \bar{F}_{mi}$$

$$+ \mu_m\frac{\partial}{\partial x_j}\left(\frac{\partial(\bar{\alpha}_m\bar{u}_{mj})}{\partial x_i} + \frac{\partial(\bar{\alpha}_m\bar{u}_{mi})}{\partial x_j}\right) - \frac{\partial(\rho_m\bar{\alpha}_m\tau_{mij})}{\partial x_j} \tag{3.50}$$

$$\tau_{mij} = \overline{u_{mi} u_{mj}} - \bar{u}_{mi}\bar{u}_{mj} \tag{3.51}$$

where τ_{mij} is the subgrid scale stress tensor, which reflects the effect of the unresolved scales on the resolved scales. The effect of the subgrid scales on the large scales can be accounted for by the standard Smagorinsky model (Smagorinsky, 1963):

$$\tau_{lij} - \tfrac{1}{3}\tau_{lkk}\delta_{ij} = -(c_s\Delta)^2|\bar{S}_l||\bar{S}_{lij}| \tag{3.52}$$

where Δ is the filter length:

$$\Delta = \sqrt[3]{r\,\delta r\,\delta\theta\,\delta z} \tag{3.53}$$

with the eddy viscosity defined as

$$\mu_{lt} = \rho_l (c_s \Delta)^2 |\bar{S}_l|$$

(3.54)

$$|\bar{S}_l| = (2\bar{S}_{lij} \bar{S}_{lij})^{1/2}$$

(3.55)

$$\bar{S}_{lij} = \frac{1}{2}\left(\frac{\partial \bar{u}_{li}}{\partial x_j} + \frac{\partial \bar{u}_{lj}}{\partial x_{li}}\right)$$

(3.56)

Here, a value of 0.1 is adopted for c_s. The effective gas viscosity can be calculated based on the following formula:

$$\mu_{g,\,eff} = \frac{\rho_g}{\rho_l}\mu_{l,eff}$$

(3.57)

The interphase coupling terms make the two-phase flow fundamentally different from the single-phase flow. F_{mi} satisfies the following relation

$$F_{li} = -F_{gi}$$

(3.58)

3.2.5 Impeller treatment

For unbaffled stirred tanks, a suitable rotating reference frame is often adopted to treat the impeller rotating. In the case of baffled stirred tanks, the contradiction between the rotating impeller and the stationary baffles is commonly treated by involving some specific numerical methods.

3.2.5.1 "Black box" model

As reviewed by Ranade (1995), most of the investigators at that time treated the impeller and its sweep region as a "black box". The experimental values of mean velocity components and turbulence quantities are imposed on the surface swept by the impeller blades as boundary conditions. The flow equations are solved in the entire vessel excluding the impeller region. The shortcoming of such a method is obvious in that the specifications of boundary conditions are strongly dependent on knowledge acquired from experimental data. The impeller design, vessel geometry, operating conditions, physicochemical properties, etc. have great influence on the boundary conditions to be specified and hence this approach cannot be employed to predict the flow generated under novel conditions without any experimental backup. This approach hence cannot be used as a design tool.

3.2.5.2 Snapshot method

In stirred tanks, once the flow has fully developed, the flow pattern becomes cyclically repeating. In this case, a snapshot of this flow can describe the flow within the impeller blades at that particular instant. The principal components of the snapshot approach consist of approximation of a time-dependent expression in terms of the spatial derivative and assumption of its negligible magnitude in the bulk region and hence its removal from the momentum equations to be solved. Ranade and co-workers employed such a method to simulate the flow generated by axial impellers

(Ranade and Dommeti, 1996), Rusthon impellers (Ranade, 1997), and gas–liquid flow (Ranade and Van den Akker, 1994) in stirred tanks. Good agreement between the predicted results and experimental data was found without requiring any empirical input of boundary conditions. However, the model predictions have not been validated over the entire flow domain for all the flow variables.

3.2.5.3 Inner–outer iteration (IO)

Daskopoulos and Harris (1996) and Brucato et al. (1994) have performed simulations using an inner–outer approach. Using this approach, the whole vessel is divided into two partly overlapping zones, as shown in Figure 3.2: an "inner" domain containing the impeller region and an "outer" one including the baffles and reaching the vessel wall and bottom. Firstly, the inner region flow is calculated in a reference frame rotating with the impeller with arbitrary boundary conditions imposed on the surface $\Sigma 1$. The trial flow field is thus obtained in the whole impeller region, including the distribution of velocity and turbulence quantities on the boundary surface $\Sigma 2$. Secondly, the variables on the surface $\Sigma 2$ are used as boundary conditions to calculate the outer flow in the inertial frame. Information on the flow field in the whole vessel, including surface $\Sigma 1$, is now available. After a few iterative calculations between the "inner" and "outer" domains, satisfactory numerical convergence will be achieved. All simulations are conducted under steady assumptions in their own reference frame. Due to the difference between the two frames, the information iteratively exchanged should be corrected for the relative motion and averaged over the azimuthal direction.

The inner–outer iteration method is an essential improvement with respect to the "black box" model for its advanced treatment of impeller region boundary conditions.

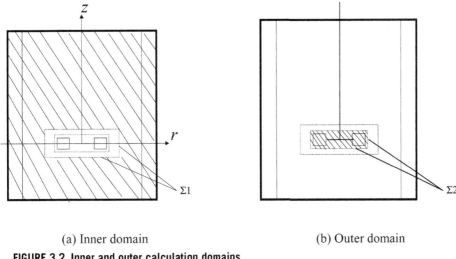

(a) Inner domain   (b) Outer domain

FIGURE 3.2 Inner and outer calculation domains.

(a) Inner domain. (b) Outer domain. Shaded area is excluded from the computation; $\Sigma 1$ and $\Sigma 2$ are control volume boundaries on which boundary conditions are iteratively imposed.

However, the information on the surfaces of the "inner" and "outer" domains is averaged over the azimuthal direction, which ignores some important flow features generated by the periodical rotation of the impeller. An improved inner–outer iteration was proposed by Wang and Mao (2002), in which the unsteady turbulent properties were not averaged over the azimuthal direction. They simulated the flow in single-phase and gas–liquid stirred tanks and found that this method can predict results with better accuracy.

3.2.5.4 Multiple reference frame (MRF)

The multiple reference frame method is analogous to inner–outer iteration. Brucato et al. (1994) and Harris et al. (1996) applied this method to simulate flow in stirred vessels. Two different reference frames are adopted to treat the rotating domain and the stationary frame respectively. The difference lies in the treatment of the interface region. The IO approach has an overlapping region, the width of which and the exact location of its boundaries are largely arbitrary. By contrast, in the MFR the "inner" and "outer" steady-state solutions are implicitly matched along a single boundary surface. The choice of this surface is not arbitrary, since it has to be assumed a priori as a surface where flow variables do not change appreciably either with angular location or with time. Since there is no overlap between the inner and outer regions, the MRF approach is computationally less intensive than the IO method.

3.2.5.5 Sliding mesh (SM)

Murthy et al. (1994) have made use of an SM approach to model impeller rotation in stirred tanks. Using this approach the computational domain is divided into two non-overlapping submeshes, one rotating with the impeller while the other one is fixed as in the MRF method. In contrast, the SM method allows the moving mesh to shear and slide relative to the stationary mesh along the interface. The coupling between computational cells along the sliding interface is accounted for by re-establishing the cell connectivity each time when sliding occurs. The two submeshes are implicitly coupled, which enhances the numerical stability. Since a transient calculation is conducted, authentic flow in stirred tanks can be studied. Unfortunately, the full time solution of flow in a stirred tank makes the computational costs greater by an order of magnitude than those required using steady-state simulations.

3.2.5.6 Methods to deal with axial flow impellers

According to the flow pattern produced, impellers are classified into two broad categories: (i) radial flow impellers and (ii) axial flow impellers. However, the simulation of the latter is obviously more difficult, because there may be some difficulties in dealing with the moving blade surface with irregular geometry, especially when "in-house" codes are used. Two methods are recommended to deal with pitched-blade impellers.

Vector distance method

The so-called "vector distance" is introduced to determine whether a node under consideration is in the solid domain of an impeller or not (Shan et al., 2008). For example, if point A is outside the impeller, as shown in Figure 3.3a, and its

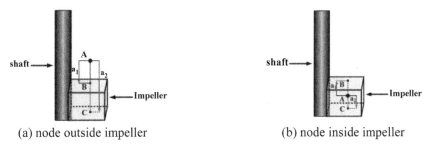

(a) node outside impeller　　　　　　(b) node inside impeller

FIGURE 3.3 Geometric rule for identifying an active node for velocity components and pressure (Shan et al., 2008).

(a) Node outside impeller. (b) Node inside impeller.

distances to two surfaces of an impeller blade are expressed by vectors $\mathbf{a}_1$ and $\mathbf{a}_2$ respectively, their dot product $\mathbf{a}_1 \bullet \mathbf{a}_2$ is positive; if point A is inside a blade, as shown in Figure 3.3b, the dot product $\mathbf{a}_1 \bullet \mathbf{a}_2$ would be negative; if A is on the surface, the dot product is equal to zero. With such a simple geometric rule, all the nodes in the liquid domain can be identified, given that all surfaces of the impeller are already specified. Thus, the smooth blade surface is now approximated by a rough one, which would produce some numerical errors in the simulation results, but these errors will decrease as the grid is refined.

Mirror fluid method (MFM)

Yang and Mao (2005) proposed the MFM and validated it in the numerical simulation of sedimentation of a solid particle in a Newtonian fluid. Using the MFM, the solid–fluid flow issue with the interface of complex geometry is resolved in a single regular domain, including a mirror fluid domain that is originally occupied by the solid particle. The advantage is that a fixed and regular grid can be applied for the entire simulation without mesh regeneration or using boundary-fitted coordinates. If a control volume containing an interface segment is assigned a source term of the shear force of the same magnitude as experienced by the real solid–fluid interface segment but in the opposite direction, the net force on the solid–fluid interface is zero but the true status of force balance is not changed. Thus, the no-slip boundary condition is enforced implicitly. This assignment can be implemented by the mirror relation, namely considering the flow in the virtual domain occupied by solid particles as the flipped mirror image of that in the real fluid domain, or in other words, by rotating the outside flow field $(\mathbf{u} - \mathbf{U})$ and pressure field by 180° around the surface segments. As shown in Figure 2.5 (see Chapter 2), point A in the real fluid corresponds to the mirror point of node B located in the solid domain. Thus, a definite one-to-one mirror relation is easily set to assign the fictitious velocity vector $(\mathbf{u}_B)$ and pressure (p_B) of node B by

$$\mathbf{u}_B = -(\mathbf{u}_A - \mathbf{U}) + \mathbf{U} = 2\mathbf{U} - \mathbf{u}_A \tag{3.59}$$

$$p_B = p_A \tag{3.60}$$

Compared with the immersed boundary method that usually adopts a linear interpolation procedure to determine the velocity at the interface, the MFM has no restriction such as the a priori assumption, and the boundary conditions are implicitly satisfied in the MFM without modifying the physical momentum balance equations. The MFM has been used to deal with pitched-blade impellers by Wang, T. et al. (2013). When the MFM is applied in numerical treatment of the impellers with complex geometries (Figure 2.6 in Chapter 2), the domain occupied by the impeller (solid dots) is assigned suitable flow parameters (e.g., velocity vector and pressure) explicitly by the mirror relation mentioned above, so that an interface segment (solid line) is eventually guaranteed to have the correct shear and normal forces on the fluid side. Consequently, the sum of the real stress contributed by the real fluid (open dots) around the impeller and the fictitious stress by the mirror fluid (solid dots) to the interface segment is kept at zero. In the MFM, integration over an incomplete cell cut by the real boundary is expanded to a whole cell covering the incomplete one and the complementary part in the solid region occupied by the impeller, as if the boundary is an artificial one immersed in the expanded domain. Thus, the whole domain, including the real fluid and the mirror fluid, can be solved together by a single set of equations. Meanwhile, the mirror fluid is regarded as having the same density and viscosity as the real fluid, so solution of the Navier–Stokes equation in the extended fluid domain is straightforward. The problem of interface boundary conditions including some jump conditions is thus replaced by the fictitious parameters specified to the mirror fluid in the solid domain.

3.2.6 Numerical details

3.2.6.1 Discretization of partial differential equations

In order to solve the governing equations numerically, some approaches such as the FEM (finite element method), FDM (finite difference method), and FVM (finite volume method) have been proposed to discretize the partial differential equations. The FVM has been used extensively in recent years because of its simple data structure. In FVM, the computational domain is divided into a number of control units or volumes of regular or irregular shapes. Computational lattices for the one-dimensional case are shown in Figure 3.4. The discretization of partial differential equations is achieved by control volume formulation with a staggered arrangement of primary variables, and then pressure–velocity coupling is dealt with using the SIMPLE

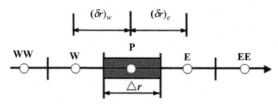

FIGURE 3.4 Computational lattices for the one-dimensional case.

algorithm (Patankar, 1980). In the LES, high-order accuracy is needed to decrease the numerical dissipation. To avoid the physical phenomena being masked, a second-order QUICK scheme (Freitas et al., 1985) is used for convective terms (as shown in Eqs. (3.61)–(3.64)) and a second-order central scheme for diffusive terms. The time integration is discretized using a Crank–Nicolson implicit scheme:

$$\phi_e = \tfrac{1}{8}(6\phi_P - \phi_E - \phi_W) + \tfrac{1}{2}\phi_E^* \qquad u_e > 0 \tag{3.61}$$

$$\phi_w = \tfrac{1}{8}(6\phi_W + 3\phi_P) - \tfrac{1}{8}\phi_{WW}^* \qquad u_w > 0 \tag{3.62}$$

$$\phi_e = \tfrac{1}{8}(6\phi_E + 3\phi_P) - \tfrac{1}{8}\phi_{EE}^* \qquad u_e < 0 \tag{3.63}$$

$$\phi_w = \tfrac{1}{8}(6\phi_P - \phi_E - \phi_W) + \tfrac{1}{2}\phi_W^* \qquad u_w < 0 \tag{3.64}$$

For two-phase flow, two continuity equations should be satisfied at the same time, and two phases share the same pressure field. Carver and Salcudean (1986) combined the two continuity equations to obtain the pressure-correction formula after each conservation equation was normalized by respective density. This combination ensures the total phase continuity equation is satisfied and the numerical stability/convergence is satisfactory. Nevertheless, the simultaneous conservation of each phase is not guaranteed theoretically.

3.2.6.2 Boundary conditions

In order to solve the closed set of governing equations, it is necessary to specify appropriate boundary conditions. For any CFD problems, it will be necessary to select an appropriate solution domain, which isolates the multiphase system being modeled from the surrounding environment. The influence of the environment on the flow within the solution domain is exerted through suitable boundary conditions. The commonly used boundary conditions include inlet, outlet, symmetry, periodic, wall, etc.

The symmetry condition is commonly used for problems of symmetric solution domains to reduce the computational requirements. In stirred tanks, when half of a tank is calculated, the symmetry condition will be enforced at the center axis below the impeller disk. At a symmetric surface, the normal velocity is set to zero and the normal gradients of all variables except the normal velocity are set to zero, expressed as

$$u_{c,r} = u_{c,q} = u_{d,r} = u_{d,q} = 0, \qquad \partial\phi / \partial r = 0 \quad (\phi \neq u_{c,r}, u_{c,q}, u_{d,r}, u_{d,q}) \tag{3.65}$$

For full tank simulations in a cylindrical coordinate system, the condition of symmetry at the axis should not be enforced, because the axis is just an internal point where fluid flows are not subject to any external restriction. Zhang et al. (2006) calculated the velocity components at the axis and used these values as the asymmetric condition in a cylindrical domain. Using this method, the nonzero velocity vectors at the axis are first updated and then the velocity component at the node serving as the

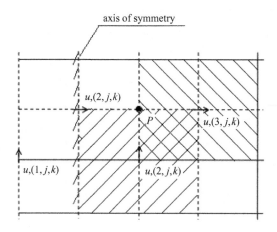

FIGURE 3.5 The staggered grid in the r–θ plane at the axis (Zhang et al., 2006).

boundary is calculated as indicated in Figures 3.5 and 3.6. For the radial and angular velocity components in a staggered grid, the velocity vector in the horizontal plane at the axis, $\mathbf{u}_{r\theta,0}$, is actually the average of all $\mathbf{u}$ vectors of all neighboring nodes at the axis in the r–θ plane. Using the unit vectors $\mathbf{i}$ and $\mathbf{j}$ in the Cartesian reference frame, the resultant velocity vector is

$$\mathbf{u}_{r\theta,0}(k) = u_{0x}(k)\mathbf{i} + u_{0y}(k)\mathbf{j}$$

$$= \frac{1}{N_\theta}\left\{ \begin{array}{l} \displaystyle\sum_j u_r(3,j,k)\left[\cos\left(\theta\left(j+\frac{1}{2}\right)\right)\mathbf{i} + \sin\left(\theta\left(j+\frac{1}{2}\right)\right)\mathbf{j}\right] + \\ \displaystyle\sum_j u_\theta(2,j,k)\left[\cos\left(\theta(j)+\frac{\pi}{2}\right)\mathbf{i} + \sin\left(\theta(j)+\frac{\pi}{2}\right)\mathbf{j}\right] \end{array} \right\} \tag{3.66}$$

with

$$u_{0x} = \frac{1}{N_\theta}\left(\sum_j u_r(3,j,k)\cos(\theta(j+\tfrac{1}{2})) + \sum_j u_\theta(2,j,k)\cos\left(\theta(j)+\frac{\pi}{2}\right)\right) \tag{3.67}$$

$$u_{0y} = \frac{1}{N_\theta}\left(\sum_j u_r(3,j,k)\sin(\theta(j+\tfrac{1}{2})) + \sum_j u_\theta(2,j,k)\sin\left(\theta(j)+\frac{\pi}{2}\right)\right) \tag{3.68}$$

where N_θ is the total grid number in the θ direction. The next step is to project $\mathbf{u}_{r\theta,0}(k)$ onto the radial coordinate axis with different orientations ($\theta(j+1/2)$ or $\theta(j)$) to get the boundary values for the subsequent solutions of u_r and u_θ:

$$u_r(2,j,k) = u_{0x}(k)\cos(\theta(j+\tfrac{1}{2})) + u_{0y}(k)\sin(\theta(j+\tfrac{1}{2})) \tag{3.69}$$

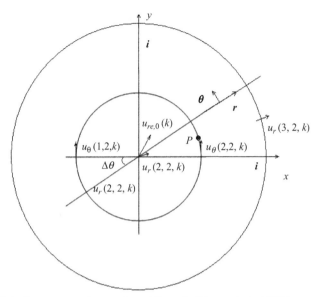

FIGURE 3.6 Velocity components u_r and u_θ at the axis (Zhang et al., 2006).

$$u_{0\theta}(j,k) = u_{0x}(k)\cos\left(\theta(j)+\frac{\pi}{2}\right) + u_{0y}(k)\sin\left(\theta(j)+\frac{\pi}{2}\right) \qquad (3.70)$$

$$u_\theta(1,j,k) = 2u_{0\theta}(j,k) - u_\theta(2,j,k) \qquad (3.71)$$

where the subscript 0 means the axis, and $u_\theta(1,j,k)$ denotes the extended boundary nodes for solving $u_\theta(i,j,k)$. It is noted that the flux across the axis can be treated the same as in the method of radial velocity components but ignoring the influence of angular velocity. For other variables,

$$\Phi(0,j,k) = \frac{1}{N_\theta}\sum\Phi(2,j,k) \qquad (3.72)$$

This update of the boundary values of velocity components is necessary to guarantee the liquid flowing freely across the central axis and the three-dimensionality of turbulent flow.

For solid surfaces such as the tank wall, bottom, baffle, shaft, impeller, and disk, a no-slip wall condition is the appropriate condition for velocity components. Since the k–ε or EASM turbulent model at high Reynolds numbers is employed, the wall function is necessary to solve the flow velocity and turbulent quantities at the nodes adjacent to the solid walls.

The treatment of the top surface of the solution domain is quite important for successful simulations, especially for modeling gas–liquid flows. For an aerated tank, the inlet and outlet boundary conditions are required for a gas stream. Commonly, the top surface is assumed to be flat and the normal gradients of all variables except the normal velocity are set to zero. In order to improve numerical stability for gas–liquid systems, Ranade (2002) proposed two other alternatives. In the first method, if the terminal rise velocities of gas bubbles are known, the top surface can be defined as an "inlet". The normal liquid velocity may be set to zero while the normal gas velocity may be set to the terminal rise velocity. The implicit assumption here is that gas bubbles escape the top surface as an inlet, and the gas volume fraction at the top surface is a free variable. There is no implicit forcing of the gas volume fraction distribution. Alternatively, the top surface of the dispersion can be modeled as a no-shear wall, which will automatically set the normal liquid velocity to zero. It will also set the normal gas velocity to zero. In order to represent escaping gas bubbles, an appropriate sink may be defined for all the computational cells attached to the top surface. Such formulations of the top surface avoid handling sharp gradients of gas volume fractions at the gas–liquid interface and are much more stable numerically.

3.3 TWO-PHASE FLOW IN STIRRED TANKS

3.3.1 Solid–liquid systems

The main purpose of the widely used solid–liquid agitated tanks in industry is to enhance the heat and mass transfer between two phases. Traditionally, flat-bottomed stirred tanks are used for liquid systems, while dished or elliptically bottomed tanks are widely used to aid particle suspension stirred by conventional impellers such as the Rushton turbine, pitched blade impeller and flat blade paddles, and typical axial flow impellers, such as Lightnin A100, A200, and A310 impellers. A number of investigations were focused on achieving empirical correlations, mostly on the distribution of solid holdups in stirred tanks and the criteria for off-bottom solid suspension. In recent years, reports on the above topics are available through experimental methods such as LDV (laser Doppler velocimetry), PIV (particle image velocimetry), and CT (computed tomography) instruments. However, experimental measurements are insufficient to provide insight into solid–liquid suspensions in stirred tanks or for design and scale-up purposes.

Numerical studies on solid–liquid suspensions have been carried out extensively during the last several decades. Many researchers did their best to develop appropriate numerical methods and accurate mathematical models. In general, there are two approaches to simulate multiphase flow: Eulerian–Eulerian and Eulerian–Lagrangian approaches. The Eulerian–Eulerian approach is often preferred for its lower computational load, faster numerical solution, and especially its capability to handle high solid loading conditions (Tamburini et al., 2009). Based on this approach, a great deal of simulation work has been reported in the literature (Gosman et al., 1992; Micale et al., 2000; Ljungqvist and Rasmuson, 2001; Wang, F. et al., 2004a; Montante and Magelli, 2005, 2007; Khopkar et al., 2006a; Guha et al., 2008; Kasat et al., 2008; Tamburini et al., 2009; Sardeshpande et al., 2010; Feng et al., 2012b). Most researchers preferred to handle turbulence using k–ε models. The main weakness of this is that it fails to predict accurately the anisotropic turbulent flow due to its assumption of isotropic turbulence. There are two other methods, direct numerical simulation (DNS; Sbrizzai et al., 2006) and large eddy simulation (LES; Derksen, 2003), which were also employed for solid–liquid flows in stirred tanks. However, the DNS or the LES with Lagrangian simulation of the dispersed phase limits their application to high solid concentrations in practical engineering cases because of the considerable computational cost in handling numerous solid particles. A two-phase explicit algebraic stress model (EASM) based on an Eulerian–Eulerian approach has been developed to simulate solid–liquid turbulent flow in a stirred tank by Feng et al. (2012b). The two-phase EASM is found to give better predictions than the k–ε–A_p model for two-phase flow.

3.3.1.1 Suspension of solid particles

In many solid–liquid stirred tanks, the state of solid suspension may have a significant influence on the product. Kraume (1992) and Bujalski et al. (1999) have observed the following states of suspension occurring in solid–liquid stirred vessels with respect to the stirrer speed (Figure 3.7). One of the most important parameters in the assessment of the performance of impellers for solid suspension is the minimum impeller speed required for the complete off-bottom suspension of the solids (N_{js}) in the vessel. Upon further increasing the impeller speed, the transport rate between the two phases increases only slightly.

Research and experience have produced several methods to assess this parameter N_{js}. They can be divided into four main groups, namely direct, indirect, theoretical, and CFD-based methods. Regarding the CFD-based methods, Hosseini et al. (2010) employed a method called the tangent-intersection method initially suggested by Mak (1992). Wang, F. et al. (2004a) adopted a method based on the axial velocity of the solid phase (u_b) predicted in the computational cells closest to the tank bottom at different impeller speeds. Kee and Tan (2002) performed 2D transient simulations of a solid–liquid stirred tank and monitored the solid volume fraction α_b for the layer of cells adjacent to the vessel floor, the so-called transient α_b profile method. Tamburini et al. (2012) summarized all the methods described above, along with their positive and negative features. At the same time, transient RANS simulations using the sliding grid algorithm were carried out to assess the adequacy of different

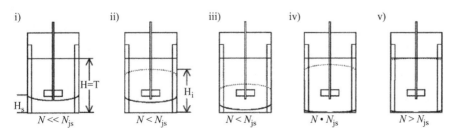

FIGURE 3.7 Solid suspension stages (solid line: static solids at the bottom, H_s; dotted line: suspended solids–liquid interface, H_i) (Bujalski et al., 1999).

methods for predicting N_{js}. An unsuspended solids criterion (USC) was introduced to judge whether the solids contained in a generic control volume should be regarded as suspended or unsuspended. Based on this criterion, the concept of impeller speed for sufficient suspension speed N_{ss} was proposed. The results suggest that it may be convenient to base the design of solid–liquid contactors on the sufficient suspension speed N_{ss} rather than on the traditional N_{js} concept (Figure 3.8).

3.3.1.2 Flow field

The attention of many authors has been focused on the entire flow field. In recent years computational fluid dynamics (CFD) has been increasingly employed as a

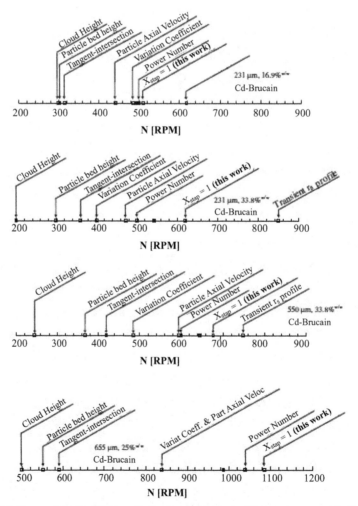

FIGURE 3.8 Comparison of all the methods examined. The solid squares represent N_{js} predicted by Zwietering's correlation (Tamburini et al., 2012).

fundamental tool to critically analyze solid–liquid flows and related phenomena. Most industrial stirred tanks operate above N_{js}. It is generally observed that the flow pattern of the solid phase is similar to that of the liquid phase whether a Rushton disk turbine (RDT; Wang, F. et al., 2004a) or a 70° pitched blade turbine downflow (PBTD; Shan et al., 2008) is used, revealing that fine solid particles follow the liquid closely (Figures 3.9 and 3.10).

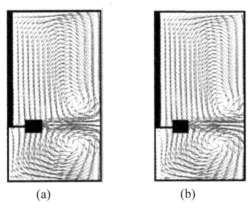

(a)   (b)

FIGURE 3.9 Velocity vectors of flow field stirred by an RDT ($T = H = 0.294$ m, $D = C = T/3$, $\alpha_{d,av} = 0.005$, $\rho_s = 2950$ kg/m³, $d_s = 232.5$ μm, $N = 300$ rpm) (Wang, F. et al., 2004a).

(a) Continuous phase, $r–z$ plane. (b) Dispersed phase, $r–z$ plane.

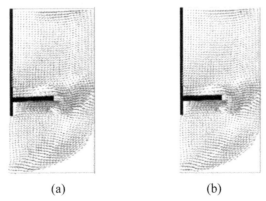

(a)   (b)

FIGURE 3.10 Vector plots of flow field stirred by a PBTD ($T = 300$ mm, $H = 420$ mm, $D = C = 160$ mm, $\alpha_{d,av} = 0.005$, $\rho_s = 1970$ kg/m3, $d_s = 80$ μm, $N = 173$ rpm) (Shan et al., 2008).

(a) Continuous phase, $r–z$ plane. (b) Dispersed phase, $r–z$ plane.

The liquid-phase mixing processes were simulated for 11 impeller rotational speeds starting from $N = 2$ to 40 rps stirred by an RDT (Kasat et al., 2008). The predicted liquid flow pattern, shown in Figure 3.11, indicates the presence of a single-loop flow pattern in the reactor. At such low impeller rotational speeds, all the solids were found to be present at the bottom of the reactor (see isosurface of solids concentration for $N = 2$ rps shown in Figure 3.12). The presence of solid bed at the bottom of the reactor offers an apparent low clearance (i.e., false bottom effect)

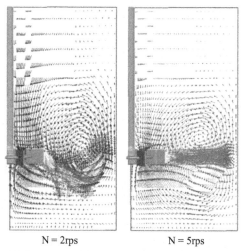

$N = 2$ rps $N = 5$ rps

FIGURE 3.11 Predicted influence of solids on liquid-phase flow pattern ($T = H = 300$ mm, $D = C = T/3$, $\alpha_{d,av} = 0.1$, $d_s = 264$ μm, $\rho_s = 2470$ kg/m^3) (Kasat et al., 2008).

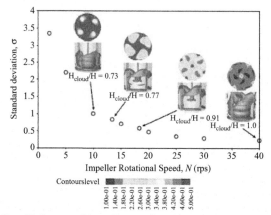

FIGURE 3.12 Predicted influence of impeller rotational speed on suspension quality ($T = H = 300$ mm, $D = C = T/3$, $\alpha_{d,av} = 0.1$, $d_s = 264$ μm, $\rho_s = 2470$ kg/m^3) (Kasat et al., 2008).

to the impeller-generated flow and therefore leads to a single-loop fluid flow pattern even with a radial flow impeller.

With an increase in the impeller rotational speed (up to $N = 5$ rps), solids start to be suspended off the vessel bottom (see Figure 3.12) and the so-called false bottom effect vanishes, resulting in the well-known two-loop flow pattern characterized (see Figure 3.12). In this pattern the rate of exchange between the two loops limits the fluid mixing efficiency. Also, as the solids are suspended in the liquid, some part of the fluid energy is dissipated at the solid–liquid interface and hence, due to less energy being available for fluid mixing, the mixing time increases to some extent.

3.3.1.3 Distribution of solid particles and cloud height

In solid–liquid systems, it is important to determine the solids distribution in the tank in accordance with the process requirements. Some processes require that the particles are just suspended off the bottom, whilst in some processes complete off-bottom solid suspension is necessary. After complete off-bottom solid suspension, the solid cloud height becomes important. The analysis of suspension conditions requires different experimental approaches, among which the visual method is the simplest. Although subjective, the visual method is better for determining off-bottom solid suspension than for complete suspension. Due to the high cost of the equipment and the technical limitations, simulation techniques such as CFD can be employed for the same purpose. To gain insight into the influence of hydrodynamics on homogeneity in such systems, detailed simulation and experimental data on the solid–liquid interaction are necessary.

With the development of CFD, numerical investigation of hydrodynamics in stirred vessels has become more and more popular, and to a certain extent it can replace experimental determination as a practical tool. The numerically simulated distribution of solid particles in a stirred tank with a radial impeller shown in Figure 3.13 suggests that the maximum solid concentration occurs at the center of the

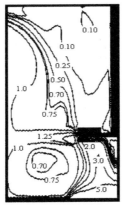

FIGURE 3.13 Contour plots of normalized concentrations of solid phase ($T = H = 0.294$ m, $D = C = T/3$, $\alpha_{d,av} = 0.005$, $\rho_s = 2950$ kg/m³, $d_s = 232.5$ μm, $N = 300$ rpm) (Wang, F. et al., 2004a).

tank bottom, and the concentration decreases gradually from the bottom to the free liquid surface. The solid concentration contour maps show a small circular region with low concentrations below the impeller plane. In the region above the impeller plane, there is also a circulation flow but no region with low solid concentrations.

Compared with a Rushton turbine, it is more difficult to construct suitable computational grids for an axial flow impeller when dealing with the surface of the impeller blades. Shan et al. (2008) conducted numerical simulation of an unbaffled stirred tank of 300 mm diameter agitated with a pitched-blade turbine downflow with a continuous phase (water) and dispersion solids ($\alpha_{av} = 0.005$, $\rho_s = 1970$ kg/m^3, $d_s = 80$ μm) using the vector distance method. From the contour profiles of the solid concentration (Figure 3.14), a relatively high concentration region exists below the impeller. The high concentration near the wall in the upper tank region can be attributed to the circumferential flow and centrifugal force. A similar observation was made using a Lagrangian simulation approach (Derksen, 2003; Ochieng and Lewis, 2006). The solids collide with the wall, losing momentum and resulting in inability of the liquid to carry them through. Consequently, the particles have a tendency to settle instead of moving along their initial trajectories. The concentration near the shaft is very low, which is caused by the central vortex. With increasing impeller speed, the concentration below the impeller also increases, while the concentrations of the regions near the free surface and the shaft decrease. However, further increase of the impeller speed above the critical suspension speed N_{js} seems to be inefficient in promoting higher homogeneity in the tank. Also, a vortex exists in the lower impeller zone near the bottom of the tank, which may be the result of the high shear stress of the continuous phase.

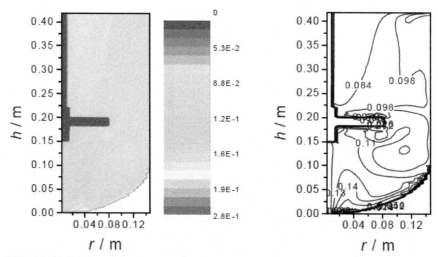

FIGURE 3.14 Solid particle concentration distribution ($T = 300$ mm, $H = 420$ mm, $D = C = 160$ mm, $\alpha_{d,av} = 0.005$, $\rho_s = 1970$ kg/m^3, $d_s = 80$ μm, $N = 173$ rpm) (Shan et al., 2008).

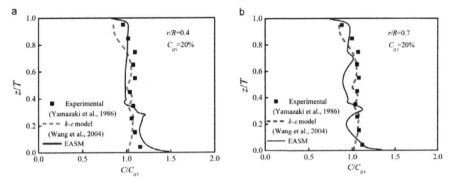

FIGURE 3.15 Axial profiles of solid concentrations at different radial positions (T = 0.3 m, N = 800 rpm, d_s = 87 μm, $\alpha_{d,av}$ = 0.20) (Feng et al., 2012b).

(a) r/R = 0.4. (b) r/R = 0.7.

A two-phase explicit algebraic stress model (EASM) based on an Eulerian–Eulerian approach has been developed to simulate solid–liquid turbulent flow in a stirred tank equipped with a Rushton turbine by Feng et al. (2012b). Comparisons of the EASM predictions with the experimental data of Yamazaki et al. (1986) and the simulated results obtained using the k–ε model of Wang, F. et al. (2004b) are shown in Figure 3.15. No apparent improvement using the present form of EASM is observed over the k–ε model. The two-phase EASM would be more sophisticated when the complex interphase force models and the effect of the dispersed phase on Reynolds stresses are incorporated into the EASM.

Some researchers derived models to predict the homogeneity as a solid cloud height. There is a distinct level (clear interface) to which most of the solid particles are lifted within the fluid at a given impeller speed. The CFD model was used to estimate the cloud height by Hosseini et al. (2010) from the computed solid concentration contours on a vertical plane, while digital photography techniques were used to obtain the normalized cloud height as a function of the impeller speed for an A310 impeller (Figure 3.16). The turbulent and fluid kinetic energy at lower impeller speeds lifted a small percentage of the solid particles from the tank bottom. However, the amount of energy imposed by the impeller was not sufficient to maintain the suspension.

3.3.1.4 Solid hydrodynamics and liquid-phase turbulence

In a solid–liquid stirred tank, if the solid concentration is too low, the hydrodynamics of the solid phase is not very different from that of the single-phase flow. Guha et al. (2008) employed two numerical methods, the LES and Eulerian simulation with the k–ε model, to evaluate the modeling results against the CARPT (computer automated radioactive particle tracking) data in dense solid–liquid suspensions (Guha et al., 2007). Feng et al. (2012b) also evaluated the EASM model with these CARPT data. Figure 3.17 plots the radial profiles of the mean velocity components and turbulent kinetic energy of the solid phase at the disk plane from the CARPT experiment, the LES, the k–ε model, and the EASM simulations. It can be seen from Figure 3.17a and b that the solid velocities

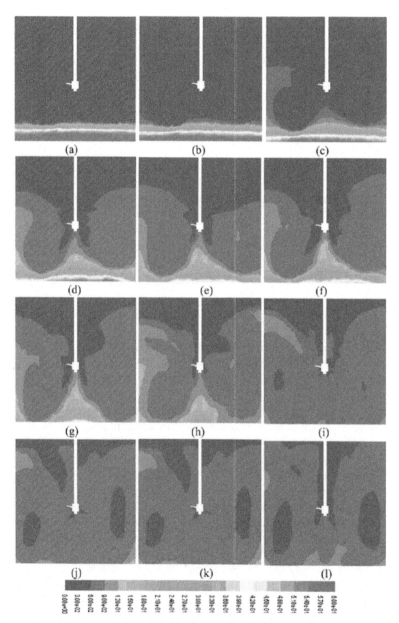

FIGURE 3.16 Solid concentration contours generated using CFD at different impeller speeds for A310 impeller ($C = T/3$, $X = 10$ wt%, $d_p = 210$ μm) (Hosseini et al., 2010).

$N =$ (a) 150, (b) 200, (c) 250, (d) 280, (e) 300, (f) 320, (g) 350, (h) 400, (i) 500, (j) 600, (k) 700, and (l) 800 rpm.

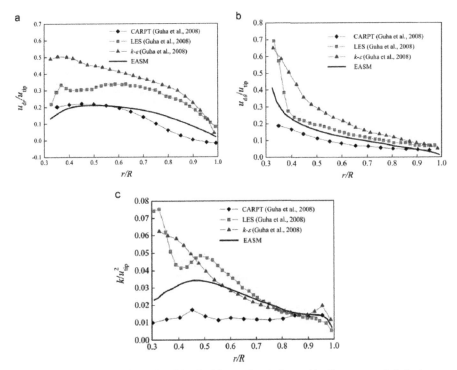

FIGURE 3.17 **Radial profiles of solid velocities and turbulence kinetic energy at disk plane** ($T = 200$ mm, $N = 1000$ rpm, $d_s = 300$ μm, $\alpha_{d,av} = 0.01$) (Feng et al., 2012b).

(a) Radial velocity. (b) Tangential velocity. (c) Turbulence kinetic energy.

predicted by the EASM are closer to the experimental data than the LES and k–ε model predictions, though all models have a good trend of profiles. Unfortunately, all models overestimate the solid tangential velocity at the impeller tip, but the discrepancy is reduced away from the impeller. The k–ε model shows the largest deviation, suggesting that the isotropic model is not adequate to treat the anisotropy in stirred tanks. Figure 3.17c shows a comparison of the solid turbulent kinetic energy found using different methods. It can be seen that the solid turbulent kinetic energy values are overpredicted by all the models. Nevertheless, the location of the peak of the turbulent kinetic energy can be well predicted by the EASM and cannot be described by the k–ε model. It should be noted that there is no additional improvement in the prediction of turbulent kinetic energy with the LES, in contrast to the k–ε model. Generally, the numerical models fail to predict solid turbulent kinetic energy even if the more accurate LES model is employed. It is conjectured that the particle–turbulence and particle–particle interactions are not adequately described by the available models. On the other hand, turbulent kinetic energy might be underestimated by the CARPT for solid–liquid flows. Rammohan et al. (2001) found that the measured values using the CARPT were about 50% lower than the LDV data from Wu and Patterson (1989).

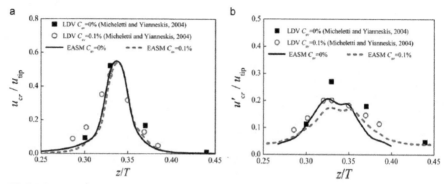

FIGURE 3.18 Axial profiles of liquid velocities at $r/R = 0.448$ with and without particles ($T = 0.0805$ m, $N = 2500$ rpm, $d_s = 186$ mm) (Feng et al., 2012b).

(a) Mean velocity. (b) r.m.s. velocity.

The liquid-phase turbulence in solid–liquid stirred tanks was also numerically investigated by Feng et al. (2012b) using the two-phase EASM, and the influence of particles on the liquid phase turbulence was discussed. Figure 3.18 shows a comparison between the EASM predictions and experimental data in terms of radial mean and r.m.s. (root mean square) fluctuation velocities of the liquid phase in conditions with and without particles. A single-phase EASM simulation was also conducted to investigate the influence of the dispersed phase on liquid-phase turbulence. Good agreement between the EASM predictions on the mean velocity and the LDV data was observed for both single-phase and solid–liquid flows. However, in contrast to the experimental observations, the presence of particles seems to have only a marginal effect on the mean velocity by numerical simulation. Regarding the predicted r.m.s. velocity, this influence is more pronounced and an approximately 15% decrease can be observed at the impeller disk plane. In order to investigate the influence of the particle concentrations on the r.m.s. velocity, two different particle concentrations ($C_{av} = 0.5\%$ and $C_{av} = 2\%$) were employed for both the experimental measurements and the EASM simulations. It is seen that the level of the r.m.s. velocity decreases with increasing particle concentration, suggesting that turbulence is suppressed by the presence of the solid phase.

3.3.2 Gas–liquid systems

The Rushton disk turbine is often employed in gas–liquid stirred tanks, because it can provide powerful shear force to break up bubbles into smaller ones. Many studies on numerical simulation of the flow in gas–liquid stirred tanks with a Rushton disk impeller, particularly the gas holdup distribution in the impeller stream, have been reported. Flooding of the tank means that gas is not well dispersed and rises up in a limited region around the shaft. To show the different gas–liquid flow patterns including flooding, the spatial distribution of gas holdup has been measured for a wide range of stirring speeds and gas holdup levels in some papers. Although the standard k–ε model is the most popular one in use, it often leads to poor results for the flow subjected to complex distributions of strain rate, e.g., in swirling flow and

curved streamline flow. Improvement of numerical methods and turbulence models
is crucial to obtain reasonable simulation of stirred tanks.

As reviewed by Sokolichin et al. (2004), for the most popularly studied system, i.e.,
air in water with bubble diameters between 2 and 8 mm, the bubble slip velocity is fairly
constant. Since, in addition, the resulting flow structure is not very sensitive to the slip
velocity, a constant slip velocity of about 25 cm/s proves to be a reasonable approximation for air/water bubble flow. This liberates us from the necessity of taking bubble coalescence or breakup into account when numerical simulation is conducted. On the other
hand, as gas–liquid mass transfer in stirred tank reactors is involved, the bubble size
distribution (BSD) must be taken into account because the local interfacial area, a, in the
volumetric mass transfer coefficient ($k_L a$) depends on BSD, which is known to vary in
agitated tanks depending on the operating conditions and locally from point to point
in the vessel. To numerically predict such processes, computational fluid dynamics
(CFD) is generally combined with accurate population balance modeling (PBM).

3.3.2.1 Flow field and energy dissipation

A two-fluid model together with the k–ε two-equation model was used to simulate
the gas–liquid turbulence flow in a stirred tank (Wang and Mao, 2002). Typical maps
for gas and liquid velocity vectors in stirred tanks are shown in Figure 3.19 (Rushton
impeller, $T = 450$ mm, $\omega = 27.8$ rad/s, $Qg = 1.67 \times 10^{-3}$ m³/s). Two large eddies are
formed away from the impeller blades, similar to those in single-phase stirred tanks.
In the right upper corner, another smaller eddy is formed due to the buoyancy of rising gas. It is also observed that the impeller discharged stream is a little inclined upward by the buoyant action of gas (Figure 3.19a). Gas and liquid behave differently
in their motion in stirred tanks. Gas bubbles out of the sparger rise up to the impeller

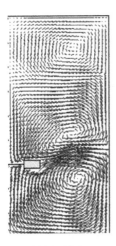
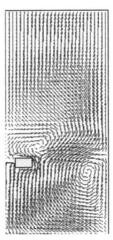

(a) Liquid phase   (b) Gas phase

**FIGURE 3.19 Velocity vector maps of gas–liquid flow in a stirred tank
(Wang and Mao, 2002).**

(a) Liquid phase. (b) Gas phase.

and are dispersed by the impeller to other regions. Compared with the eddies of the liquid phase, they are smaller and located at different positions (Figure 3.19b).

It seems that the swirl modification of the mathematical model is necessary to better match the simulation results with the experimental data. A new swirl number, R_s, is proposed for the gas–liquid flow in a stirred tank, and the k–ε model is modified accordingly by introducing R_s into the energy dissipation equation (Zhang et al., 2009). Figure 3.20 shows a comparison between the predicted results and experimental data for the resultant velocity u along a vertical line at $r = 73$ mm in a stirred tank with a diameter of 288 mm driven by a Rushton impeller. It is seen that the swirl modification model provides more reasonable results.

The large eddy simulation (LES) model has shown great potential in understanding fluid flow behaviors in recent years. Several groups (Wang et al., 1998; Apte et al., 2003; Derksen, 2003; Afshari et al., 2004; Shotorban and Mashayek, 2005) have attempted to apply large eddy simulation to two-phase flow, but in all of these papers Lagrangian approaches were employed to model the dispersed phase. Regarding engineering applications, the Eulerian formulation may be preferred. An Eulerian–Eulerian two-fluid model was proposed for gas–liquid flows using the LES for both gas and liquid phases in a three-dimensional frame by Zhang et al. (2008). The instantaneous velocity vector maps of the gas and liquid phases in the r–z plane located midway between two blades are presented in Figure 3.21 for a tank with a diameter of 288 mm. It is obvious that the flow pattern in the tank is dynamic and very complex, and such fine structures could not be well predicted by the k–ε model. There are many small vortices in both the gas and the liquid flow fields. Furthermore, the flow in the tank is not symmetrical as presumed in most of the literature. Figure 3.22 shows the profiles of gas resultant velocities at different radial positions compared with the experimental data. In comparison to the predictions of the k–ε model, the predictions from the LES are closer to the experimental data, especially in positions near the impeller tips. This is because the LES simulation is more powerful in capturing the anisotropic nature of turbulence in the impeller region and the

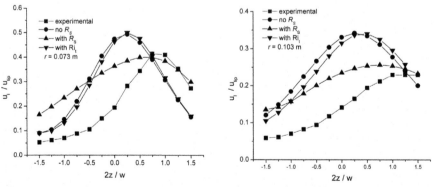

FIGURE 3.20 Predictions of mean liquid velocity with swirl modifications compared with the experimental data (Zhang et al., 2009).

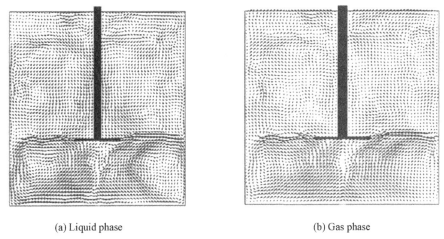

(a) Liquid phase   (b) Gas phase

FIGURE 3.21 Instantaneous velocity fields in the *r–z* plane using the LES (Zhang et al., 2008).

(a) Liquid phase. (b) Gas phase.

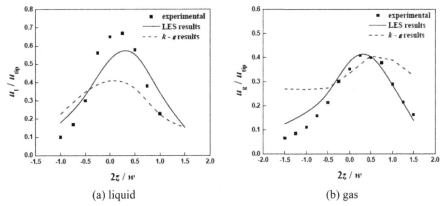

(a) liquid   (b) gas

FIGURE 3.22 Predicted velocity profiles using LES (Zhang et al., 2008) compared with the experimental data (Wang, W. J. et al., 2006) and the standard *k–ε* model prediction (Wang, W. J. et al., 2006) at different vertical positions ($\omega = 62.8$ rad/s).

(a) Liquid. (b) Gas.

impeller stream, much superior to the k–ε model based on the assumption of isotropic turbulent flow.

   Energy dissipation is an important property in a stirred tank because it provides a direct measure of energy input into the system by the impeller. Figure 3.23 presents the distribution of the energy dissipation per mass of fluid in the plane midway between two successive blades throughout the tank. The energy dissipation in the liquid phase is mainly focused on the impeller swept region and the impeller discharged

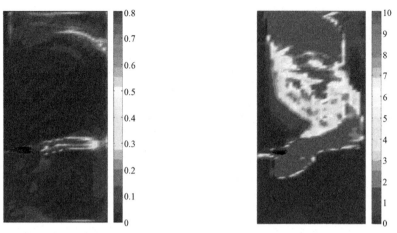

FIGURE 3.23 Energy dissipation (m²/s³) predicted by LES in the *r–z* plane between two impeller blades (Zhang et al., 2008).

stream. Very high dissipation rates occur at the blade edges and near the wakes of the blades. The region with high dissipation rates is smaller than that in single-phase flows, suggesting that the existence of the bubbles reduces the production of turbulence. It is interesting to note that the region of high energy dissipation of gas is rather large, both in the impeller outflows and in the upper part of the upper bulk circulation zone, due to the buoyancy of gas bubbles.

3.3.2.2 Gas holdup and flooding

There are different flow regimes in a gas–liquid stirred tank, which are closely related to the rate of aeration and the pumping power of the stirring impeller, as shown in Figure 3.24 (Paglianti et al., 2000). A stirred tank reactor is expected to operate in the completely dispersed regime or at least the loaded regime (the bubbles are dispersed by the impeller into the upper bulk region). As flooding occurs, the buoyancy of the gas overwhelms the pumping capacity of the impellers, and the bubbles rise in a nearly vertical line from the sparger to the surface.

A gas–liquid two-phase k–ε model was used to simulate the hydrodynamic characteristics of a stirred tank with two six-bladed turbines and four baffles, combined with the multiple size group model to determine bubble size distribution, by Wang, H. et al. (2013). A gas pattern usually referred to as "flooding" was revealed in their study, as shown in Figure 3.25 (Wang, H. et al., 2013). The buoyancy of the gas overwhelmed the pumping capacity of the impellers, and the gas rose in a vertical path from the sparger to the surface. The impeller did not adequately disperse gas to the walls for gas-consuming operations under these conditions.

Paglianti et al. (2000) plotted the Froude number $Fr = N^2D/g$ against the flow number $Fl_g = Q_g/ND^3$ at the flooding/loading transition to correlate the critical condition for the transition between the loading and flooding regimes. Numerical simulation was conducted by Wang, W. J. et al. (2006) to predict this transition. Point A

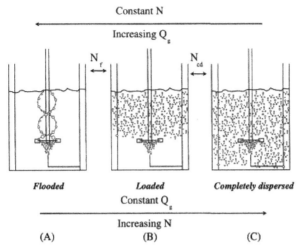

Constant N

Increasing Q$_g$

N_f N_{cd}

Flooded   Loaded   Completely dispersed

Constant Q$_g$

Increasing N

(A)   (B)   (C)

FIGURE 3.24 Bulk flow patterns with increasing stirring speed (Paglianti et al., 2000).

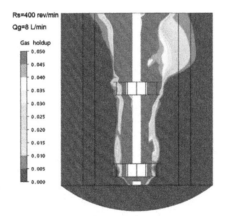

Rs=400 rev/min

Qg=8 L/min

Gas holdup

0.050
0.045
0.040
0.035
0.030
0.025
0.020
0.015
0.010
0.005
0.000

FIGURE 3.25 Model predictions of a gas pattern usually referred to as "flooding" ($T = 180$ mm, $H = 200$ mm, $R_s = 400$ rev/min, $Q_g = 8$ L/min) (Wang, H. et al., 2013).

in Figure 3.26 is the transitional state between flooding and the loaded regime, and point B is the critical condition for the transition from complete dispersion to the flooding regime because of the increase of gas flow rate at a constant stirring speed. The simulation data are in very good agreement with the data from Nienow et al. (1985) and Paglianti et al. (2000). A line based on the correlation is also plotted as a reference:

$$Fr = \frac{1}{30}\left(\frac{T}{D}\right)^{3.5} Fl_g \qquad (3.73)$$

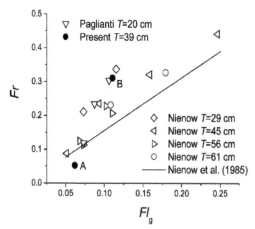

FIGURE 3.26 Comparison of simulation data for the flooding/loading transition with experimental data and correction (Wang, W. J. et al., 2006).

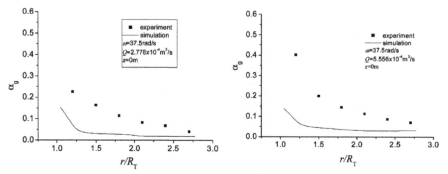

FIGURE 3.27 Comparison between simulation and experiment in impeller discharge stream (Wang, W. J. et al., 2006).

However, the significant correlation between Fr and Fl_g in Figure 3.26 seems to demonstrate a typical nonlinear relationship rather than a linear one.

The measured gas holdup profiles are presented and compared with the numerical predictions of Wang, W. J. et al. (2006). Figure 3.27 shows the radial profiles of gas holdup in the impeller stream at increasing impeller speed, revealing that the holdup decreases in the radial direction and assumes a rather high level near the impeller swept region. However, the simulation underpredicts the gas holdup largely, but has the same trend. This is attributed to the standard k–ε turbulence model used, which cannot account for the anisotropic nature of turbulence in the impeller stream. In the bulk regions, the simulation matches the experimental data well because turbulent gas–liquid flow in the bulk flow region is more likely to be isotropic, and the k–ε-A_p model describes such flow reasonably well (Figure 3.28).

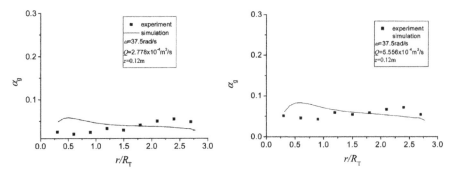

FIGURE 3.28 Comparison between simulation and experiment in bulk flow region (Wang, W. J. et al., 2006).

3.3.2.3 Bubble size distribution and mass transfer

The correct evaluation of the local bubble size distribution (BSD) plays an important role in drag force evaluation and in phase coupling in general. The drag force in gas–liquid stirrer tanks can be written as follows (Ishii and Zuber, 1979; Tomiyama, 2004):

$$F_{D,LG} = C_{D,LG} \frac{3}{4} \rho_L \frac{\alpha_G}{d_b} |u_G - u_L| (u_G - u_L) \qquad (3.74)$$

where $C_{D,LG}$ is the drag coefficient and d_b is the bubble diameter, which can be calculated as the area-averaged bubble size or Sauter diameter, d_{32}. When the BSD is very wide, an interesting alternative is to represent the population of bubbles as constituted by several dispersed phases, each characterized by specific volume fraction, velocity, and size. The closure needed for the evaluation of the drag coefficient $C_{D,LG}$ can be formulated in terms of the terminal velocity of the bubbles U_∞ by employing, for example, the following equation:

$$C_{D,LG} = \frac{4d_b(\rho_L - \rho_G)g}{3\rho_L U_\infty^2} \qquad (3.75)$$

where the bubble size can again be evaluated through the local area-averaged bubble size (d_{32}).

An important aspect that often needs to be modeled is the BSD change due to bubble coalescence and breakage, with a possibly significant impact on local values of velocity profiles, gas volume fraction, and interfacial area. A full 2D population balance model (PBM), in which the bubble size L and gas composition ϕ represent the internal coordinates, is thus preferable. The gas–liquid system is described via a number density function (NDF), defined so that the following quantity:

$$n(t, \mathbf{x}, \phi, L)d\phi dL \qquad (3.76)$$

represents the number of bubbles per unit volume at time t with size bounded between L and $L + dL$ and composition between ϕ and $\phi + d\phi$. Composition here is described in terms of the absolute number of moles of chemical component (i.e., oxygen) contained within the bubble. The evolution of this NDF is described using a population balance equation (PBE) as follows:

$$\frac{\partial n}{\partial t} + \nabla \cdot (\mathbf{u}_L n) + \frac{\partial}{\partial L}(Gn) + \frac{\partial}{\partial \phi}(\dot{\phi}n) = h \tag{3.77}$$

where $\mathbf{u}_L$ is the velocity of bubbles with size L, G is the continuous rate of change of bubble size due to mass transfer, $\dot{\phi}$ is the mass transfer rate, and h is the rate of change of the NDF due to discontinuous events such as coalescence and breakup. Standard submodels were used for all these terms (for details see Buffo et al., 2012). The bubble velocity was calculated using the classical multi-fluid model. So far only drag, lift, and virtual mass forces have been considered and for drag an empirical correlation based on the apparent swarm bubble terminal velocity was employed. The mass transfer rate was calculated using Higbie's penetration model (and consequently the rate of change of bubble size) by taking the average penetration time equal to the Kolmogorov time scale. Quadrature method of moments (QMOM) has been chosen to solve the PBE combined with CFD, to predict the local bubble size distribution in stirred gas–liquid reactors (for details see Petitti et al., 2010).

An approach combining CFD and PBM for simulation of stirred gas–liquid reactors was proposed by Petitti et al. (2010). The population balance was solved by resorting to QMOM and the effects of coalescence and breakup on the BSD were considered. The flow field was described with a two-fluid Eulerian–Eulerian model with some simplification hypotheses. A uniform bubble terminal velocity was considered, and the drag force was evaluated on the basis of the local mean Sauter diameter (rather than the entire BSD). A detailed comparison between the measured (symbols) and the predicted (continuous line) local BSDs at the 10 measuring points for these operating conditions is shown in Figure 3.29. The reconstruction of the local predicted BSD from the computed moments has been performed assuming a BSD given by the sum of two log-normal distributions, whose parameters were evaluated from the six moments, and then compared with the experimental distributions in terms of volume densities. On the whole, the model is able to describe qualitatively and quantitatively the experimental trend of bubble diameters at the measurement points. The smaller bubbles are located in the stream exiting from the impeller (R9), where turbulence is higher and the effect of break-up is more intense. On the contrary, the largest bubbles are in the recirculation zones (R4, R12), where coalescence prevails. However, when it comes to detailed comparisons of the BSD, it is clearly evident that the predicted BSDs are characterized by longer tails than the experimental ones (highlighting a possible underestimation of higher order moments).

Mass transfer has been numerically investigated in a gas–liquid stirred reactor by considering the absorption of oxygen in the liquid phase (water) by Petitti et al. (2013). A multivariate population balance model (MPBM), coupled with an Eulerian multi-fluid approach, was employed to describe the spatial and temporal evolution of the bubble sizes and composition distributions. The mass transfer coefficient k_L was

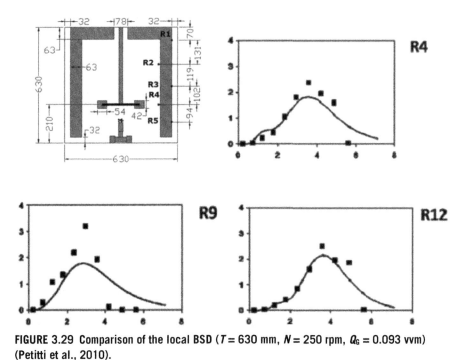

FIGURE 3.29 Comparison of the local BSD ($T = 630$ mm, $N = 250$ rpm, $Q_G = 0.093$ vvm) (Petitti et al., 2010).

evaluated with the expression used by Lamont and Scott (1970), based on the local value of turbulent dissipation rate:

$$k_L = \frac{2}{\sqrt{\pi}} D^{1/2} \left(\frac{\varepsilon_m \rho_L}{\mu_L} \right)^{1/4}$$ (3.78)

where D is the oxygen molecular diffusivity, ε_m is the turbulent dissipation rate, ρ_L is the density of the continuous phase, and μ_L is the viscosity of the continuous phase.

As a consequence of the oxygen mass transfer process from gas phase to liquid phase, an additional equation for the oxygen concentration in the liquid must be solved:

$$\frac{\partial(\alpha_c \rho_c \psi_c)}{\partial t} + \nabla \cdot (\alpha_c \rho_c \bar{U}_c \psi_c) - \nabla \cdot (\alpha_c D_c \nabla \psi_c) = S_{\psi_c}$$ (3.79)

where ψ_c is the concentration of oxygen in the continuous liquid phase, while the term appearing on the right-hand side is the source term accounting for the oxygen transfer from the gas phase. The effective diffusivity of oxygen in the liquid phase is calculated by considering both the molecular and turbulent contributions with a turbulent Schmidt number equal to 0.70. The source term is finally calculated from the mass transfer term as follows:

$$S_{\psi_c} = \int_{\Omega_L} \int_{\phi_b} n(L, \phi_b) k_L \pi L^2 \left(H_{O_2} \frac{\phi_b}{k_v L^3} - \psi_c \right) dL d\phi$$ (3.80)

The PBM is solved in terms of the moments of the NDF, defined as follows:

$$M_{k,l} = \int_0^{+\infty} \int_0^{+\infty} n(L,\phi_b)L^k\phi_b^l dLd\phi_b \tag{3.81}$$

Lower-order moments are very important since they represent specific physical properties of multiphase systems. For example, $M_{0,0}$ is the total bubble number density, $M_{1,0}$ represents the total bubble length density, and $M_{2,0}$ is related to the bubble specific surface area a through the area shape factor k_a:

$$a = k_a M_{2,0} = k_a \int_0^{+\infty} \int_0^{+\infty} n(L,\phi_b)L^2 dLd\phi_b \tag{3.82}$$

The changes in the bubble size distribution in the vessel strongly affect the specific surface area of the gas bubbles, reported in Figure 3.30a for stirring and gassing rates of 250 rpm and 0.052 vvm and in Figure 3.30b for 155 rpm and 0.018 vvm respectively. As is seen, not only does the specific surface area change when the operating conditions are changed, its location also changes throughout the vessel, and it generally increases by increasing the stirring rate. It is interesting to notice that the regions where the turbulent dissipation rate approaches its maximum (near the impeller) are also the regions where the specific surface area is the largest. These regions are also characterized by the highest values of mass transfer coefficient, k_L, which is reported in Figure 3.31 for the same operating conditions. The synergistic effect of these two variables on the final global mass transfer coefficient, $k_L a$, can be observed in Figure 3.32, where the reported contour plots show changes for this parameter of orders of magnitude.

A comparison of time evolution of the normalized oxygen concentration in the liquid as predicted by CQMOM and as measured is shown in Figure 3.33 for the four operating conditions, and the predictions are found to be in good agreement with experiments, with a unique set of theoretical constants appearing in the different submodels for coalescence, breakage, and mass transfer.

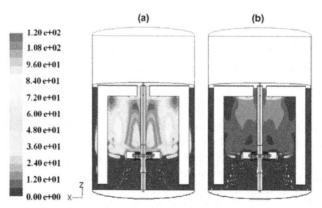

FIGURE 3.30 Contours of the specific surface area of bubbles at 250 rpm and 0.052 vvm (left, a), and at 155 rpm and 0.018 vvm (right, b) (Petitti et al., 2013).

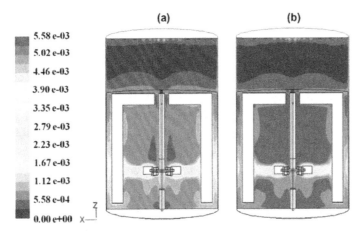

FIGURE 3.31 Contours of mass transfer coefficient (k_L) at 250 rpm and 0.052 vvm (left, a), and at 155 rpm and 0.018 vvm (right, b) (Petitti et al., 2013).

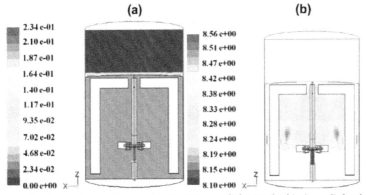

FIGURE 3.32 Contours of oxygen concentration (mol·m^{-3}) in the liquid phase (left, a) and the gas phase (right, b) at 250 rpm and 0.052 vvm (Petitti et al., 2013).

3.3.2.4 Surface aerated stirred tank

In stirred vessels, it is known that gas becomes entrained from the free surface of the liquid. This phenomenon is called surface aeration. An attractive feature of surface aeration is that it can bring about gas–liquid contact, by generating gas bubbles directly from the vessel head space and entraining them in the liquid (as shown in Figure 3.34). Thus, gas dispersion can be created without the sparger–compressor system. There are many reactions, such as hydrogenation, alkylation, chlorination, oxidation, ethoxylation, etc., in which the per-pass conversion of gas is small. In such instances, it is desirable to recycle the unreacted gas back to the reactor because the gas may be highly toxic or may pose safety problems. The recycling of unreacted gas from the head space can be achieved internally with the help of surface aerators. This eliminates the need for a recycling gas compressor. This not only means savings,

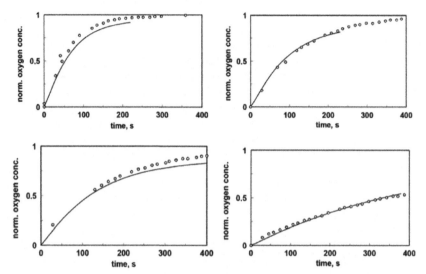

FIGURE 3.33 Comparison of predicted temporal evolution of dimensionless oxygen concentration in liquid with experimental data (Petitti et al., 2013).

From left to right and top to bottom: 250 rpm, 0.093 vvm; 250 rpm, 0.052 vvm; 220 rpm, 0.041 vvm; 155 rpm, 0.018 vvm.

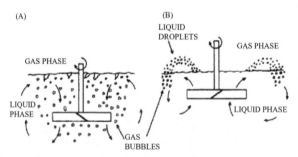

FIGURE 3.34 Schematic representation of surface aeration.

(A) G/L surface aerators. (B) L/G surface aerators.

but the operation of the system is also safer, relatively maintenance free, and more reliable. Moreover, the gas is automatically recycled from the head space, forming a dead-end system.

A two-fluid approach combined with the k–ε turbulence model is often used for gas–liquid flow simulation in a stirred tank, including the case of surface aeration. The key factor to characterize the efficiency of surface aeration is the rate of gas entrainment from the head space above the liquid surface. However, it is difficult to measure this parameter, because gas is entrapped in the liquid phase from the surface and some bubbles may escape back into the gas phase via the same interface.

A general method is to measure the gas concentration variation in the liquid phase and derive the rate of surface aeration in combination with mass conservation of several components. Topiwala (1972) and Veljkovic et al. (1991) obtained the surface aerating rates and established empirical correlations. Wang, A. H. et al. (2004) used the Na_2SO_3 fed-batch method to measure the aeration rate for a stirred tank with different surface aerating impellers. Matsumura et al. (1982) provided a useful correlation for the total entrainment rate:

$$V_s = 7.15 \times 10^{-6} N^{1.90} D^{3.95} T^{-2.5} \sigma^{-2.40} \mu^{-0.15}$$

(3.83)

So far few fundamental equations for local surface aeration rates are available in the literature, though they are essential for numerical simulation of a surface aerating stirred tank. Sun et al. (2006) suggested such a basic equation for local surface entrainment. The mechanism of surface aeration is related to several hydrodynamic characteristics, such as surface shear rate, maximum bubble diameter, turbulent eddy frequency, turbulent length scale, local energy dissipation and so on, namely

$$u_s = f(\tau, d_{max}, \eta, L, P_v)$$

(3.84)

From general fluid mechanics, the agitating power is

$$P_v = \varepsilon \rho_1$$

(3.85)

The maximum bubble size may be estimated by

$$d_{max} = 0.725 \left(\frac{\sigma}{\rho_1} \right)^{3/5} \varepsilon^{-2/5}$$

(3.86)

Uhl and Gray (1967) suggested a relation of shear rate with agitating power and bubble size:

$$\tau \propto \rho_1 \left(P_v \frac{d_{max}}{\rho_1} \right)^{2/3}$$

(3.87)

Moreover, the Kolmogorov theory of isotropic turbulence indicates that the pulsating frequency of turbulent eddies and their length scale are given respectively as (Frost and Moulden, 1977)

$$\eta = \left(\frac{\varepsilon}{v} \right)^{1/2}, \qquad L = \left(\frac{v^3}{\varepsilon} \right)^{1/4}$$

(3.88)

Equations (3.84)–(3.88) reveal that the local energy dissipation rate is an important factor related to the parameters mentioned above, thus implying an intuitive assumption that local surface aeration rate is a complex function of local energy dissipation rate. Using the Kolmogorov theory, the equilibrium turbulent feature is determined by ε and v only when the Reynolds number is large enough, and the characteristic velocity is

$$u = (u\varepsilon)^{1/4}$$

(3.89)

As a first approximation to a real aeration rate equation, it is assumed that the gas entrainment velocity is proportional to the turbulent velocity scale (Sun et al., 2006):

$$u_s = k(v_l \varepsilon_l)^{1/4} \tag{3.90}$$

with the coefficient k to be determined or estimated from experimental data.

This constitutive equation could be used in numerical simulation of surface aeration as the boundary condition at the free surface. Another constraint is also necessary that gas is entrained according to Eq. (3.89) only when the axial velocity component of the liquid is negative (downwards). As a comparison, Eq. (3.83) is also used as the local rate of uniform surface aeration in the simulation. When the co-axial surface baffle is involved in the simulation, the rotating speed measured in the experiment is specified to the surface baffle. The baffle at the free surface is a simple representation of the patented one for intensifying surface aeration as described by Yu et al. (2002). The no-slip conditions are applied to the surface underneath the baffle. The open holes in the baffle are treated as the free surface for surface aeration according to Eq. (3.90) or (3.83).

Observations in the gas and liquid velocity vector maps and the gas holdup contour map (Figures 3.35 and 3.36) reveal that gas is discharged from the impeller region into the impeller stream, and two vortices are formed close to the blade tip. The gas holdup contours show that there is a high holdup region near the blade tip. Another recirculation zone near the surface is formed from the interaction between the vertical upflow of gas, and gas is sucked from the liquid surface.

As for the liquid flow, two vortices are formed in the upper and lower bulk regions respectively, and they are rather symmetrical with respect to the impeller plane. A

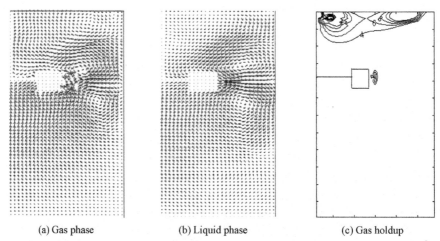

(a) Gas phase   (b) Liquid phase   (c) Gas holdup

FIGURE 3.35 Gas and liquid flow fields and gas holdup distribution in the r–z plane (with $T/2$ Rushton impeller without surface baffle at $\omega = 31.4$ rad/s) (Sun et al., 2006).

(a) Gas phase. (b) Liquid phase. (c) Gas holdup.

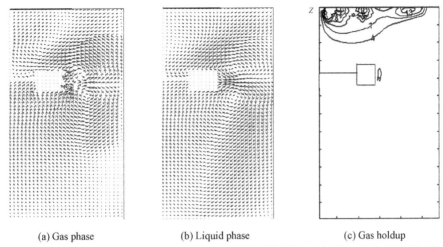

(a) Gas phase　　　　　　(b) Liquid phase　　　　　　(c) Gas holdup

FIGURE 3.36 Gas and liquid flow fields and gas holdup distribution in the *r–z* plane (with *T/2* Rushton impeller with surface baffle at *ω* = 31.4 rad/s) (Sun et al., 2006).

(a) Gas phase. (b) Liquid phase. (c) Gas holdup.

third liquid vortex is formed close to the impeller shaft at the surface as has also been observed experimentally. This vortex causes gas bubbles to accumulate there, creating a high void region. The role of the surface baffle becomes obvious when compared with the configuration without the surface baffle. The lower gas eddy becomes weaker and closer to the impeller plane, and the gas holdup in the lower bulk region is lower. However, owing to the presence of the surface baffle, the increase of gas holdup in the upper bulk region and the total gas holdup is obvious. More air is entrained from the opening in the surface baffle and the third gas holdup peak occurs at the surface.

The contour lines are concentrated in the upper bulk region above the impeller. Another region with dense contour lines is at the impeller blade tip, but rather fewer lines appear in the lower bulk region. This suggests that intensive whole-tank circulation of the liquid phase is necessary for uniform distribution of the entrained gas in the whole tank. The turbulence model needs to be upgraded to more sophisticated versions to account for the swirling and anisotropic nature of turbulent two-phase flow in stirred tanks. Furthermore, the gas bubble size distribution is considered to be a major problem to be resolved in numerical programs for greater improvement in the accuracy of simulation.

3.3.3 Liquid–liquid systems

Liquid–liquid dispersions in agitated vessels are widely encountered in the chemical and metallurgical industry, such as suspension polymerization, chemical reaction, and solvent extraction. In liquid–liquid systems, agitation plays an important role, which controls breakage, coalescence, and also suspension of drops. The purpose of this operation is to increase the interfacial area by intensifying the dispersion of one liquid into another and to consequently enhance interphase heat/mass transfer and chemical reactions.

The fluid dynamic interaction between two liquid phases plays a significant role in determining the properties of the dispersion, but is not fully understood. Experimental study of flow structure for detailed information of flow and turbulence is difficult as intrusive measurement techniques may disturb the flow, while the dispersed phase limits the optical accessibility of non-intrusive measurement techniques (e.g., LDV and PIV). Therefore, computational fluid dynamics (CFD) seems to be a convenient and inquiring tool to explore the hydrodynamic details of both phases in liquid–liquid dispersion systems. Numerical simulations of liquid–liquid flows in stirred tanks are commonly conducted based on the Eulerian–Eulerian approach with the k–ε turbulence model. Compared with other two-phase flows, the difficulties are mainly due to the additional complexity generated by the drop deformation, breakage and coalescence, and the occurrence of circulation inside drops. Models of interphase drag force and drop size are crucial for successful simulation of liquid–liquid flows.

3.3.3.1 Interphase drag force

For multiphase flow in stirred tanks, the interphase drag force plays a significant role, in contrast to other interphase forces. The choice of drag coefficient correlation should be prudent because it can affect the prediction of flow field and especially holdups.

Laurenzi et al. (2009) assumed the droplets behaved as rigid non-interacting spheres with a fixed size rising in a stagnant liquid. The classical drag correlation was adopted in their simulation as (Clift et al., 1978)

$$
\begin{aligned}
C_{\mathrm{D}} &= \frac{24(1+0.15Re_{\mathrm{d}}^{0.678})}{Re_{\mathrm{d}}}, & Re_{\mathrm{d}} &< 1000 \\
C_{\mathrm{D}} &= 0.44, & Re_{\mathrm{d}} &\geq 1000
\end{aligned}
\tag{3.91}
$$

A drag correlation with the deformation of droplets taken into account is as follows (Ishii and Zuber, 1979):

$$
C_{\mathrm{D}} = \frac{24}{Re_{\mathrm{d}}}(1+0.1Re_{\mathrm{d}}^{0.75})
\tag{3.92}
$$

$$
Re_{\mathrm{d}} = \frac{d_{\mathrm{d}}\left|\mathbf{u}_{\mathrm{d}} - \mathbf{u}_{\mathrm{c}}\right|\rho_{\mathrm{c}}}{\mu_{\mathrm{m}}}
\tag{3.93}
$$

$$
\mu_{\mathrm{m}} = \mu_{\mathrm{c,lam}}\left(1 - \frac{\alpha_{\mathrm{d}}}{\alpha_{\mathrm{m}}}\right)^{-2.5\alpha_{\mathrm{m}}\frac{\mu_{\mathrm{d,lam}}+0.4\mu_{\mathrm{c,lam}}}{\mu_{\mathrm{d,lam}}+\mu_{\mathrm{c,lam}}}}
\tag{3.94}
$$

Another correlation considering both wall effect and droplet deformation was given by Barnea and Mizrahi (1975):

$$C_{\mathrm{D}} = (1+\alpha_{\mathrm{d}}^{1/3})\left(0.63+\frac{4.8}{\sqrt{Re_{\mathrm{d}}}}\right)^2 \tag{3.95}$$

$$Re_{\mathrm{d}} = \frac{d_{\mathrm{d}}|\mathbf{u}_{\mathrm{d}}-\mathbf{u}_{\mathrm{c}}|\rho_{\mathrm{c}}}{\mu_{\mathrm{m}}} \tag{3.96}$$

$$\mu_{\mathrm{m}} = \mu_{\mathrm{c}}K_{\mathrm{b}}\frac{\frac{2}{3}K_{\mathrm{b}}+\frac{\mu_{\mathrm{d}}}{\mu_{\mathrm{c}}}}{K_{\mathrm{b}}+\frac{\mu_{\mathrm{d}}}{\mu_{\mathrm{c}}}} \tag{3.97}$$

$$K_{\mathrm{a}} = \frac{\mu_{\mathrm{c}}+2.5\mu_{\mathrm{d}}}{2.5\mu_{\mathrm{c}}+2.5\mu_{\mathrm{d}}} \tag{3.98}$$

$$K_{\mathrm{b}} = \exp\left(\frac{5\alpha_{\mathrm{d}}K_{\mathrm{a}}}{3(1-\alpha_{\mathrm{d}})}\right) \tag{3.99}$$

Cheng et al. (2011) examined these drag correlations in terms of the dispersed phase holdup profiles, as shown in Figure 3.37. It can be seen that the drag models have a certain influence on the dispersed phase holdup distributions. Both the correlations of Ishii and Zuber (1979) and of Barnea and Mizrahi (1975) capture exactly the same profiles, which indicate that the droplet deformation plays a much more important role than the wall effect in calculation of the dispersed phase holdup distribution. Also, these two correlations show better agreement with experimental data than the classical Clift model (Clift et al., 1978) and thus are more suitable for modeling liquid–liquid flow in stirred tanks.

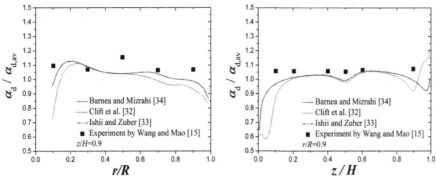

FIGURE 3.37 Comparison of the simulated dispersed phase holdup distributions using different drag correlations with experimental data (Cheng et al., 2011).

3.3.3.2 Evaluation of drop size

The drop size is an important parameter and controls the increase of drop drag coefficient due to turbulence. The prevailing drop size distribution in a liquid–liquid stirred reactor is controlled by many parameters, such as the dispersed phase holdup, impeller speed, and reactor configuration. There are several methods to evaluate the drop sizes in liquid–liquid systems in the open literature: uniform drop size, empirical correlations, and population balance equation (PBE).

Uniform drop size

The simplest way to evaluate the drop size is to use a uniform drop diameter throughout the vessel. Laurenzi et al. (2009) assumed the drop behaved as rigid non-interacting spheres with a fixed size, and three sizes of 50, 200 and 1000 μm were employed for the purpose of assessing the influence of drop sizes on numerical results. As can be observed in Figure 3.38, the effect of the droplet size is significant: when d_d is assumed to be equal to 1000 μm, the two liquids remain completely separated, while for the other two cases oil entrainment is predicted. By contrast, the prediction with a drop size of 50 μm matches the experimental observation more closely. Cheng et al. (2011) also employed six drop sizes to investigate the effect of droplet sizes. Their findings were similar: the larger the drop size was, the more the dispersed phase holdup distribution deviated from the experimental results, as shown in Figure 3.39. However, the prediction of hydrodynamics was hardly affected by the drop size, mainly because the differences between physical properties of two liquid phases are not large.

Empirical correlations

Although the method of uniform drop size is convenient and can sometimes give a quantitative prediction, there is a wide distribution of drop sizes in a liquid–liquid

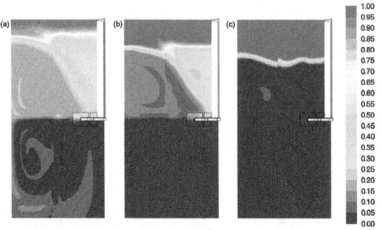

FIGURE 3.38 Maps of computed oil holdup at $N = 3.33$ s^{-1} (Laurenzi et al., 2009).

(a) $d_d = 50$ μm. (b) $d_d = 200$ μm. (c) $d_d = 1000$ μm.

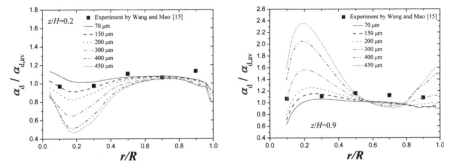

FIGURE 3.39 Comparison of simulated dispersed phase holdup distributions at different drop sizes with experimental data (Cheng et al., 2011).

stirred tank, which complicates the problem of selecting a suitable drop diameter if a uniform monodispersed drop size is used. Also, the effect of the drop size on the dispersed phase holdup distribution is significant. Such a method seems to use the drop size as an adjustable parameter, which is not recommended in CFD simulations. A better way is to evaluate the mean drop size from empirical correlations. Hinze (1955) was the first to show that the maximum stable equilibrium drop size could be related to the maximum local energy dissipation rate in a stirred vessel by the following relationship:

$$d_{\max} = K_1 (\varepsilon_T)_{\max}^{-0.4} \left(\frac{\sigma}{\rho_c} \right)^{0.6} \qquad (3.100)$$

It was also assumed that at equilibrium $d_{\max}$ was proportional to d_{32} and the maximum local energy dissipation rate was proportional to the mean energy dissipation rate, so that the correlation could be rearranged into the form of the impeller Weber number:

$$d_{32} / D = K_2 We^{-0.6}$$
$$We = \frac{\rho_c N^2 D^3}{\sigma} \qquad (3.101)$$

where D is the impeller diameter and K_1 and K_2 are dimensionless constants. In all the above developments, the dispersed phase volume fraction is considered to be sufficiently small. However, for cases of high dispersed phase concentrations, coalescence is considered to occur in the region away from the impeller, possibly with damping of the turbulence in the continuous phase due to the presence of the dispersed phase, and thus the effect of the dispersed phase volume fraction on the drop size should be taken into account in the correlation. A well-developed empirical correlation is given by

$$\frac{d_{32}}{D} = A(1 + \gamma \alpha_{d,av})(We)^{-0.6} \qquad (3.102)$$

This correlation has been found to work quite well by many workers. Many reported correlations show some differences in the values of constants A and γ. For a batch process, constants A and γ are 0.06 and 9 reported by Calderbank (1958), 0.051 and 3.14 by Brown and Pitt (1970), 0.047 and 2.5 by van Heuven and Beek (1971), and 0.058 and 5.4 by Mlynek and Resnick (1972).

For different operating conditions, such empirical correlations can give different mean drop sizes, which are more appropriate compared to the uniform drop size method. However, for a given operating condition, the drop size also varies from one location to another. An evaluation of local drop sizes seems to be more appropriate. Wang and Mao (2005) estimated the local value of the drop diameter using an empirical correlation as follows:

$$d = 10^{(-2.316+0.672\alpha_d)} \, v_{c,lam}^{0.0722} \, \varepsilon^{-0.914} \left(\frac{\sigma g}{\rho_c}\right)^{0.196} \tag{3.103}$$

The drop size distribution was obtained by the local value of energy dissipation rate. They simulated the liquid–liquid flow and the dispersed phase holdup distribution in stirred tanks, as shown in Figure 3.40. At the impeller speed of 350 rpm, the dispersed phase seems to accumulate around the impeller shaft at the top of the tank, which is in accordance with the experimental observations. When the impeller speed is increased, the distribution gradually tends to be homogeneous.

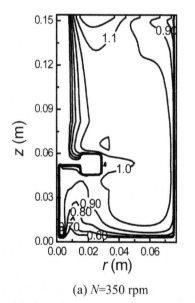

(a) N=350 rpm

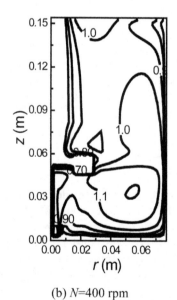

(b) N=400 rpm

FIGURE 3.40 Predicted contour plots of normalized holdup of the dispersed phase (Wang and Mao, 2005).

(a) N = 350 rpm. (b) N = 400 rpm.

Population balance equation

The empirical correlation of local drop sizes is not so proper when the RANS turbulence modeling approach is employed as this method commonly underestimates the energy dissipation rate, especially at the impeller region. The more accurate method to evaluate drop size distribution is using the population balance equation (PBE). The PBE is a set of mathematical tools that enable one either to predict the time evolution of the drop size distribution (DSD) or to determine specific information, such as breakage frequency and daughter size distribution or collision frequency and coalescence efficiency, from analysis of time-variant drop size data. In addition to liquid–liquid systems, the PBE method has been widely applied to other chemical processes such as crystallization, grinding, interphase heat and mass transfer, multiphase reaction, floatation, etc.

The most general form of the PBE, applicable for a Reynolds averaged flow system, can be written as

$$\nabla[\rho \mathbf{u} n(\upsilon; \mathbf{x}, t)] - \nabla[\Gamma_{\mathrm{eff}} \nabla[n(\upsilon; \mathbf{x}, t)]] = \rho h(\upsilon; \mathbf{x}, t) \tag{3.104}$$

where

$$h(\upsilon; \mathbf{x}, t) = B^{\mathrm{a}}(\upsilon; \mathbf{x}, t) - D^{\mathrm{a}}(\upsilon; \mathbf{x}, t) + B^{\mathrm{b}}(\upsilon; \mathbf{x}, t) - D^{\mathrm{b}}(\upsilon; \mathbf{x}, t) \tag{3.105}$$

The source terms refer to birth and death rates for a drop of a specific diameter or volume. Figure 3.41 shows a general scheme for the events taking place. $B^{\mathrm{a}}(\upsilon; \mathbf{x}, t)$ and $D^{\mathrm{a}}(\upsilon; \mathbf{x}, t)$ represent the birth and death rates due to coalescence, $B^{\mathrm{b}}(\upsilon; \mathbf{x}, t)$ and $D^{\mathrm{b}}(\upsilon; \mathbf{x}, t)$ represent the birth and death rates due to breakage, which are expressed as

$$B^{\mathrm{a}}(\upsilon; \mathbf{x}, t) = \frac{1}{2} \int_{0}^{\upsilon} F(\upsilon - \varepsilon, \varepsilon) n(\upsilon - \varepsilon; \mathbf{x}, t) n(\varepsilon; \mathbf{x}, t) \mathrm{d}\varepsilon \tag{3.106}$$

$$D^{\mathrm{a}}(\upsilon; \mathbf{x}, t) = n(\upsilon; \mathbf{x}, t) \int_{0}^{+\infty} F(\upsilon, \varepsilon) n(\varepsilon; \mathbf{x}, t) \mathrm{d}\varepsilon \tag{3.107}$$

$$B^{\mathrm{b}}(\upsilon; \mathbf{x}, t) = \int_{\upsilon}^{+\infty} g(\varepsilon) \beta(\upsilon, \varepsilon) n(\varepsilon; \mathbf{x}, t) \mathrm{d}\varepsilon \tag{3.108}$$

$$D^{\mathrm{b}}(\upsilon; \mathbf{x}, t) = g(\upsilon) n(\upsilon; \mathbf{x}, t) \tag{3.109}$$

where $F(\upsilon, \varepsilon)$ is the coalescence rate between the drop volume of υ and ε, $g(\upsilon)$ is the breakage rate, and $\beta(\upsilon, \varepsilon)$ is the daughter drop size distribution. The volume-based expressions are often transformed into a diameter-based function via $\upsilon = k_v L^3$ and $k_v = \pi/6$ as the volume factor for a sphere, while all drops are assumed as ideal spheres.

The coalescence rate is given by a product of collision frequency of the drops and the collision efficiency:

$$F(L, \xi) = h(L, \xi) \lambda(L, \xi) \tag{3.110}$$

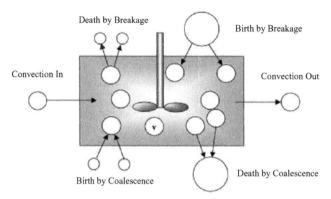

FIGURE 3.41 Population balance events for drops of volume v.

The collision frequency equation and collision efficiency equation are commonly given as (Coulaloglou and Tavlarides, 1977):

$$h(L,\xi) = c_{1,c} \frac{\varepsilon^{1/3}}{1+\varphi_d}(L+\xi)^2(L^{2/3}+\xi^{2/3})^{1/2} \tag{3.111}$$

$$\lambda(L,\xi) = \exp\left(-c_{2,c}\frac{\mu_c\rho_c\varepsilon}{\sigma^2(1+\varphi)^3}\left(\frac{L\xi}{L+\xi}\right)^4\right) \tag{3.112}$$

As for the breakage rate, many authors assume that drop breakage occurs due to drop–eddy collisions. Coulaloglou and Tavlarides (1977) gave the breakage rate equation as

$$g(L) = c_{1,b}\frac{\varepsilon^{1/3}}{(1+\varphi_d)L^{2/3}}\exp\left(-c_{2,b}\frac{\sigma(1+\varphi_d)^2}{\rho\varepsilon^{2/3}L^{5/3}}\right) \tag{3.113}$$

To determine the daughter drop distribution, it is necessary to specify the number of daughter drops. This is usually done by giving a fixed value. Most authors assumed the binary breakage of a mother drop into two daughter drops with a maximum probability of forming two equally sized daughter drops. If diameter is the variable used in the population balance equations, the beta function can be written as

$$\beta(L,\xi) = \frac{90L^2}{\xi^3}\left(\frac{L^3}{\xi^3}\right)^2\left(1-\frac{L^3}{\xi^3}\right)^2 \tag{3.114}$$

Usually, analytical solution of the PBE cannot be obtained except in some very simple situations. Thus, the PBE equations are solved using numerical methods. Because some of the source terms such as coalescence and breakage rates cannot be expressed as a function of the number density, some methods are often adopted to close the PBE equations. There are mainly three approaches: class method (CM), Monte Carlo method (MCM), and moment method (MM), among which the CM and MM methods are widely encountered.

Many authors have employed PBE technology to study the drop size distribution in liquid–liquid stirred tanks. One of the main problems encountered in their simulations is how to evaluate the flow and turbulent parameters in the breakage and coalescence models. Alopaeus et al. (1999) conducted a simulation of the population balances for liquid–liquid systems in a non-ideal stirred tank using a class method. A multi-block stirred tank model was employed to obtain the local turbulent energy dissipation and fluid flow. They considered that such a model had an attractive feature, that it can predict the non-uniformity of dispersions and some scale-up phenomena. Schmelter (2008) derived a mathematical model describing the different phenomena influencing the drop size distribution in a stirred tank. The turbulent flow field was simulated by the RANS method together with the k–ε turbulence model. The behavior of drops was modeled by a Reynolds-averaged PBE approach in which the coalescence and breakage phenomena were taken into account by means of integral terms. However, only single-phase flow fields were considered in his work.

Although PBE technology is able to give a reasonable drop size distribution and has been widely used in scientific research, there have been few practical industrial applications conducted using this method. Firstly, solution of the PBE increases the computational load by several fold. On the other hand, the available drop size distribution information in the present cases is unfortunately not adequate to calculate the parameters appearing in the coalescence and break-up kernels. Furthermore, the drop population balances cannot be applied in a satisfactory manner as the turbulent kinetic energy dissipation rate cannot be predicted correctly by the RANS approach, thus causing a large deviation of drop break-up rates. High-efficiency numerical methods and more accurate description of energy dissipation rates should be developed in future.

3.4 **THREE-PHASE FLOW IN STIRRED TANKS**

Compared with two-phase flow, three-phase flow is more complex due to the presence of a third phase. In experimental work, the distribution of phase holdup and the rate of heat/mass transfer between phases are difficult to be determined. For numerical simulations, the interaction between the two dispersed phases and the contribution of dispersed phases on the turbulence of the continuous phase make the numerical solution of the governing equations even more challenging. Gas–liquid–solid three-phase systems have often been reported in the literature, but other three-phase systems have rarely been approached numerically.

3.4.1 **Liquid–liquid–solid systems**

Liquid–liquid–solid three-phase stirred tanks are popular in process industries. Typical applications include reactive flocculation, solid catalyzed liquid–liquid reactions, etc. Understanding of the hydrodynamic characteristics, such as suspension of solid particles, dispersion of the dispersed liquid phases, and their spatial distribution in a stirred tank is critical for determination of the rates of heat/mass transfer, and therefore of great importance for reliable design and scale-up of such chemical reactors (Wang, F. et al., 2006).

3.4.1.1 Experimental measurements

For liquid–liquid–solid three-phase flow in stirred tanks, it is desirable to obtain information on the state of dispersion of the dispersed phases. Wang, F. et al. (2006) provided experimental measurements of axial and radial variations of phase holdups of the two dispersed phases in a lab-scale stirred tank under different operating conditions using a sample withdrawal method. Tap water, n-hexane, and glass beads were chosen as the liquid, liquid, and solid phases respectively. The measurement results normalized with the respective phase volume fraction are presented in Figure 3.42. It can be seen that at $N = 300$ rpm, the larger local holdup of solids appears close to the tank bottom, and the higher local holdup of oil at the top surface, indicating that both solid and oil phases are not sufficiently dispersed at such low impeller speeds. Increasing the stirring speed can obviously promote the dispersion of oil phase in the continuous phase. It is also observed that the local holdup of the oil phase below the impeller is larger than that in the upper sections, which can be explained by the fact that some droplets adhere to the surface of solid particles as a result of better wettability to oil. This suggests that more ways of phase interaction may appear when a new phase is introduced.

Mass transfer between two phases is an important process in multiphase systems. In a liquid–liquid–solid stirred vessel, the introduction of a solid phase is influential in the breakup and coalescence of droplets and the circulation in a droplet. Furthermore, suspension of the solid phase also affects the mass transfer between two liquid phases. Fang et al. (2005) examined the typical liquid–liquid extraction system n-butanol–deionized water–succinic acid and chose glass beads of various diameters as the inert solid phase. The effects of operating conditions on the volumetric mass transfer coefficients were studied. At lower agitation speeds, the introduction of a solid phase is harmful to the mass transfer between liquid–liquid phases. But with increasing agitation speed, the presence of solids reinforces turbulence intensity because of separation of eddies generated by the relative motion between solid particles and liquid droplets, which is beneficial to liquid–liquid mass transfer. Also, the faster agitating speed is better for uniform solid suspension, resulting in larger mass transfer areas. The mass

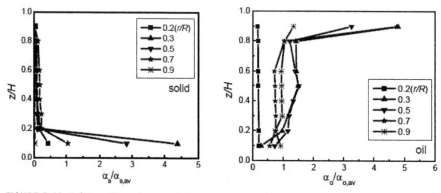

FIGURE 3.42 Axial profiles of normalized holdup of solid and oil phases at $\alpha_{s,av} = 0.10$, $\alpha_{o,av} = 0.10$, and $N = 300$ rpm (left: solid; right: oil) (Wang, F. et al., 2006).

transfer coefficients also differ with different impeller types. Generally, radial impellers perform better than axial ones. However, if solid particles are not suspended uniformly enough, the enhanced drag force is an obstacle to turbulence, which is also destructive for mass transfer. Furthermore, the influence of particle diameters on $k_L a$ is negative when the diameter exceeds 100 μm (Fang et al., 2005; Wang, F. et al., 2006).

3.4.1.2 Numerical simulation

For numerical simulation of liquid–liquid–solid three-phase flow, the Eulerian multi-fluid model is preferred. Wang, F. et al. (2006) treated water, oil, and solid phases as different continua, interpenetrating and interacting with each other everywhere in the computational domain under consideration. The oil and solid phases are in the form of spherical dispersed droplets and particles respectively. The effect of breakup and coalescence of droplets is neglected. The pressure field is shared by all three phases, which are exerted respectively by the pressure gradient multiplied by respective volume fraction. The motion of each phase is governed by the respective mass and momentum conservation equations.

The RANS version of governing equations for three-phase flow is similar to that of two-phase flows. Reynolds stresses are always modeled by introducing the Boussinesq hypothesis. Although different from two-phase flows, there are still interactions between the dispersed droplets and solid particles in liquid–liquid–solid three-phase flow, which have to be modeled. However, this factor has not always been included in literature reports. In the model developed by Padial et al. (2000) for a gas–liquid–solid three-phase bubble column, the drag between solid particles and gas bubbles was modeled identically to drag between liquid and gas bubbles based on the notion that the particles in the vicinity of gas bubbles tend to follow the liquid. In the simulation of Michele and Hempel (2002), the momentum exchange terms between the dispersed gas and solid phases were expressed as

$$F_{\text{g-s}, i} = -F_{\text{s-g}, i} = \frac{3\alpha_g \alpha_s C_{\text{g-s}} |\mathbf{u}_s - \mathbf{u}_g| (u_{si} - u_{gi})}{4d_p} \tag{3.115}$$

The combination of $C_{\text{g-s}} |\mathbf{u}_s - \mathbf{u}_g|$ was defined to be a fitting parameter determined by fitting model predictions to measured local solid holdups. Since the two dispersed phases are presumed to be continua as mentioned above, it is reasonable to expect the drag between solid particles and oil droplets to behave in a similar way as the drag force between the continuous and the dispersed phases (Wang, F. et al., 2006):

$$F_{\text{o-s,drag}, i} = -F_{\text{s-o,drag}, i} = \frac{3\rho_o \alpha_o \alpha_s C_{\text{o-s}} |\mathbf{u}_s - \mathbf{u}_o| (u_{si} - u_{oi})}{4d_s} \tag{3.116}$$

Numerical simulation of such a liquid–liquid–solid system seems to be successful. In the flow fields of both the continuous and dispersed phases (Figure 3.43a–c), the well-documented flow patterns generated by a disk turbine in a stirred tank are clearly illustrated. Two large ring eddies exist, above and below the impeller plane respectively, and a high-velocity radial impeller stream is also predicted. Overall,

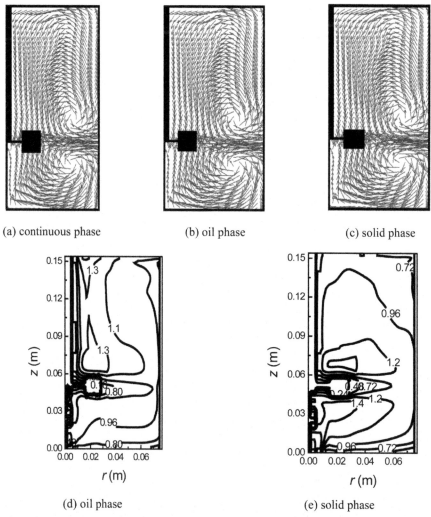

(a) continuous phase   (b) oil phase   (c) solid phase

(d) oil phase   (e) solid phase

FIGURE 3.43 Velocity vector plots of the continuous, oil, and solid phases and contour plots of normalized holdups of two dispersed phases in the r–z plane at $\alpha_{s,av} = 0.10$, $\alpha_{s,av} = 0.30$, and $N = 400$ rpm (Wang, F. et al., 2006).

(a) Continuous phase. (b) Oil phase. (c) Solid phase. (d) Oil phase. (e) Solid phase.

the flow fields of the three phases are very similar to each other in most parts of the domain. The velocity field of the dispersed oil phase shows a trend of drifting upwards at lower impeller speeds. This might be due to the fact that the oil phase with lower density accumulates easily in the top section of the tank. The time-averaged flow field of the solid phase reveals a small recirculation zone above the center of the tank bottom, implying that the solid particles tend to settle down in this zone.

The calculated normalized local holdups of the two dispersed phases are presented as contour plots in Figure 3.43d and e. It is clear that the distributions of oil and solid phases are both less homogeneous at low impeller speeds. The maximum oil phase holdup occurs in the center of the free surface because of the ring vortices in the upper bulk zone, in qualitative agreement with experimental investigations. The distributions of both dispersed phases become more uniform at higher agitation speeds. The maximum solid concentration is located at the center of the tank bottom due to the density difference and the ring eddy at the bottom, as confirmed above.

3.4.2 Gas–liquid–liquid systems

Gas–liquid–liquid systems are commonly encountered in catalytic hydrogenation, hydroformylation, carbonylation, and extraction processes. Yu et al. (2000) studied the characteristics of three-phase dispersion and gas–liquid mass transfer in an air–water–kerosene system in a self-aspirating stirred tank reactor equipped with a draft tube. The critical speed for gas aspiration and oil dispersion, rate of gas aspiration, gas–liquid mass transfer coefficient, and power consumption for some types of impeller were discussed in detail. Recently Cheng et al. (2012) determined experimentally the macromixing time in a gas–liquid–liquid stirred tank.

The critical speed for oil dispersion (N_o) means that the oil drops appear in the continuous phase (water). This denotes the ability of the impeller to disperse the oil phase. N_o is dependent on impeller diameter and immersion depth. A disk turbine (labeled as DT) with double rings fixed over and beneath the blades respectively proved to be more suitable for oil phase dispersion because of its smaller value of N_o. Both rings help to form a pressure gradient, which helps to suck in the oil phase. The critical speed for gas intake (N_g) is also related to impeller diameter and immersion depth. White and De Villiers (1977) suggested a nondimensional number:

$$C = \frac{N_g^2 D^2}{gS}$$

(3.117)

where S is the immersion depth of the impeller. From the experimental results, the DT and the mineral flotation impeller performed better in gas aspiration. In addition, this impeller did a good job in increasing the gas aspiration rate (Q_g).

The oil phase always has a large volume fraction in industrial reactors. It is revealed that the value of N_o decreases with increasing oil phase fraction, because a larger amount of oil phase makes the initial interface between the oil and water phases closer to the impeller, so that the initial production of oil drops becomes easier. The oil phase fraction will not influence the depth of the interface between the liquid and gas phases. That is why the oil fraction correlates negatively with N_g. However, the oil phase fraction has much influence on the gas aspiration rate. With increasing oil phase fraction, the system viscosity is also enhanced, which reinforces the stability of the gas cavity generated behind the blades. The existence of a gas cavity reduces the pressure gradient, and results in a decrease of the amount of gas inspiration.

3.4.3 Liquid–liquid–liquid systems

A combination of two aqueous phases (mixture of acetone and aqueous solution of ammonium sulfate) and an oil phase (cyclohexane) or an aqueous phase and two oil phases constitutes a liquid–liquid–liquid system. Yu et al. (2007) studied the dispersion and phase separation of the triple-liquid-phase stirred tank with a CCD camera system. The results show that different types of impeller had different dispersion abilities. In this specific system, the height of the upper phase to the tank bottom was suggested to indicate the dispersion ability. Generally, radial impellers performed better than axial ones as a result of the greater shear generated by vertical blades. For axial impellers, the downward flow type was better than the upward flow counterpart. Similar to the critical suspension speed in a liquid–solid phase system, Skelland and Ramsay (1987) defined the minimum agitator speed needed to obtain complete liquid–liquid dispersion, i.e., the rotational speed just sufficient to completely disperse one liquid into the other. In the experiments of Yu et al. (2007), the middle liquid and the lowest liquid were dispersed with each other firstly with an augmented impeller speed. The impeller speed when the upper liquid was also completely dispersed into the other two liquid phases was defined as the critical dispersion speed. The phase volume ratio had a great influence on the dispersion pattern. The increase of middle phase volume ratio had little effect on the critical impeller speed, which tended to increase with an increase of middle to bottom phase volume ratio.

The phase separation process of the triple-liquid-phase system is classified into types A and B, based on the differences in the status of phase dispersion. A mathematic model was proposed to predict the type A phase separation process (Figure 3.44). Cockbain and McRoberts (1953) measured the rate of coalescence of a single oil/water drop at the oil–water interface and found that the stability of drops

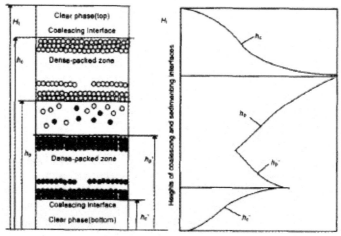

FIGURE 3.44 Schematic diagram of type A phase separation process (Yu et al., 2007).

of equal size under the same conditions was different. The distribution for the drop lifetime (defined as the time interval for the drop to coalesce) is

$$N / N_0 = e^{-kt}$$

(3.118)

in which N is the number of drops remaining uncoalesced at time t, N_0 is the total number of drops, and k is the coalescence constant. Gillespie and Rideal (1956) concluded with an equation for the lifetime distribution of equal size at the oil–water interface by detailed analysis of the probability of rupture of the film between the drop and the interface, namely

$$N / N_0 = e^{-k(t-t_0)^{1.5}}$$

(3.119)

in which t_0 is the time of film drainage. Another entirely empirical mathematical function was proposed by Yu and Mao (2004) to replace Eqs. (3.118) and (3.119):

$$N / N_0 = (1 + k_1 t^{k_2}) e^{-k_3 t^{k_4}}$$

(3.120)

in which k_1, k_2, k_3, and k_4 are the fitting constants and have no clear physical meaning. Substituting N and N_0 in Eq. (3.120) with height parameters and phase fraction, Eq. (3.120) becomes

$$\frac{H_t \varepsilon_T - (H_t - h_C)}{H_t \varepsilon_T} = (1 + k_1 t^{k_2}) e^{-k_3 t^{k_4}}$$

(3.121)

and the height of coalescence interface h_C can be deduced from Eq. (3.121) as

$$h_C = H_t (1 - \varepsilon_T) + H_t \varepsilon_T (1 + k_1 t^{k_2}) e^{-k_3 t^{k_4}}$$

(3.122)

The percentage of moving drops in the whole system is very small. So an assumption can be made that all the uncoagulative drops are concentrated in the dense-packed zone. Thus,

$$\frac{(h_C - h_P) \varepsilon_P}{H_t \varepsilon_T} = (1 + k_1 t^{k_2}) e^{-k_3 t^{k_4}}$$

(3.123)

The relation of the interface of the dense-packed zone with time can be found:

$$h_P = H_t (1 - \varepsilon_T) + \left(1 - \frac{1}{\varepsilon_P}\right) H_t \varepsilon_T (1 + k_1 t^{k_2}) e^{-k_3 t^{k_4}}$$

(3.124)

And the results show that the prediction is consistent with experimental observations (Figure 3.45).

3.4.4 Gas–liquid–solid systems

Stirred reactors involving gas, liquid, and solid phases are very common in chemical and allied industries. The solid phase may act as a catalyst or a resultant or undergo chemical reactions. Typical applications include catalytic hydrogenation,

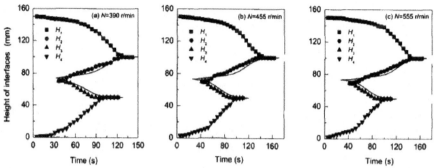

FIGURE 3.45 Modeling and experimental results for three-liquid-phase separation (dispersion by PBTU impeller) (Yu et al., 2007).

Fischer–Tropsch synthesis, oxidation of *p*-xylene to terephthalic acid, production of polymers using suspension polymerization, oxidative leaching of ores, and many other economically important reactions (Murthy et al., 2007). In these reactors, the agitator plays a dual role of keeping the solids suspended, while dispersing the gas uniformly as bubbles. The design and scale-up of mechanically agitated gas–liquid–solid reactors are generally based on semi-empirical methods. Because of additional complexities associated with three-phase systems, few published papers are available on the mathematical models and CFD simulations of three-phase systems (Murthy et al., 2007; Panneerselvam et al., 2008).

CFD simulations have been carried out to study solid suspension and gas dispersion in gas–liquid–solid mechanically agitated contactors using an Eulerian–Eulerian multi-fluid approach along with the standard k–ε turbulence model (Murthy et al., 2007; Panneerselvam et al., 2008). There are various interchange forces such as drag, lift force, and added mass force during momentum exchange between different phases. But the main interaction force is due to the drag force caused by the slip between different phases. Khopkar et al. (2003, 2005) studied the influence of different interphase forces and reported that the effect of the virtual mass force was not significant in the bulk region of stirred vessels and the magnitude of the Basset force was also much smaller than that of the interphase drag force. Further they reported that turbulent dispersion terms were significant only in the impeller discharge stream. Similarly Ljungqvist and Rasmuson (2001) also found very little influence of the virtual mass force and the lift force on the simulated solid holdup profiles. Hence, based on their recommendations and also to reduce the computational time, only the interphase drag force is usually considered in numerical simulation and the only non-drag force considered is the turbulent dispersion force.

The drag force between the liquid and solid phases is represented by the following equation:

$$F_{D,LS} = C_{D,LS} \frac{3}{4} \rho_L \frac{\varphi_S}{d_P} |u_S - u_L| (u_S - u_L) \tag{3.125}$$

In the work of Murthy et al. (2007), the drag coefficient proposed by Pinelli et al. (2001) was used as follows:

$$\frac{C_{D0}}{C_{D,LS}} = \left[0.4\tanh\left(\frac{16\lambda}{d_p} - 1\right) + 0.6 \right]^2 \tag{3.126}$$

However, in the work of Panneerselvam et al. (2008), the drag coefficient proposed by Brucato et al. (1998b) was used as follows:

$$\frac{C_{D,LS} - C_{D0}}{C_{D0}} = 8.67 \times 10^{-4} \left(\frac{d_p}{\lambda}\right)^3 \tag{3.127}$$

where d_p is the particle size, λ is the Kolmogorov length scale, and C_{D0} is the drag coefficient in a stagnant liquid, which is given as

$$C_{D0} = \frac{24}{Re}(1 + 0.15Re_p^{0.687}) \tag{3.128}$$

The drag force between the gas and liquid phases is represented by the following equation:

$$F_{D,LG} = C_{D,LS}\frac{3}{4}\rho_L\frac{\varphi_G}{d_b}|u_G - u_L|(u_G - u_L) \tag{3.129}$$

where the drag coefficient exerted by the dispersed gas phase on the liquid phase is obtained using the modified Brocade drag model (Khopkar et al., 2006a, b), which accounts for the interphase drag by microscale turbulence and is given by

$$\frac{C_{D,LG} - C_D}{C_D} = 6.5 \times 10^{-6} \left(\frac{d_p}{\lambda}\right)^3 \tag{3.130}$$

where C_D is the drag coefficient of a single bubble in a stagnant liquid, given by

$$C_D = \max\left(\frac{24}{Re_b}(1 + 0.15Re_b^{0.687}), \frac{8}{3}\frac{Eo}{Eo+4}\right) \tag{3.131}$$

where Eo is the Eotvos number and Re_b is the bubble Reynolds number, given by

$$Re_b = \frac{|u_L - u_G|d_b}{v_L}, \quad Eo = \frac{g(\rho_L - \rho_G)d_b^2}{\sigma} \tag{3.132}$$

Furthermore, the turbulent dispersion force is considered and the following equation derived by de Bertodano (1992) is used:

$$F_{TD} = -C_{TD}\rho_L k_L \nabla\varphi_L \tag{3.133}$$

where C_{TD} is the turbulent dispersion coefficient and the recommended value for turbulent dispersion coefficient C_{TD} is in the range of 0.1–1.0. In the literature (Cheung

Table 3.1 Experimental (Chapman et al., 1983) and Predicted Values (Murthy et al., 2007) of Overall Gas Holdup

No.	Impeller speed (s⁻¹)	Gas sparging rate (m³/s)	% Gas hold-up (ε_G)	
			Experimental	Predicted
1	3.3	0.57	1.7	1.5
2	4	1.14	2.8	2.5
3	5	2.32	5.6	5

Table 3.2 Effect of Impeller Type on Solid Concentration Distribution (d_p = 180 μm, ρ_p = 2520 kg/m³, 6.6 wt%, V_G= 2 mm/s; Murthy et al., 2007)

No.	Impeller type	Critical impeller speed for off-bottom suspension (N_{CS}) (s⁻¹)	
		Experimental	Simulated
1	PBTD	6.5	6.5
2	DT	9.2	9.5
3	PBTU	11.1	11.3

et al., 2007; Lucas et al., 2007), the most widely used value for turbulent dispersion coefficient is 0.1.

The three-phase CFD predictions of Murthy et al. (2007) were compared with experimental data (Chapman et al., 1983; Rewatkar et al., 1991; Zhu and Wu, 2002) to understand the distribution of solids over a wide range of solid loading (0.34–15 wt%) for different impeller designs, including Rushton turbine (RT) and pitched blade down and upflow turbines (PBT45), solid particle sizes (120–1000 μm) and various superficial gas velocities (0–10 mm/s), as listed in Tables 3.1 and 3.2. The comparison is quite reasonable. CFD simulations have also been performed to study the effects of impeller design, impeller speed, particle size, and superficial gas velocity on prediction of critical impeller speed and cloud height for the just suspended condition in mechanically agitated gas–liquid–solid reactor by Panneerselvam et al. (2008), as shown in Figures 3.46 and 3.47.

3.5 SUMMARY AND PERSPECTIVE

3.5.1 Summary

Up to now, two-phase flows (solid–liquid, liquid–liquid and gas–liquid) in stirred tanks have been successfully predicted using numerical methods. The Eulerian–Eulerian model (or two-fluid model) is the dominant approach in such simulations

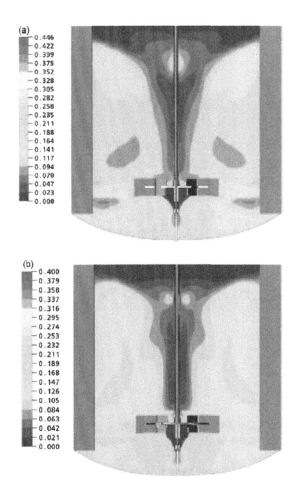

FIGURE 3.46 Effect of particle size on solid concentration distribution for Rushton turbine using CFD simulations (ρ_p = 4200 kg/m^3, 30 wt%) at the critical impeller speed (Panneerselvam et al., 2008).

(a) 125 μm. (b) 180 μm.

because of the lower computational effort required, faster numerical resolution, and especially its capability to handle high dispersed phase loading conditions. Different types of turbulence models including steady models and transient models have been developed and used. In most cases, the particle size is dealt with as a uniform drop size. Empirical correlations and even the population balance equation (PBE) have also been accepted to evaluate the particle sizes in the simulation. These have greatly improved the accuracy of numerical simulations. The distribution of solids over a wide range of solid loadings, the dispersion of bubbles for various superficial gas velocities, and evolving drop characteristics have been well understood. Knowledge of the two-phase macromixing process has been relatively well developed and more

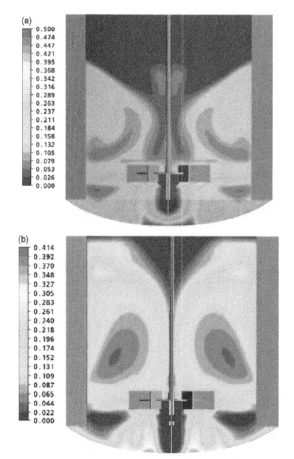

FIGURE 3.47 Effect of air flow rate on solid concentration distribution for Rushton turbine using CFD simulations (d_p = 230 μm, ρ_p = 4200 kg/m³, 30 wt%) at the critical impeller speed (Panneerselvam et al., 2008).

(a) 0. (b) 0.5 vvm.

attempts should be made to model this process. On the whole, the CFD approach has become a reliable tool for design of two-phase stirred tanks in addition to the stage-wise test-and-trial approach.

Compared with two-phase flows, three-phase flows in stirred reactors are more complex due to the presence of the third phase. For numerical simulations, the interaction between two dispersed phases and the contribution of the dispersed phases on the turbulence of the continuous phase make the numerical solution of the governing equations even more challenging. Fewer reports on the numerical simulation of three-phase flows in stirred tanks can be found and most of these have been validated against the experimental results of overall or macro performance. There are two reasons for this: (1) the complexity of three-phase flows lead to the difficulty in constructing three-phase flow models, and the interphase force models being used are those prevalently used in two-phase flows; (2) up to now, there are no mature experimental techniques to measure the local distributions of each phase in three-phase flows.

3.5.2 **Perspective**

The CFD approach has been a reliable tool for design of lab-scale two-phase stirred tanks. In fact, flow under relevant industrial operating conditions is more complex and chaotic (dense dispersed-phase holdup and large liquid flow rates, large bubbles and vortices). Under these conditions, the mechanism of momentum transfer is not well understood because in such denser dispersed-phase holdups, the levels of shear rate and turbulence in the vicinity of any particle (bubble, drop, or solid particle) are very different from a single isolated particle. The estimation of a drag coefficient for a particle in a swarm is so essential that work on the motion and interactions of the particle swarm should be emphasized using experimental and CFD approaches. Furthermore, fast and accurate prediction of industrial stirred tanks needs the development of appropriate turbulent models, efficient algorithms, and massive parallel techniques.

In order to develop reliable numerical simulation of three-phase stirred tanks, the following targets need to be resolved:

1. **Measurement techniques of local phase holdups.** It is important to measure the local distributions of two dispersed phases because the knowledge of the local distributions can help to establish a new model for such a three-phase stirred tank and these data are also essential in verifying the new model. Various methods that had been reported in the literature for measurement in a two-phase stirred vessel have particular shortcomings and cannot be used directly for three-phase measurement. The vision probe developed by Honkanen et al. (2010) has been utilized in experiments for four three-phase industrial applications. Mobile phone technology has encouraged the production of small camera sensors with several megapixel resolution, which also opens up a new possibility for the development of such vision fiber-optical probes.

2. **Expressions of drag coefficient for a particle in the presence of a particle swarm.** The current drag correlations are mostly based on stationary or shear flow. There are two ways to obtain such expressions: (a) Correlating the experimental data directly against particle swarm. However, this approach is unrealistic considering the defects and limitations of existing techniques. (b) Using a theoretical or numerical simulation method. Firstly, the drag force of a single particle in the continuous phase is considered, then the interaction between the particles. Thus, the relationship between expressions of the drag coefficient for a single particle and for a particle swarm can be obtained.

3. **Numerical techniques.** The k–ε model assumes turbulence isotropy. Future work should include further improvements of RANS models such as EASM and RSM for both good prediction ability and computational efficiency. Also, faster and more accurate LES models need be developed to verify the improved

version of the RANS models, particularly for Eulerian–Eulerian simulation of diversified stirred multiphase systems. Thus, efforts should also be devoted to finding efficient algorithms and massive parallel techniques.

4. **The full model of stirred reactors.** The mass/heat transfer and complex chemical reactions need to be incorporated into the CFD simulation of stirred tanks, so as to produce a full numerical program to simulate the performance of industrial stirred reactors. The coupling of multiphase flow with macromixing/micromixing, transport phenomena, and chemical reactions is the key issue to be resolved before achieving this ultimate goal of chemical reaction engineering.

NOMENCLATURE

A	area, or coefficient in the discretized equation	m^2, –
a_{ij}	Reynolds stress anisotropy tensor	m^2/s^2
B_c	bottom clearance	m
C	clearance of the impeller plane off the bottom of the stirred tank	m
C_A	added mass coefficient, or concentration of substance A	–, mol/m^3 or kg/m^3
C_b	empirical constant in extra production of turbulent kinetic energy	–
C_D	drag coefficient of a swarm	–
C_d	drag coefficient of a single bubble, drop or particle	–
C_f	empirical parameter in drag force	
C_l	lift coefficient	–
C_p	concentration of product	mol/m^3
C_{TD}	coefficient of turbulent dispersion force	–
C_V	coefficient of virtual mass force	
C_W	coefficient of the wall force	–
C_μ	model constant of k–ε model	–
c_0	nodes adjacent to outlet boundary	
c_k	parameters in EASM	
c_ε	parameters in EASM	
c_s	parameters in Smagorinsky model	
c_w	parameters in EASM	
D_L	molecular diffusivity of gas in liquid	m^2/s
D, d	diameter	m
du, dv, dw	pressure correction coefficient	$m/(Pa \cdot s)$
Eo	Eotvos number	–
Eo_d	modified Eotvos number	–
F	force	N
G	production of turbulent kinetic energy	$kg/(m \cdot s^3)$
G_e	extra production of turbulent kinetic energy	$kg/(m \cdot s^3)$
g	acceleration due to gravity	m/s^2

H	height of fluid in the reactor	m
h	monitoring height	m
h_f	feed position	m
K	correction parameter	
k	turbulent kinetic energy	m^2/s^2
$k_L a$	gas transfer coefficient	s^{-1}
L_p	height of packing	m
Mo	Morton number	–
N	number	
N	rotation speed	rpm
N_c	production to dissipation ratio	
n	number density	
P	pressure	Pa
Re	Reynolds number	–
Re_d	Reynolds number of the dispersed phase, $Re_d = d_p\lvert\mathbf{u}_d - \mathbf{u}_c\rvert\rho_c/\mu_c$	
r	radial position	m
r_0	radial coordinate at the axis	m
S	source term in the discretized equation	
S_b	area of the bubble	m^2
S_{ij}	normalized mean strain rate tensor	
T	diameter of the stirred tank	m
T_c	top clearance	m
T_L	integral time scale of turbulence	s
T_{kij}	viscous stress tensor in phase k	kg/(m·s²)
t	time	s
t_c	circulation time	s
$u, U, \mathbf{u}$	velocity	m/s
V_b	volume of the bubble	m^3
V_s	total entrainment rate	$m\cdot s^{-1}$
v	velocity	m/s
v_s	local entrainment rate	m/s
w	velocity	m/s
We	Weber number	–
W_{ij}	normalized absolute rotation rate tensor	
x, y, z	Cartesian coordinate	m

Greek letters

α	holdup of phase	–
β	coefficient in EASM	β
Δ	filter length	Δ
δ_{ij}	Kronecker delta	
ε	turbulent energy dissipation	m^2/s^3
ζ_r	ratio of characteristic time of turbulence in liquid and characteristic time scale of bubble necessary to cross the containing energy eddies	–

η_r	ratio of characteristic time scale of turbulence seen by gas phase and characteristic time scale of bubble entrainment by liquid motion	–
η_{1-5}	five independent invariants	
θ	azimuthal coordinate	°
Γ	diffusion coefficient	m²/s
λ	Kolmogorov length scale	m
μ	viscosity	Pa·s
Φ, ϕ	general variables	
ν	kinematic viscosity	m²/s
ρ	density	kg/m³
ΔH	heat of reaction	kJ/mol
σ	surface tension	N/m
σ_k	model constant of $k–\varepsilon$ model	–
σ_t	turbulent Schmidt number	–
σ_ε	model constant of $k–\varepsilon$ model	–
τ_{ij}	Reynolds stress tensor	m²/s²
τ_l	mean eddy lifetime	s
τ_p	particle response time	s
Ω_i	rotation rate vector of non-inertial frame	rad/s
ω_{ij}	normalized mean rotation rate tensor	

Subscripts

A	added force
ap	apparent
BI	bubble induced
b	bubble
c	continuous phase or column
D	drag force
d	drag force or downcomer
eff	effective
FOU	first order upwind
g	gas
H_2	hydrogen
i, j, k	directions of coordinate
k	phase
l	liquid or lift
m	mixture or phase
p	dispersion phase or particle
r	radial direction or riser
SI	shear induced
s	separator
slip	slip velocity
t	terminal velocity
θ	tangential direction

V	virtual
VM	virtual mass
z	axial direction

Superscripts

dc	deferred correction
n	value in the step n
norm	normalized quantity
T	matrix transposition
t, turb	turbulent
*	predicted value in last iteration
'	correction in this iteration or turbulent fluctuation

REFERENCES

Afshari, A., Shotorban, B., Mashayek, F., Shih, T. I. P., & Jaberi, F. A. (2004). Development and validation of a multi-block flow solver for large eddy simulation of turbulent flows in complex geometries. *Proceedings of 42nd AIAA aerospace sciences meeting and exhibit* (pp. 1492–1500), Reno, Nevada.

Alopaeus, V., Koskinen, J., & Keskinen, K. I. (1999). Simulation of the population balances for liquid–liquid systems in a nonideal stirred tank. Part 1. Description and qualitative validation of the model. *Chem. Eng. Sci.*, *54*(24), 5887–5899.

Apte, S. V., Gorokhovski, M., & Moin, P. (2003). LES of atomizing spray with stochastic modeling of secondary breakup. *Int. J. Multiphase Flow*, *29*, 1503–1522.

Bakker, A., & van den Akker, H. E. A. (1994). A computational model for the gas–liquid flow in stirred reactors. *Chem. Eng. Res. Des.*, *72*(A4), 594–606.

Barnea, E., & Mizrahi, J. (1975). A generalised approach to the fluid dynamics of particulate systems part 2: Sedimentation and fluidisation of clouds of spherical liquid drops. *Can. J. Chem. Eng.*, *53*(5), 461–468.

Brown, D. E., & Pitt, K. (1970). Drop break-up in a stirred liquid–liquid contactor. *Proc. Chemeca.*, *70*, 87–90.

Brucato, A., Ciofalo, M., Grisafi, F., & Micale, G. (1994). Complete numerical solution of flow fields in baffled stirred vessels: The inner–outer approach. *Proceeding of 8th European conference on mixing* (pp. 155–162), Cambridge, UK.

Brucato, A., Grisafi, F., & Montante, G. (1998 a). Particle drag coefficients in turbulent fluids. *Chem. Eng. Sci.*, *53*(18), 3295–3314.

Brucato, A., Ciofalo, M., Grisafi, F., & Micale, G. (1998 b). Numerical prediction of flow fields in baffled stirred vessels: A comparison of alternative modeling approaches. *Chem. Eng. Sci.*, *53*, 3653–3684.

Buffo, A., Vanni, M., & Marchisio, D. L. (2012). Multidimensional population balance model for the simulation of turbulent gas–liquid systems in stirred tank reactors. *Chem. Eng. Sci.*, *70*, 31–40.

Bujalski, W., Takenaka, K., Paolini, S., Jahoda, M., Paglianti, A., Takahashi, A., Nienow, A. W., & Etchells, A. W. (1999). Suspensions and liquid homogenisation in high solids concentration stirred chemical reactors. *Trans. IChemE*, *77*, 241–247.

Calderbank, P. H. (1958). Physical rate processes in industrial fermentation. Part I: The interfacial area in gas–liquid contacting with mechanical agitation. *Trans. Inst. Chem. Eng.*, *36*, 443–463.

Carver, M. B., & Salcudean, M. (1986). Three-dimensional numerical modeling of phase distribution of two-fluid flow in elbows and return bends. *Numer. Heat Transfer A.*, *10*(3), 229–251.

Chapman, C. M., Nienow, A. W., Cooke, M., & Middleton, J. C. (1983). Particle–gas–liquid mixing in stirred vessels, part III: Three phase mixing. *Chem. Eng. Res. Des.*, *60*, 167–181.

Cheng, D., Cheng, J. C., Yong, Y. M., Yang, C., & Mao, Z. -S. (2011). CFD prediction of the critical agitation speed for complete dispersion in liquid–liquid stirred reactors. *Chem. Eng. Technol.*, *34*(12), 1–13.

Cheng, D., Cheng, J. C., Li, X. Y., Wang, X., Yang, C., & Mao, Z. -S. (2012). Experimental study on gas–liquid–liquid macro-mixing in a stirred tank. *Chem. Eng. Sci.*, *75*, 256–266.

Cheung, S. C. P., Yeoh, G. H., & Tu, J. Y. (2007). On the modeling of population balance in isothermal vertical bubbly flows – Average bubble number density approach. *Chem. Eng. Process*, *46*, 742–756.

Clift, R., Grace, J. R., & Weber, M. E. (1978). *Bubbles, drops and particles*. New York: Academic Press.

Cockbain, E. G., & McRoberts, T. S. (1953). The stability of elementary emulsion drops and emulsions. *J. Colloid Sci.*, *8*, 440–451.

Coulaloglou, C. A., & Tavlarides, L. L. (1977). Description of interaction processes in agitated liquid–liquid dispersions. *Chem. Eng. Sci.*, *32*(11), 1289–1297.

Daskopoulos, P., & Harris, C. K. (1996). Three dimensional CFD simulations of turbulent flow in baffled stirred tanks: An assessment of the current position. *I. Chem. E. Symposium Series*, *140*, 1–113.

de Bertodano, L.M. A. (1992). *Turbulent bubbly two-phase flow in a triangular duct*. Ph.D. Thesis, Rensselaer Polytechnic Institute, Troy, New York.

Deen, N. G., Solberg, T., & Hjertager, B. H. (2002). Flow generated by an aerated Rushton impeller: Two-phase PIV experiments and numerical simulations. *Can. J. Chem. Eng.*, *80*(4), 638–652.

Derksen, J. J. (2003). Numerical simulation of solids suspension in a stirred tank. *AIChE J.*, *49*(11), 2700–2714.

Fang, J., Yang, C., Yu, G. Z., & Mao, Z. -S. (2005). Preliminary investigation on interphase mass transfer in agitated liquid–liquid–solid dispersion. *Chinese J. Process Eng.*, *5*, 125–130 (in Chinese).

Feng, X., Cheng, J. C., Li, X. Y., Yang, C., & Mao, Z. -S. (2012 a). Numerical simulation of turbulent flow in a baffled stirred tank with an explicit algebraic stress model. *Chem. Eng. Sci.*, *69*(1), 30–44.

Feng, X., Li, X. Y., Cheng, J. C., Yang, C., & Mao, Z. -S. (2012 b). Numerical simulation of solid–liquid turbulent flow in a stirred tank with a two-phase explicit algebraic stress model. *Chem. Eng. Sci.*, *82*(12), 272–284.

Freitas, C. T., Street, R. L., Frindikakis, A. N., & Keseff, J. R. (1985). Numerical simulation of three-dimensional flow in a cavity. *Int. J. Numer. Meth. Fluids*, *5*, 561–575.

Frost, W., & Moulden, T. (1977). *Handbook of turbulence.* Volume 1 – *Fundamentals and applications.* New York: Plenum Press.

Gatski, T., & Speziale, C. (1993). On explicit algebraic stress models for complex turbulent flows. *J. Fluid Mech., 254*(9), 59–78.

Gillespie, T., & Rideal, E. K. (1956). Coalescence of drops at oil–water interface. *Trans. Faraday Soc., 52,* 173–183.

Gosman, A. D., Lekakou, C., Politis, S., Issa, R. I., & Looney, M. K. (1992). Multidimensional modeling of turbulent two-phase flows in stirred vessels. *AIChE J., 38*(12), 1946–1956.

Grienberger, J., & Hofmann, H. (1992). Investigations and modelling of bubble columns. *Chem. Eng. Sci., 47*(9–11), 2215–2220.

Guha, D., Ramachandran, P. A., & Dudukovic, M. P. (2007). Flow field of suspended solids in a stirred tank reactor by Lagrangian tracking. *Chem. Eng. Sci., 62,* 6143–6154.

Guha, D., Ramachandran, P. A., Dudukovic, M. P., & Derksen, J. J. (2008). Evaluation of large eddy simulation and Euler–Euler CFD models for solids flow dynamics in a stirred tank reactor. *AIChE J., 54*(3), 766–778.

Harris, C. K., Roekaerts, D., & Rosendal, F. J. J. (1996). Computational fluid dynamics for chemical reactor engineering. *Chem. Eng. Sci., 51,* 1569–1594.

Hinze, J. O. (1955). Fundamentals of the hydrodynamic mechanism of splitting in dispersion processes. *AIChE J., 1*(3), 289–295.

Honkanen, M., Eloranta, H., & Saarenrinne, P. (2010). Digital imaging measurement of dense multiphase flows in industrial processes. *Flow Measur. Instrum., 21*(1), 25–32.

Hosseini, S., Patel, D., Ein-Mozaffari, F., & Mehrvar, M. (2010). Study of solid–liquid mixing in agitated tanks through computational fluid dynamics modeling. *Ind. Eng. Chem. Res., 49,* 4426–4435.

Ishii, M., & Zuber, N. (1979). Drag coefficient and relative velocity in bubbly, droplet or particulate flows. *AIChE J., 25*(5), 843–855.

Joshi, J. B., Nere, N., Rane, C. V., Murthy, B. N., Mathpati, C. S., Patwardhan, A. W., & Ranade, V. V. (2011). CFD simulation of stirred tanks: Comparison of turbulence models (Part I: Radial flow impellers, Part II: Axial flow impellers, multiple impellers and multiphase dispersions). *Can. J. Chem. Eng., 89*(23–82), 754–816.

Kasat, G. R., Khopkar, A. R., Ranade, V. V., & Pandit, A. B. (2008). CFD simulation of liquid-phase mixing in solid–liquid stirred reactor. *Chem. Eng. Sci., 63,* 3877–3885.

Kataoka, I., Besnard, D. C., & Serizawa, A. (1992). Basic equation of turbulence and modeling of interfacial transfer terms in gas–liquid two-phase flow. *Chem. Eng. Commun., 118,* 221–236.

Kee, R. C. S., & Tan, R. B. H. (2002). CFD simulation of solids suspension in mixing vessels. *Can. J. Chem. Eng., 80,* 721–726.

Khopkar, A. R., & Ranade, V. V. (2006). CFD simulation of gas–liquid stirred vessel: VC, S33, and L33 flow regimes. *AIChE J., 52*(5), 1654–1672.

Khopkar, A. R., Aubin, J., Xureb, C., Le Sauze, N., Bertrand, J., & Ranade, V. V. (2003). Gas–liquid flow generated by a pitched blade turbine: PIV measurements and CFD simulations. *Ind. Eng. Chem. Res., 42,* 5318–5332.

Khopkar, A. R., Rammohan, A., Ranade, V. V., & Dudukovic, M. P. (2005). Gas–liquid flow generated by a Rushton turbine in stirred vessel: CARPT/CT measurements and CFD simulations. *Chem. Eng. Sci., 60,* 2215–2222.

Khopkar, A. R., Kasat, G. R., Pandit, A. B., & Ranade, V. V. (2006 a). Computational fluid dynamics simulation of the solid suspension in a stirred slurry reactor. *Ind. Eng. Chem. Res., 45*(12), 4416–4428.

Khopkar, A. R., Kasat, G. R., Pandit, A. B., & Ranade, V. V. (2006 b). CFD simulation of mixing in tall gas–liquid stirred vessel: Role of local flow patterns. *Chem. Eng. Sci., 61,* 2921–2929.

Kraume, M. (1992). Mixing time in stirred suspensions. *Chem. Eng. Technol., 15,* 313–318.

Lamont, J. C., & Scott, D. S. (1970). An eddy cell model of mass transfer into the surface of a turbulent liquid. *AIChE J., 16,* 513–519.

Lane, G. L., Schwarz, M. P., & Evans, G. M. (2005). Numerical modelling of gas–liquid flow in stirred tanks. *Chem. Eng. Sci., 60*(8–9), 2203–2214.

Launder, B. E., Reece, G. J., & Rodi, W. (1975). Progress in development of a Reynolds-stress turbulence closure. *J. Fluid Mech., 68,* 537–566.

Laurenzi, F., Coroneo, M., Montante, G., Paglianti, A., & Magelli, F. (2009). Experimental and computational analysis of immiscible liquid–liquid dispersions in stirred vessels. *Chem. Eng. Res. Des., 87*(4), 507–514.

Li, X. Y., Yang, C., Yang, S. F., & Li, G. Z. (2012). Fiber-optical sensors: Basics and applications in multiphase reactors. *Sensors, 12,* 12519–12544.

Ljungqvist, M., & Rasmuson, A. (2001). Numerical simulation of the two-phase flow in an axially stirred vessel. *Chem. Eng. Res. Des., 79*(5), 533–546.

Lucas, D., Kreppera, E., & Prasserb, H. M. (2007). Use of models for lift, wall and turbulent dispersion forces acting on bubbles for poly-disperse flows. *Chem. Eng. Sci., 62,* 4146–4157.

Mak, A. T. C. (1992). Solid–liquid mixing in mechanically agitated vessels. *Ph. D. Thesis.* University of London.

Matsumura, M., Hideo, S., & Tamitoshi, Y. (1982). Gas entrainment in a new entraining fermenter. *J. Ferment. Technol., 60*(5), 457–467.

Micale, G., Montante, G., Grisafi, F., Brucato, A., & Godfrey, J. (2000). CFD simulation of particle distribution in stirred vessels. *Chem. Eng. Res. Des., 78,* 435–444.

Micale, G., Grisafi, F., Rizzuti, L., & Brucato, A. (2004). CFD simulation of particle suspension height in stirred vessels. *Chem. Eng. Res. Des., 82*(9), 1204–1213.

Michele, V., & Hempel, D. (2002). Liquid flow and phase holdup-measurement and CFD modeling for two- and three-phase bubble columns. *Chem. Eng. Sci., 57,* 1899–1908.

Mlynek, Y., & Resnick, W. (1972). Drop sizes in an agitated liquid–liquid system. *AIChE J., 18*(1), 122–127.

Montante, G., & Magelli, F. (2005). Modelling of solids distribution in stirred tanks: Analysis of simulation strategies and comparison with experimental data. *Int. J. Comput. Fluid Dyn., 19,* 253–262.

Montante, G., & Magelli, F. (2007). Mixed solids distribution in stirred vessels: Experiments and computational fluid dynamics simulations. *Ind. Eng. Chem. Res., 46,* 2885–2891.

Murthy, B. N., & Joshi, J. B. (2008). Assessment of standard k–ε, RSM and LES turbulence models in a baffled stirred vessel agitated by various impeller designs. *Chem. Eng. Sci., 63,* 5468–5495.

Murthy, B. N., Ghadge, R. S., & Joshi, J. B. (2007). CFD simulations of gas–liquid–solid stirred reactor: Prediction of critical impeller speed for solid suspension. *Chem. Eng. Sci., 62,* 7184–7195.

Murthy, J. Y., Mathur, S. R., & Choudhary, D. (1994). CFD Simulation of flows in stirred tank reactors using a sliding mesh technique. *Proceedings of 8th European conference on mixing,* 341–348, Cambridge, UK.

Nienow, A., Warmoeskerken, M., Smith, J., & Konno, M. (1985). On the flooding–loading transition and the complete dispersal condition in aerated vessels agitated by a Rushton turbine. *Proceedings of the 5th European conference on mixing* (pp. 143–154), Bedford, UK.

Ochieng, A., & Lewis, A. E. (2006). Nickel solids concentration distribution in a stirred tank. *Miner. Eng.*, *19*(2), 180–189.

Padial, N., Vander Heyden, W., Rauenzahn, R., & Yarbro, S. (2000). Three-dimensional simulation of a three-phase draft-tube bubble column. *Chem. Eng. Sci.*, *55*, 3261–3273.

Paglianti, A., Pintus, S., & Giona, M. (2000). Time-series analysis approach for the identification of flooding/loading transition in gas–liquid stirred tank reactors. *Chem. Eng. Sci.*, *55*(23), 5793–5802.

Panneerselvam, R., Savithri, S., & Surender, G. D. (2008). CFD modeling of gas–liquid–solid mechanically agitated contactor. *Chem. Eng. Res. Des.*, *86*, 1331–1344.

Patankar, S. V. (1980). *Numerical heat transfer and fluid flow*. New York: McGraw-Hill.

Paul, E., Atiemo-Obeng, A., & Kresta, S. (2004). *Handbook of industrial mixing: Science and practice*. New York: John Wiley.

Petitti, M., Nasuti, A., Marchisio, D. L., Vanni, M., Baldi, G., Mancini, N., & Podenzani, F. (2010). Bubble size distribution modeling in stirred gas–liquid reactors with QMOM augmented by a new correction algorithm. *AIChE J.*, *228*, 1182–1194.

Petitti, M., Vanni, M., Marchisio, D. L., Buffo, A., & Podenzani, F. (2013). Simulation of coalescence, break-up and mass transfer in a gas–liquid stirred tank with CQMOM. *Chem. Eng. J.*, *228*, 1182–1194.

Pinelli, D., Nocentini, M., & Magelli, F. (2001). Solids distribution in stirred slurry reactors: Influence of some mixer configurations and limits to the applicability of a simple model for predictions. *Chem. Eng. Commun.*, *188*(1), 91–107.

Pope, S. (1975). A more general effective-viscosity hypothesis. *J. Fluid Mech.*, *72*(2), 331–340.

Rammohan, A. R., Kemoun, A., Al-Dahhan, M. H., & Dudukovic, M. P. (2001). Characterization of single phase flows in stirred tanks via computer automated radioactive particle tracking (CARPT). *Chem. Eng. Res. Des.*, *79*, 831–844.

Ranade, V. V. (1995). Computational fluid dynamics for reactor engineering. *Rev. Chem. Eng.*, *11*, 229–289.

Ranade, V. V. (1997). An efficient computational model for simulating flow in stirred vessels: A case of Rushton turbine. *Chem. Eng. Sci.*, *52*(24), 4473–4484.

Ranade, V. V. (2002). *Computational flow modeling for chemical reactor engineering*. New York: Academic Press.

Ranade, V. V., & Dommeti, S. M. S. (1996). Computational snapshot of flow generated by axial impellers in baffled stirred vessels. *Chem. Eng. Res. Des.*, *74*(4), 476–484.

Ranade, V. V., & Van den Akker, H. E. A. (1994). A computational snapshot of gas–liquid flow in baffled stirred reactors. *Chem. Eng. Sci.*, *49*(24B), 5175–5192.

Ranade, V. V., Perrard, M., Xuereb, C., Le Sauze, N., & Bertrand, J. (2001). Influence of gas flow rate on the structure of trailing vortices of a Rushton turbine: PIV measurements and CFD simulations. *Chem. Eng. Res. Des.*, *79*(A8), 957–964.

Rewatkar, V. B., Raghava Rao, K. S. M. S., & Joshi, J. B. (1991). Critical impeller speed for solid suspension in mechanical agitated three-phase reactors. 1. Experimental part. *Ind. Eng. Chem. Res.*, *30*, 1770–1784.

Sardeshpande, M. V., Juvekar, V. A., & Ranade, V. V. (2010). Hysteresis in cloud heights during solid suspension in stirred tank reactor: Experiments and CFD simulations. *AIChE J.*, *56*, 2795–2804.

Sbrizzai, F., Lavezzo, V., Verzicco, R., Campolo, M., & Soldati, A. (2006). Direct numerical simulation of turbulent particle dispersion in an unbaffled stirred-tank reactor. *Chem. Eng. Sci.*, *61*, 2843–2851.

Schmelter, S. (2008). Modeling, analysis, and numerical solution of stirred liquid–liquid dispersions. *Comput. Method. Appl. Mech. Eng.*, *197*(49–50), 4125–4131.

Schwarz, M. P., & Turner, W. J. (1989). Applicability of the standard k–ε turbulence model to gas-stirred baths. *Math. Comput. Model.*, *12*(3), 273–279.

Shan, X. G., Yu, G. Z., Yang, C., Mao, Z. -S., & Zhang, W. G. (2008). Numerical simulation of liquid–solid flow in an unbaffled stirred tank with a pitched-blade turbine downflow. *Ind. Eng. Chem. Res.*, *47*(9), 2926–2940.

Shotorban, B., & Mashayek, F. (2005). Modeling subgrid-scale effects by approximate deconvolution. *Phys. Fluids*, *17*, 1–4.

Skelland, A., & Ramsay, G. (1987). Minimum agitator speeds for complete liquid–liquid dispersion. *Ind. Eng. Chem. Res.*, *26*, 77–81.

Smagorinsky, J. (1963). General circulation experiments with the primitive equations: I. The basic experiment. *Monthly Weather Rev.*, *91*(3), 99–165.

Sokolichin, A., Eigenberger, G., & Lapin, A. (2004). Simulation of buoyancy driven bubbly flow: Established simplifications and open questions. *AIChE J.*, *50*(1), 24–45.

Sun, H. Y., Mao, Z. -S., & Yu, G. Z. (2006). Experimental and numerical study of gas hold-up in surface aerated stirred tanks. *Chem. Eng. Sci.*, *61*(12), 4098–4110.

Tamburini, A., Cipollina, A., Micale, G., Ciofalo, M., & Brucato, A. (2009). Dense solid–liquid off-bottom suspension dynamics: Simulation and experiment. *Chem. Eng. Res. Des.*, *87*, 587–597.

Tamburini, A., Cipollina, A., Micale, G., Brucato, A., & Ciofalo, M. (2012). CFD simulations of dense solid–liquid suspensions in baffled stirred tanks: Prediction of the minimum impeller speed for complete suspension. *Chem. Eng. J.*, *193–194*, 234–255.

Tomiyama, A. (2004). Drag, lift and virtual mass forces acting on a single bubble. *3rd international symposium on two-phase flow modelling and experimentation* (pp. 22–24), Pisa, Italy.

Topiwala, H. (1972). Surface aeration in a laboratory fermenter at high power inputs. *J. Ferment. Technol.*, *50*, 668–675.

Uhl, V., & Gray, J. (1967). *Mixing: Theory and practice* (2). New York: Academic Press.

van Heuven, J. W., & Beek, W. J. (1971). Power input, drop size and minimum stirrer speed for liquid–liquid dispersions in stirred vessels. *Solvent extraction: Proceedings of the international solvent extraction conference*, *51*, 70–81, London, UK.

Veljkovic, V. B., Bicok, K. M., & Simonovi, D. M. (1991). Mechanism, onset and intensity of surface aeration in geometrically similar, sparged, agitated vessels. *Can. J. Chem. Eng.*, *69*(4), 916–926.

Wang, A. H., Yu, G. Z., & Mao, Z. -S. (2004). Characteristics of gas–liquid mass transfer of surface aerator with self-rotating floating baffle. *Petrol. Process. Petrochem.*, *35*, 51–55 (in Chinese).

Wang, F., & Mao, Z. -S. (2005). Numerical and experimental investigation of liquid–liquid two-phase flow in stirred tanks. *Ind. Eng. Chem. Res.*, *44*(15), 5776–5787.

Wang, F., Wang, W. J., & Mao, Z. -S. (2004 a). Numerical study of solid–liquid two-phase flow in stirred tanks with Rushton impeller – (I) Formulation and simulation of flow field. *Chinese J. Chem. Eng.*, *12*(5), 599–609.

Wang, F., Mao, Z. -S., & Shen, X. Q. (2004 b). Numerical study of solid–liquid two-phase flow in stirred tanks with Rushton impeller – (II) Prediction of critical impeller speed. *Chinese J. Chem. Eng.*, *12*(5), 610–614.

Wang, F., Mao, Z. -S., Wang, Y., & Yang, C. (2006). Measurement of phase holdups in liquid–liquid–solid three-phase stirred tanks and CFD simulation. *Chem. Eng. Sci.*, *61*, 7535–7550.

Wang, H., Jia, X., Wang, X., Zhou, Z., Wen, J., & Zhang, J. (2013). CFD modeling of hydrodynamic characteristics of a gas–liquid two-phase stirred tank. *Appl. Math. Model.*, *38*(1), 63–92.

Wang, Q. Z., Squires, K. D., & Simonin, O. (1998). Large eddy simulation of turbulent gas–solid flows in a vertical channel and evaluation of second-order models. *Int. J. Heat Fluid Flow*, *19*, 505–511.

Wang, T., Cheng, J. C., Li, X. Y., Yang, C., & Mao, Z. -S. (2013). Numerical simulation of a pitched-blade turbine stirred tank with mirror fluid method. *Can. J. Chem. Eng.*, *91*(5), 902–914.

Wang, W. J., & Mao, Z. -S. (2002). Numerical simulation of gas–liquid flow in a stirred tank with a Rushton impeller. *Chinese J. Chem. Eng.*, *10*(4), 385–395.

Wang, W. J., Mao, Z. -S., & Yang, C. (2006). Experimental and numerical investigation on gas holdup and flooding in an aerated stirred tank with Rushton impeller. *Ind. Eng. Chem. Res.*, *45*(3), 1141–1151.

Wallin, S., & Johansson, A. (2000). An explicit algebraic Reynolds stress model for incompressible and compressible turbulent flows. *J. Fluid Mech.*, *403*, 89–132.

White, D., & De Villiers, J. (1977). Rates of induced aeration in agitated vessels. *Chem. Eng. J.*, *14*, 113–118.

Wu, H., & Patterson, G. (1989). Laser-Doppler measurements of turbulent-flow parameters in a stirred mixer. *Chem. Eng. Sci.*, *44*, 2207–2221.

Yamazaki, H., Tojo, K., & Miyanami, K. (1986). Concentration profiles of solids suspended in a stirred tank. *Powder Technol.*, *48*, 205–216.

Yang, C., & Mao, Z. -S. (2005). Mirror fluid method for numerical simulation of sedimentation of a solid particle in a Newtonian fluid. *Phys. Rev. E.*, *71*, 036704.

Yu, G. Z., & Mao, Z. -S. (2004). Sedimentation and coalescence profiles in liquid–liquid batch settling experiments. *Chem. Eng. Technol.*, *27*(4), 407–413.

Yu, G. Z., Wang, R., & Mao, Z. -S. (2000). Dispersion characteristics and gas–liquid mass transfer in a gas–liquid–liquid self-aspirating reactor. *Petrol. Process. Petrochem.*, *31*, 54–59 (in Chinese).

Yu, G. Z., Mao, Z. -S., & Wang, R. (2002). A novel surface aeration configuration for improving gas–liquid mass transfer. *Chinese J. Chem. Eng.*, *10*(1), 39–44.

Yu, Q., Yu, G. Z., Yang, C., & Mao, Z. -S. (2007). Dispersion and phase separation characteristics of liquid–liquid–liquid systems. *Chinese J. Process Eng.*, *7*, 229–234.

Zhang, Y. H., Yang, C., & Mao, Z. -S. (2006). Large eddy simulation of liquid flow in a stirred tank with improved inner–outer iterative algorithm. *Chinese J. Chem. Eng.*, *14*(3), 321–329.

Zhang, Y. H., Yang, C., & Mao, Z. -S. (2008). Large eddy simulation of the gas–liquid flow in a stirred tank. *AIChE J.*, *54*(8), 1963–1974.

Zhang, Y. H., Yong, Y. M., Mao, Z. -S., Yang, C., Sun, H. Y., & Wang, H. L. (2009). Numerical simulation of gas–liquid flow in a stirred tank with swirl modification. *Chem. Eng. Technol.*, *32*(8), 1266–1273.

Zhang, Y., Bai, Y., & Wang, H. (2013). CFD analysis of inter-phase forces in a bubble stirred vessel. *Chem. Eng. Res. Des.*, *91*(1), 29–35.

Zhou, L. X., Chen, T., & Liao, C. M. (1994). A unified second-order moment two-phase turbulence model for simulating gas-particle flows. *ASME, J. Fluids Eng. Division*, *185*, 307–313.

Zhu, Y., & Wu, J. (2002). Critical impeller speed for suspending solids in aerated agitation tanks. *Can. J. Chem. Eng.*, *80*, 1–6.

Airlift loop reactors

4

4.1 INTRODUCTION

The advantages of airlift loop reactors (ALRs) include no moving parts, low power consumption, high mass and heat transfer characteristics, good solid suspension, homogeneous shear, and rapid mixing (Chisti, 1998; Petersen and Margaritis, 2001). The ALR is widely used in biotechnological processes such as fermentation (Pollard et al., 1998), wastewater treatment (Heijnen et al., 1990), and cell production. In all these applications, the prediction of gas-induced circulation of liquid in the reactor is a key step in reactor design (Chisti and Moo-Young, 1993). There are two types of ALRs: (1) the external loop airlift reactor (EALR), where the liquid circulation in the riser and downcomer takes place in two separate compartments, and (2) the internal airlift loop reactor (IALR) in which either a vertical baffle or a draft tube divides the reactor into a riser and a downcomer. The EALR usually reaches nearly total gas disengagement at the top separator, which gives rise to a higher difference in density or hydrostatic pressure. This results in lower gas recirculation (thus lower mass transfer) with higher liquid circulation so that liquid mixing is enhanced. The flow in these reactors is quite ordered in a cyclic pattern like a loop beginning from top to bottom.

It is widely accepted that the design and scale-up of ALRs still remain difficult due to the nature of complex multiphase flow, though considerable research advances have been made. With the development of computer hardware and software, and the advances in numerical algorithms and theories of fluid dynamics, computational fluid dynamics (CFD) has been popularly adopted in the design of chemical reactors in recent years. The first step in CFD simulations is to adopt the appropriate theoretical model, which means deducing the formulation that can represent the key physical phenomena in the reactors. However, the limiting steps in model development are the formulation of appropriate boundary conditions, closure laws determining turbulent effects, interfacial transfer fluxes, and bubble coalescence and breakage processes. The multi-fluid model is found to represent a tradeoff between accuracy and computational requirements for practical applications. The governing set of equations consists of the continuity and momentum equations for $N + 1$ phases, i.e., one phase corresponds to the liquid phase and the remaining N phases are gas bubble groups. It is well known that multiphase flow simulations are numerically unstable with both 2D and 3D models because of the application of low-accuracy discretization, the large density difference between the phases, ill-conditioned implementation of interfacial closure in the limit of zero void or liquid fraction, and large gradient or discontinuity in flow fields that may occur in these flow systems. Therefore, further studies on these issues are highly recommended. In the past two decades, there have been substantial developments in both experimental descriptions and modeling of bubbly flows in ALRs. However, physical understanding of local spatial and temporal scales is still very limited, and modeling has so far been restricted to very low gas fractions (Jakobsen et al., 2005).

In this chapter, both mathematical models of transport phenomena (including momentum transfer, mass/heat transfer, and chemical reactions) and numerical methods in CFD developed in recent years to predict the bubbly flow in ALRs are described. The performance of closure in these mechanistic models is evaluated. The experimental and theoretical progress in ALRs including hydrodynamics, macromixing and micromixing is addressed, and guidelines for the design and scale-up of ALRs are also suggested.

4.2 FLOW REGIME IDENTIFICATION

It is widely accepted that the hydrodynamics in a gas–liquid or gas–liquid–solid reactor is characterized by different flow regimes, namely homogeneous (bubbly flow), transitional, and heterogeneous (churn-turbulent flow) regimes, mainly depending on the superficial gas velocity. Another regime, i.e., slug flow, is also observed in small-diameter laboratory columns when large bubbles are restricted spatially by the column wall and draft tube. However, this regime does not occur in large industrial reactors (Vial et al., 2001). The homogeneous regime is encountered at low gas velocities and is characterized by a narrow bubble size distribution, a uniform radial gas holdup profile, and a minor bubble–bubble interaction. Consequently, coalescence and breakup phenomena can be neglected in this regime. It is well known that homogeneous conditions in a bubble column are generally present up to a superficial gas velocity of 3 cm/s in the air–water system if an efficient gas distributor is used.

However, in ALRs, the homogeneous regime can be extended up to higher superficial gas velocities because of the stabilizing effect of the circulating liquid velocity on the stability of gas–liquid flow. At higher gas velocities, the flow becomes unstable and the homogeneous regime cannot be maintained. Large bubbles are formed by coalescence, and these move with higher rise velocities than smaller bubbles. This flow pattern is referred to as a heterogeneous regime and is characterized by a wide bubble size distribution and a non-uniform radial gas holdup profile. The presence of large bubbles is evidence of churn-turbulent flow and can be used in regime discrimination. These two patterns are separated by a regime called the transition regime. It is noteworthy that when a poor gas distributor is used, heterogeneous conditions prevail at all gas flow rates. The bubble–bubble interactions are different in different flow regimes, and it is therefore extremely important to understand different hydrodynamics and flow regime transitions for the purpose of reactor design, operation, control, and scale-up.

It is widely accepted that the hydrodynamics, heat and mass transfer, and mixing behavior are quite different in different regimes, and flow regime identification can be realized based on variations in bubble size distributions. Typical flow patterns in a vertical pipe are shown in Figure 4.1 (the virtual side projections of three-dimensional gas fractions are shown on the left, while the two-dimensional cross-sectional diagram distributions of gas fractions are shown on the right) and the parameters for these examples are summarized in Table 4.1.

Both experimental studies and numerical simulations show that large bubbles rise up mainly in the central region of the column with large rise velocities, while

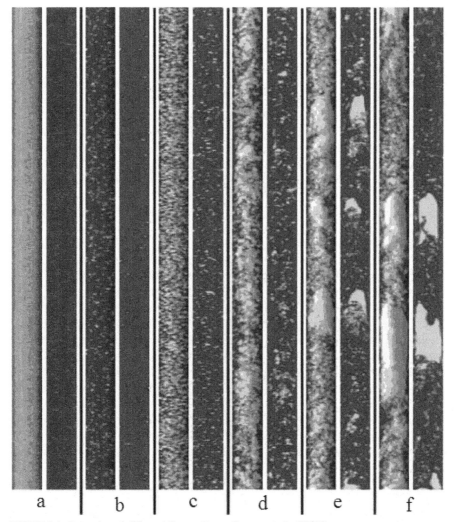

FIGURE 4.1 Examples of different flow patterns (Lucas et al., 2005).

Table 4.1 Parameters for the Different Flow Patterns Shown in Figure 4.1 (Lucas et al., 2005)

Test no.	u_l (m/s)	u_g (m/s)	Flow pattern
a	4.047	0.2190	Finely dispersed
b	0.405	0.0096	Bubbly flow with a wall peak
c	0.405	0.0574	Bubbly flow in the transition region
d	1.017	0.2190	Bubbly flow with a centre (core) peak
e	1.017	0.3420	Bubbly flow with a bimodal bubble size distribution
f	1.017	0.5340	Slug flow

small bubbles form a peak near the pipe wall and the radial profile of the volume fraction is much more uniform. Generally, the gas holdup increases with superficial gas velocity. However, this trend of increase shows different characteristics in the homogeneous and heterogeneous regimes: the average gas holdup increases almost linearly with increasing superficial gas velocity in the homogeneous regime, while the increase is less pronounced in the heterogeneous regime.

Much research has been undertaken to study the regime transition in bubbly flows and many advances have been made in the last decades. Several experimental approaches to identify the flow regime transition have been proposed. These fall generally into two types (Wang et al., 2007): one is based on the sharp variation in gas holdup (Vial et al., 2001) or drift flux (Thorat and Joshi, 2004) with respect to the superficial gas velocity, or the dynamic gas disengagement (DGD) technique (Li and Prakash, 2000); the other type is based on analysis of the dynamic signal by signal processing methods such as statistical analysis (Zhang et al., 1997), fractal analysis (Drahoš et al., 1992), chaotic analysis (Luewisutthichat et al., 1997), spectral analysis (Drahoš et al., 1991), and time–frequency analysis (Bakshi et al., 1995). In recent years, the regime prediction methods based on theoretical analysis, such as linear analysis (Thorat and Joshi, 2004) and the population balance model (PBM; Wang et al., 2005b), have also been developed. Example charts of flow regimes in a bubble column and an airlift loop reactor respectively, identified by the method of variation of gas volume fraction versus superficial velocity, are shown in Figure 4.2 (Vial et al., 2001). In the homogeneous regime, the slope of gas holdup versus superficial gas velocity is greater than or closer to 1 (SI unit, s/m) and the line passes through the origin. However, the slope is much less than 1 in the heterogeneous regime (Thorat and Joshi, 2004). It can be seen in Figure 4.2a with a multiple orifice sparger that the limit of the linear region, corresponding to the upper bound of the homogeneous region, is reached at a superficial gas velocity of 3 cm/s, and the heterogeneous regime begins when the value of the superficial gas velocity is higher than 9 cm/s, when the curve exhibits a minimum. As noted above and shown in Figure 4.2b, transition takes place at higher superficial gas velocities due to the presence of global liquid circulation in ALRs. Consequently, the limit of the homogeneous region is reached at a gas flow rate of about 5 cm/s, and heterogeneous conditions seem to prevail when

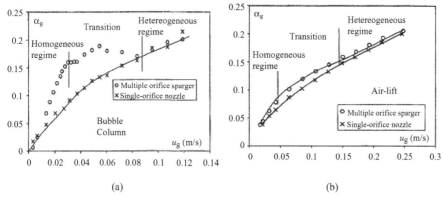

FIGURE 4.2 Gas holdup data in bubble columns and airlift loop reactors (Vial et al., 2001).

(a) Bubble column. (b) Airlift loop reactor.

the superficial gas velocity is greater than 15 cm/s. However, accurate determination of the transition is very difficult by simply observing the smallish slope evolution when the transition occurs, and some other models in this case, for instance the drift-flux analysis of the Zuber and Findlay (1965) model, should be used instead.

Compared to bubble columns, the global liquid circulation in an ALR has some special features. Firstly, it is preferable to homogenously suspend catalyst particles, which makes it possible to operate ALRs at low superficial gas velocities without solid sedimentation. Secondly, the gas holdup and mass transfer coefficient in an ALR are a little lower than those in a bubble column. Thirdly, variations of gas holdup and volumetric mass transfer coefficient with system properties and operating conditions, such as superficial gas velocity, pressure, temperature, liquid viscosity, and solid concentration, are less pronounced in an ALR than those in the bubble column. The latter occurs because, when the gas holdup increases due to variation in the operating conditions, e.g., with an increase in superficial gas velocity, the liquid circulation velocity also increases, which has a restraining influence on the increase of the gas holdup and volumetric mass transfer coefficient. Thus, significant differences in hydrodynamics between these two types of reactors are observed.

4.3 MATHEMATICAL MODELS AND NUMERICAL METHODS

Gas-induced liquid circulation in ALRs is a critical aspect of their design and operation, and how to predict and control the circulation is a crucial issue. In the past, many researchers studied IALRs using CFD to describe the hydrodynamics and have achieved significant success (e.g., Mudde and Van Den Akker, 2001; Vial et al., 2002; van Baten et al., 2003; Blažej et al., 2004a; Wu and Merchuk, 2004; Talvy et al., 2007a; Huang et al., 2010). Basically, two different modeling approaches are mostly used to predict the hydrodynamics of gas–liquid two-phase flow, i.e., the Eulerian–Eulerian (E-E) approach and the Eulerian–Lagrangian

(E-L) approach. In the Eulerian–Lagrangian representation, the individual bubbles are tracked separately while the liquid is treated as a continuum. The bubble motion is described by Newton's second law, offering the advantage of simple implementation for the forces exerted on individual bubbles. In the Eulerian–Eulerian approach, which is also referred to as the two-fluid model, both phases are treated as interpenetrating continua of fluids. The ensemble-averaged mass and momentum conservation equations are used to describe the time-dependent motion of both phases. The ensemble-averaged interaction terms, describing the effects of drag, virtual mass, wall lubrication, and lift forces, appear in the momentum balances of both phases. In contrast to the Eulerian–Lagrangian approach, no individual bubbles are considered. Both methods have some advantages and disadvantages in their respective ranges of applicability. Bubble–bubble interactions can be handled in an easier manner in the Eulerian–Lagrangian approach than those in the Eulerian–Eulerian method. However, Eulerian–Lagrangian simulations contain several severe drawbacks. Because of the necessity of individually tracking each bubble, these simulations require high-performance computers with large computer memory and storage space, and therefore are expensive and time-consuming. The advantage of the Eulerian–Eulerian approach over the Eulerian–Lagrangian approach becomes obvious when the void fraction of the dispersed phase is relatively high. The use of the Eulerian–Eulerian approach in these situations is much more suitable and practical. Therefore, the Eulerian–Eulerian multi-fluid model will be addressed in the following sections.

4.3.1 Eulerian–Eulerian two-fluid model

As a tradeoff between accuracy and computational efficiency in practical applications (Jakobsen et al., 2005), a two-fluid model is commonly adopted in slurry bubbly flow due to the difficulty of getting a convergent solution of gas–liquid–solid three-phase flows. For simulation of bubbly flow with transport and reaction, the Favre averaging two-fluid model has commonly been used (Ayed et al., 2007; Talvy et al., 2007a, b). The mass and momentum conservation equations can be written as

$$\frac{\partial \rho_k}{\partial t} + \frac{\partial}{\partial x_j}(\alpha_k \rho_k u_{ki}) = 0 \tag{4.1}$$

$$\frac{\partial \rho_k u_{ki}}{\partial t} + \frac{\partial}{\partial x_j}(\alpha_k \rho_k u_{ki} u_{ki}) = F_{ki} + \frac{\partial}{\partial x_j}(\alpha_k \tau_{kij}) - \frac{\partial}{\partial x_j}(\rho_k \alpha_k \overline{u'_{ki} u'_{kj}}) \tag{4.2}$$

where subscript k refers to liquid (l) or gas (g) phases; α_k, ρ_k and $\overline{u'_{ki} u'_{kj}}$ represent holdup, density, and the Reynolds stress tensor of phase k respectively; F_{ki} is the interphase momentum exchange between liquid and gas, which is a sum of familiar interphase forces including drag, added mass effect, lift, wall lubrication, and turbulent dispersion forces. Using the Boussinesq gradient transport hypothesis, the effect of fluctuating velocity correlation, which is called Reynolds stress, can be modeled following the practice of single-fluid flow as

$$\overline{u'_{ki}u'_{kj}} = -v_{kt}\left(\frac{\partial u_{ki}}{\partial x_j} + \frac{\partial u_{kj}}{\partial x_i}\right) + \frac{2}{3}k_k\delta_{ij} \tag{4.3}$$

The shear stress tensor is usually related to the mean velocity gradient using a Boussinesq hypothesis:

$$\tau_{kij} = \mu_{keff}(\nabla u_k + (\nabla u_k)^T) \tag{4.4}$$

It has been accepted in CFD that the only open questions are the models of the interphase forces and turbulence (Sokolichin et al., 2004). The closures of interfacial momentum exchange and turbulence models will be discussed in subsequent sections. A unified formulation for the continuity and momentum equations can be formulated as follows:

$$\frac{(\rho_k\alpha_k\phi)}{\partial t} + \nabla\cdot(\rho_k\alpha_k\mathbf{u}\phi) = \nabla\cdot(\alpha_k\Gamma_{\phi eff}\nabla\phi) + S_\phi \tag{4.5}$$

In the Cartesian coordinate system, the unified form can be expressed as

$$\frac{\partial(\rho_k\alpha_k\phi)}{\partial t} + \frac{\partial(\rho_k\alpha_k u_{kx}\phi)}{\partial x} + \frac{\partial(\rho_k\alpha_k u_{ky}\phi)}{\partial y} + \frac{\partial(\rho_k\alpha_k u_{kz}\phi)}{\partial z} = \frac{\partial}{\partial x}\left(\alpha_k\Gamma_{\phi eff}\frac{\partial\phi}{\partial x}\right)$$
$$+ \frac{\partial}{\partial y}\left(\alpha_k\Gamma_{\phi eff}\frac{\partial\phi}{\partial y}\right) + \frac{\partial}{\partial z}\left(\alpha_k\Gamma_{\phi eff}\frac{\partial\phi}{\partial z}\right) + S_\phi \tag{4.6}$$

However, the general formulation for the momentum and continuity equations in a cylindrical coordinate system can be written as

$$\frac{(\rho_k\alpha_k\phi)}{\partial t} + \frac{1}{r}\frac{\partial}{\partial r}(\rho_k\alpha_k ru_{kr}\phi) + \frac{1}{r}\frac{\partial}{\partial\theta}(\rho_k\alpha_k u_{k\theta}\phi) + \frac{\partial}{\partial z}(\rho_k\alpha_k u_{kz}\phi)$$
$$= \frac{1}{r}\frac{\partial}{\partial r}\left(\alpha_k\Gamma_{\phi eff}r\frac{\partial\phi}{\partial r}\right) + \frac{1}{r}\frac{\partial}{\partial\theta}\left(\frac{\alpha_k\Gamma_{\phi eff}}{r}\frac{\partial\phi}{\partial\theta}\right) + \frac{\partial}{\partial z}\left(\alpha_k\Gamma_{\phi eff}\frac{\partial\phi}{\partial z}\right) + S_\phi \tag{4.7}$$

The source terms and diffusion coefficients in the respective equations and coordinate systems are summarized in Table 4.2.

For Newtonian, viscous, and incompressible fluids in both phases, the densities of the liquid and gas are constant, and the local volume fractions are limited by

$$\alpha_1 + \alpha_g = 1.0 \tag{4.8}$$

4.3.2 Closure of interfacial forces

In multi-fluid models, different phases interact via interfacial forces and pressure. All these contributions are included as source terms in the corresponding momentum conservation equations. The interfacial coupling force, F_k, is assumed to be a

Table 4.2 Source Terms and Diffusion Coefficients for Different Coordinate Systems

Equation		ϕ	$\Gamma_{\phi,\text{eff}}$	Source term expression
Continuity		1	0	0
Cartesian coordinate system	Momentum source in x direction	u_x	$\mu_{k\text{eff}}$	$-\alpha_k \dfrac{\partial P}{\partial x} + \rho_k \alpha_k g_x + F_{kx} - \dfrac{2}{3}\rho_k \dfrac{\partial(\alpha_k k)}{\partial x} + \dfrac{\partial}{\partial x}\left(\alpha_k \mu_{k\text{eff}}\dfrac{\partial u_x}{\partial x}\right)$ $+ \dfrac{\partial}{\partial y}\left(\alpha_k \mu_{k\text{eff}}\dfrac{\partial u_y}{\partial x}\right) + \dfrac{\partial}{\partial z}\left(\alpha_k \mu_{k\text{eff}}\dfrac{\partial u_z}{\partial x}\right)$
	Momentum source in y direction	u_y	$\mu_{k\text{eff}}$	$-\alpha_k \dfrac{\partial P}{\partial y} + \rho_k \alpha_k g_y + F_{ky} - \dfrac{2}{3}\rho_k \dfrac{\partial(\alpha_k k)}{\partial y} + \dfrac{\partial}{\partial x}\left(\alpha_k \mu_{k\text{eff}}\dfrac{\partial u_x}{\partial y}\right)$ $+ \dfrac{\partial}{\partial y}\left(\alpha_k \mu_{k\text{eff}}\dfrac{\partial u_y}{\partial y}\right) + \dfrac{\partial}{\partial z}\left(\alpha_k \mu_{k\text{eff}}\dfrac{\partial u_z}{\partial y}\right)$
	Momentum source in z direction	u_z	$\mu_{k\text{eff}}$	$-\alpha_k \dfrac{\partial P}{\partial z} + \rho_k \alpha_k g_z + F_{kz} - \dfrac{2}{3}\rho_k \dfrac{\partial(\alpha_k k)}{\partial z} + \dfrac{\partial}{\partial x}\left(\alpha_k \mu_{k\text{eff}}\dfrac{\partial u_x}{\partial z}\right)$ $+ \dfrac{\partial}{\partial y}\left(\alpha_k \mu_{k\text{eff}}\dfrac{\partial u_y}{\partial z}\right) + \dfrac{\partial}{\partial z}\left(\alpha_k \mu_{k\text{eff}}\dfrac{\partial u_z}{\partial z}\right)$
Cylindrical coordinate system	Momentum source in r direction	u_r	$\mu_{k\text{eff}}$	$F_{kr} + \dfrac{1}{r}\dfrac{\partial}{\partial r}\left(r\alpha_k \mu_{k\text{eff}}\dfrac{\partial u_{kr}}{\partial r}\right) + \dfrac{1}{r}\dfrac{\partial}{\partial \theta}\left(r\alpha_k \mu_{k\text{eff}}\dfrac{\partial}{\partial r}\left(\dfrac{u_{k\theta}}{r}\right)\right) +$ $\dfrac{\partial}{\partial z}\left(\alpha_k \mu_{k\text{eff}}\dfrac{\partial u_{kz}}{\partial r}\right) - \dfrac{2\alpha_k \mu_{k\text{eff}}}{r^2}\dfrac{\partial u_{k\theta}}{\partial \theta} - \dfrac{2\alpha_k \mu_{k\text{eff}} u_{kr}}{r^2} +$ $\dfrac{\rho_k \alpha_k u_{k\theta}^2}{r} - \alpha_k \dfrac{\partial p}{\partial r} - \rho_k \dfrac{2}{3}\dfrac{\partial(\alpha_k k)}{\partial r}$
	Momentum source in θ direction	u_θ	$\mu_{k\text{eff}}$	$F_{k\theta} + \alpha_k \mu_{k\text{eff}}\dfrac{\partial}{\partial r}\left(\dfrac{u_{k\theta}}{r}\right) - \dfrac{1}{r}\dfrac{\partial}{\partial r}(\alpha_k \mu_{k\text{eff}} u_{k\theta}) - \dfrac{\rho_k \alpha_k u_{kr} u_{k\theta}}{r}$ $+ \dfrac{\alpha_k \mu_{k\text{eff}}}{r^2}\dfrac{\partial u_{kr}}{\partial \theta} + \dfrac{1}{r}\dfrac{\partial}{\partial r}\left(\alpha_k \mu_{k\text{eff}}\dfrac{\partial u_{kr}}{\partial \theta}\right) + \dfrac{1}{r}\dfrac{\partial}{\partial \theta}\left(\dfrac{\alpha_k \mu_{k\text{eff}}}{r}\dfrac{\partial u_{k\theta}}{\partial \theta}\right)$ $+ \dfrac{1}{r}\dfrac{\partial}{\partial \theta}\left(2\alpha_k \mu_{k\text{eff}}\dfrac{u_{kr}}{r}\right) + \dfrac{\partial}{\partial z}\left(\dfrac{\alpha_k \mu_{k\text{eff}}}{r}\dfrac{\partial u_{kz}}{\partial \theta}\right) - \dfrac{\alpha_k}{r}\dfrac{\partial p}{\partial \theta} -$ $\rho_k \dfrac{2}{3}\dfrac{1}{r}\dfrac{\partial(\alpha_k k)}{\partial \theta}$
	Momentum source in z direction	u_z	$\mu_{k\text{eff}}$	$F_{kz} + \dfrac{1}{r}\dfrac{\partial}{\partial r}\left(r\alpha_k \mu_{k\text{eff}}\dfrac{\partial u_{kr}}{\partial z}\right) + \dfrac{1}{r}\dfrac{\partial}{\partial \theta}\left(\alpha_k \mu_{k\text{eff}}\dfrac{\partial u_{k\theta}}{\partial z}\right) +$ $\dfrac{\partial}{\partial z}\left(\alpha_k \mu_{k\text{eff}}\dfrac{\partial u_{kz}}{\partial z}\right) - \alpha_k \dfrac{\partial p}{\partial z} - \rho_k \alpha_k g - \rho_k \dfrac{2}{3}\dfrac{\partial(\alpha_k k)}{\partial z}$

linear combination of several underling body forces, such as drag, turbulent dispersion force, lift force, wall lubrication force, and virtual mass force. Mao (2008) pointed out that some of these forces have a sound physical basis, but some do not; some forces have been correlated with simple formulations with sufficient accuracy, but some demand further attention. Sokolichin et al. (2004) argued that the pressure and the drag are the most relevant forces of all, and without the action of the pressure force a bubble released in stagnant water would not rise, while without the effect of the drag force the released bubble would accelerate infinitely. The purpose of this section is to summarize the interfacial forces commonly included in the simulations.

The forces acting on a bubble are usually related to some nondimensional numbers, i.e., Reynolds number, Eötvös number, and Morton number (Lucas et al., 2007), defined as

$$Re = \frac{\rho_1 u d_b}{\mu_1} \tag{4.9}$$

$$Eo = \frac{g(\rho_1 - \rho_g)d_b^2}{\sigma} \tag{4.10}$$

$$Mo = \frac{g\mu_1^4(\rho_1 - \rho_g)}{\rho_1 \sigma^3} \tag{4.11}$$

4.3.2.1 Pressure force and gravity force
In the frame of a multi-fluid model, the pressure force resulting from the global pressure gradient can be written as

$$\mathbf{F}_p = \alpha_g \nabla p \tag{4.12}$$

The influence of gravity, which constitutes only part of $\mathbf{F}_p$, can be computed by the following relation:

$$\mathbf{F}_p = \alpha_g \rho_1 \mathbf{g} \tag{4.13}$$

4.3.2.2 Drag force
A steadily moving bubble accelerates part of the surrounding liquid and in turn its motion will be hindered by the surrounding liquid. This interphase interaction is called drag force. In the early stage, many researchers (e.g., Schwarz and Turner, 1988; Azzaro et al., 1992; Becker et al., 1994; Deng et al., 1996; Lin et al., 1997b; Lapin et al., 2001) used a very simple expression for the drag:

$$\mathbf{F}_d = 5 \times 10^4 \alpha_1 \alpha_g (\mathbf{u}_g - \mathbf{u}_1) \tag{4.14}$$

which is deduced from a mean bubble slip velocity of about 0.2 m/s and agrees well with the experimental data for air bubbles in tap water. This velocity is close to the velocities measured for bubbles in the size range 1–10 mm in bubbly flow, and this simple correlation may thus often provide a better estimation than more complex ones. An essential advantage of this formulation lies in the fact that it requires very little computational expense.

Usually, the drag force on a bubble can be calculated by

$$\mathbf{F}_d = \tfrac{1}{2}\rho_l C_d A |\mathbf{u}_g - \mathbf{u}_l|(\mathbf{u}_l - \mathbf{u}_g) \tag{4.15}$$

where A is its projection area and can be written as

$$A = \pi d_b^2 / 4 \tag{4.16}$$

The number of bubbles per unit volume can be determined as

$$N_b = \frac{6\alpha_g}{\pi d_b^3} \tag{4.17}$$

So the force acting on N_b bubbles per unit volume can be found:

$$\mathbf{F}_d = \frac{3}{4}\frac{C_d}{d_b}\rho_l \alpha_g |\mathbf{u}_g - \mathbf{u}_l|(\mathbf{u}_l - \mathbf{u}_g) \tag{4.18}$$

When the dispersed phase is present in its pure form, the above equation gives a very large drag while it should be solved only with the single-phase momentum equations at such locations. In order to ensure that the interfacial force vanishes, the following formulation is used instead (Chen, P., 2004):

$$\mathbf{F}_d = \frac{3}{4}\frac{C_d}{d_b}\rho_l \alpha_l \alpha_g |\mathbf{u}_g - \mathbf{u}_l|(\mathbf{u}_l - \mathbf{u}_g) \tag{4.19}$$

The aforementioned equation can be taken as a correction to consider the effect of the adjacent bubbles to some extent. This formulation has a wide range of application and has been used by many authors (Kerdouss et al., 2006; Khopkar and Ranade, 2006; Khopkar et al., 2006; Huang et al., 2010).

The derivation of reliable empirical correlations for the drag coefficient C_d is complicated by the fact that direct measurement of the drag force acting on a gas bubble is possible only for the terminal rise velocity of a single air bubble in a stagnant liquid (Sokolichin et al., 2004). Since the experimental information available to date is insufficient to determine unambiguously the drag coefficient as an exact function of related variables, the literature abounds with various correlations with slight differences.

Tomiyama (1998b) proposed the following set of corrections as the bubble drag coefficient for a pure, slightly contaminated or contaminated system respectively:

$$C_d = \max\left\{\min\left[\frac{16}{Re}(1+0.15Re^{0.687}), \frac{48}{Re}\right], \frac{8}{3}\frac{Eo}{Eo+4}\right\} \tag{4.20}$$

$$C_d = \max\left\{\min\left[\frac{24}{Re}(1+0.15Re^{0.687}), \frac{72}{Re}\right], \frac{8}{3}\frac{Eo}{Eo+4}\right\} \tag{4.21}$$

$$C_{\mathrm{d}} = \max\left[\frac{24}{Re}(1+0.15Re^{0.687}), \frac{8}{3}\frac{Eo}{Eo+4}\right] \tag{4.22}$$

As noted by Tomiyama (1998b), the air–tap water system may correspond to contaminated or slightly contaminated water. Water carefully distilled two or more times belongs to the pure liquid system. These correlations represent experimental data very well for $10^{-2} < Eo < 10^3$, $10^{-3} < Re < 10^6$, and $10^{-14} < Mo < 10^7$.

Morsi and Alexander (1972) proposed an expression related to the Reynolds number:

$$C_{\mathrm{d}} = a_1 + \frac{a_2}{Re} + \frac{a_3}{Re^2} \tag{4.23}$$

where a_1, a_2, and a_3 are constants, which are summarized in Table 4.3. This model has been adopted by Chandavimol (2003) to predict the bubbly flow in a stirred tank.

There is another relation adopted by many researchers (e.g., Kuo and Wallis, 1988; Lahey et al., 1993; Boisson and Malin, 1996; Ilegbusi et al., 1998; de Matos et al., 2004):

$$C_{\mathrm{d}} = \begin{cases} \dfrac{24}{Re}, & \text{if } Re < 0.49 \\[2mm] \dfrac{20.68}{Re^{0.643}}, & \text{if } 0.49 < Re < 100 \\[2mm] \dfrac{6.3}{Re^{0.385}}, & \text{if } Re < 100,\ We \le 8,\ Re \le \dfrac{2065.1}{We^{2.6}} \\[2mm] \dfrac{We}{3}, & \text{if } Re < 100,\ We \le 8,\ Re > \dfrac{2065.1}{We^{2.6}} \\[2mm] \dfrac{8}{3}, & \text{if } Re > 100,\ We > 8 \end{cases} \tag{4.24}$$

Table 4.3 Constants for the Drag Coefficient of Eq. (4.23)

Re	a_1	a_2	a_3
$Re < 0.1$	0	24	0
$0.1 < Re < 1.0$	3.69	22.73	0.0903
$1.0 < Re < 10.0$	1.222	29.17	−3.889
$10.0 < Re < 100.0$	0.6167	46.5	−116.67
$100.0 < Re < 1000.0$	0.3644	98.33	−2778.0
$1000.0 < Re < 5000.0$	0.357	148.62	−47,500.0
$5000.0 < Re < 10,000.0$	0.46	−490.55	578,700.0
$10,000.0 < Re < 50,000.0$	0.5191	−1662.5	5,416,700.0

Some other frequently used correlations for the drag coefficient in bubbly flow are presented as follows (Ishii and Zuber, 1979; Karamanev and Nikolov, 1992; Delnoij et al., 1997; Jakobsen et al., 1997):

$$C_d = \begin{cases} \dfrac{24}{Re}(1+0.15Re^{0.687}), & \text{if } Re < 1000 \\ 0.44, & \text{if } Re > 1000 \end{cases} \tag{4.25}$$

$$C_d = \begin{cases} \dfrac{24}{Re}(1+0.15Re^{0.687}) + \dfrac{0.413}{(1+16.3Re^{-1.09})}, & \text{if } Re < 135 \\ 0.95, & \text{if } Re \geq 135 \end{cases} \tag{4.26}$$

$$C_d = \tfrac{2}{3}\sqrt{Eo} \tag{4.27}$$

$$C_d = 0.622/(0.235+1/Eo) \tag{4.28}$$

The drag coefficient depends on flow regime and liquid properties. As a classical topic, the drag coefficient for a gas bubble in liquid has been extensively studied through the years. The sensitivity of the simulation results to the chosen drag coefficient is discussed later. Additionally, the knowledge on single bubbles in a quiescent fluid is relatively satisfactory, but other situations (turbulence, shear, rotation, oscillation, and unsteady motion) have not been understood sufficiently. We should note that the correlation has its own scope of application and the details were usually ignored by many researchers in practice.

However, the sensitivity study of drag formulation and drag coefficient correlations should be examined, and screening of the models of interphase forces should be done before the numerical simulations. Huang et al. (2008) compared the different drag formulations in an IALR and found that, despite considerable differences in functional forms, all the drag formulations resulted in good predictions comparable to experimental data. It should be noted that the drag coefficient correlation used in Eqs. (4.18) and (4.19) is modeled by Eq. (4.26). It is surprising that even the simple drag formulation of Eq. (4.14) has good accuracy, almost the same as those of complex ones. That is, the good results predicted with the simple drag formulation indicate that the predicted results are not sensitive with respect to a fixed slip velocity in bubbly flow. The different results of different drag formulations are compared with experimental data in Figures 4.3 and 4.4. The drag predicted by Eq. (4.14), which has a constant slip velocity of 0.2 m/s, slightly underpredicts the overall gas holdup in the reactor when the superficial gas velocity is greater than 0.04 m/s.

For Eqs. (4.23), (4.25), and (4.26), the drag coefficients are dependent on Reynolds numbers, though they have different forms. As compared in Figure 4.5, the predicted results of Eqs. (4.23) and (4.26) are negligible though there is a tremendous difference between them. For Eq. (4.26), the first part was proposed by Turton and Levenspiel (1986) for isolated and spherical bubbles, and the second part was put forward by Karamanev and Nikolov (1992) for isolated bubbles (including implicitly the effect of deformation).

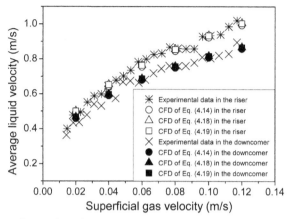

FIGURE 4.3 Comparisons of predicted average liquid velocities using different drag formulations with experimental data (Huang et al., 2008).

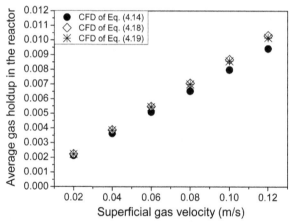

FIGURE 4.4 Comparisons of predicted gas holdups using different drag formulations with experimental data (Huang et al., 2008).

The predicted results using the same drag formulation of Eq. (4.19) but different frequently used drag coefficient correlations in bubbly flow are illustrated in Figures 4.6 and 4.7. It is shown that the drag coefficient correlation plays a dominant role in the simulations, and the higher the superficial gas velocity, the greater the importance. It is clear that the drag coefficient correlation of Eq. (4.26) gives the best agreement with experimental data and all the simulations agree very well at low superficial gas velocities, while underestimation occurs at high superficial gas velocities. The deviation between the CFD simulations and experiments may be due to bubble breakup and coalescence, and the flow field changes from the homogeneous flow to the transition regime flow or the heterogeneous flow at high superficial

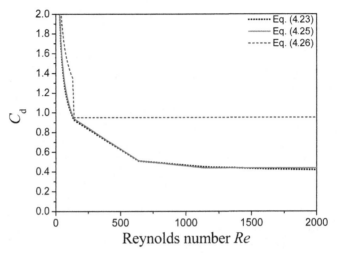

FIGURE 4.5 Predicted drag coefficients using different correlations versus Reynolds numbers.

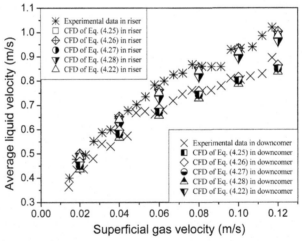

FIGURE 4.6 Comparisons of predicted average liquid velocities using different drag coefficient correlations with experimental data (Huang et al., 2008).

gas velocities. It can be concluded that the influence of drag coefficient correlations is greater than that of the drag formulations.

Recently, the influence of turbulence on drag has entered the scope of research related to CFD. Khopkar and Ranade (2006) argued that bubbles underwent significantly higher turbulence than that of a bubble moving through a stagnant liquid. Some attempts have been made to understand the influence of prevailing turbulence on drag coefficients in stirred tanks. The characteristic spatial scale of prevailing

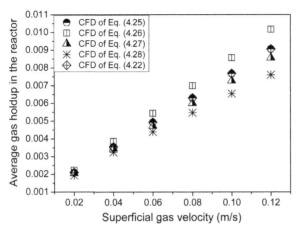

FIGURE 4.7 Comparisons of predicted gas holdups using different drag coefficient correlations (Huang et al., 2008).

turbulence is related to the drag coefficients in these models, which is a promising improvement.

Brucato et al. (1998) proposed a correlation of C_d based on the Kolmogorov length scale of turbulence and the ratio of the bubble size:

$$\frac{C_d - C_{d0}}{C_{d0}} = k_{ct} \left(\frac{d_b}{\lambda} \right)^3 \tag{4.29}$$

where C_d is the drag coefficient in turbulent flow and C_{d0} that in a stagnant liquid, k_{ct} is a correlation constant and was recommended to be 6.5×10^{-6}, and λ is the Kolmogorov length scale based on the volume-averaged energy dissipation rate:

$$\lambda = \left(\frac{v^3}{\varepsilon} \right)^{0.25} \tag{4.30}$$

where v is the kinematical viscosity of the continuous phase.

Bakker and Van den Akker (1994) attempted to relate the influence of turbulence on drag coefficient by using a modified Reynolds number in a usual correlation developed for the stagnant liquid. The modified Reynolds number is expressed as

$$Re^* = \frac{\rho_l |\mathbf{u}_b - \mathbf{u}_l| d_b}{\mu_l + \frac{2}{9} \mu_{t,l}} \tag{4.31}$$

It is noteworthy that the effective viscosity is calculated by adding some fraction of turbulent viscosity to molecular viscosity. The fraction here is introduced to account for the effect of turbulence in reducing the slip velocity.

Huang et al. (2008) investigated the influence of turbulence on the drag coefficient correlation in an IALR, and the results are illustrated in Figures 4.8 and 4.9

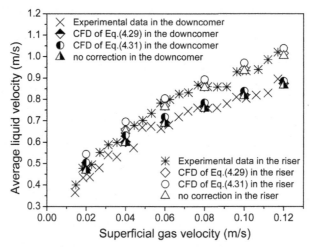

FIGURE 4.8 Comparisons of predicted average liquid velocities using different turbulent correction relations with experimental data (Huang et al., 2008).

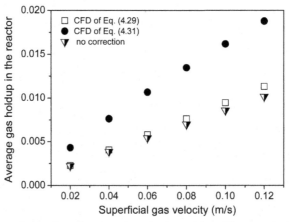

FIGURE 4.9 Comparisons of predicted total gas holdups using different turbulent correction relations (Huang et al., 2008).

with the drag formulation of Eq. (4.18) and the drag coefficient correlation of Eq. (4.26). It is found that using either Eq. (4.29) or Eq. (4.31) increases the actual drag coefficient, which results in the increase of gas holdup and liquid velocity compared to the no-turbulence correction. It is seen that the difference between these two methods is significant. The underprediction can be alleviated a little by the first correction method and the results agree well with experimental data. However, the second correction method gives overprediction in all the tested cases, and it needs further improvement to consider this effect in an IALR.

In practical multiphase processes, the bubbles are often small so as to generate a large specific surface area, which will result in high interphase mass and heat transfer rates. Therefore, the interactions between the bubbles must be considered in the simulations when the gas voidage is high. It is necessary to establish the correlations including the interactions between bubbles in a swarm. Unfortunately, the work on bubble swarms is not sufficient as compared with its counterpart for single bubbles (Mao, 2008; Mao and Yang, 2009).

There are two categories of extension for the drag coefficient from a single bubble (C_d) to a swarm (C_D). The first one argues that the drag increases with increasing gas voidage (e.g., Richardson and Zaki, 1954), and the second one believes that the situation is just the reverse. However, which one is right is still not yet clear (Schlueter and Raebiger, 1998; Sokolichin et al., 2004). Jakobsen et al. (1997) proposed two formulations to represent the reduced drag caused by cluster formation, bubbles in line or in wakes, etc. These two equations are purely empirical and dependent on bubble diameters, which fit the experimental data and have no theoretical basis, as follows:

$$C_D = 0.2 + \frac{(d_b - 0.0015)^2}{0.42(d_b - 0.0015)^2 + 2.7 \times 10^{-6}} \tag{4.32}$$

$$C_D = \left(1 + \frac{10^{-5}}{(d_b + 0.002)^2}\right) \times \left(\frac{(d_b - 0.0015)^2}{0.42(d_b - 0.0015)^2 + 5.7 \times 10^{-6}} + 0.4\right) \tag{4.33}$$

Similarly, Tomiyama (2004) also proposed a method to account for the effect of adjacent bubbles:

$$C_D = \frac{8}{3} \frac{Eo(1 - E^2)}{E^{2/3} Eo + 16(1 - E^2)E^{4/3}} F(E)^{-2} \tag{4.34}$$

$$E = \frac{1}{1 + 0.163 Eo^{0.757}} \tag{4.35}$$

$$F(E) = \frac{\sin^{-1} \sqrt{(1 - E^2)} - E\sqrt{(1 - E^2)}}{1 - E^2} \tag{4.36}$$

However, most of the correlations for bubble swarms are some kind of extension with gas voidage incorporated into the drag coefficient correlation of a single bubble. Pan et al. (1999, 2000) developed a drag correlation from that of Tomiyama (1998a), taking into account the effect of gas content by defining the Reynolds number and a correction coefficient. The drag coefficient can be expressed as

$$C_D = \max\left[\frac{24}{Re_\alpha}(1 + 0.15 Re_\alpha^{0.687}), \frac{8}{3}\frac{Eo}{Eo + 4} f(\alpha_g)\right] \tag{4.37}$$

$$Re_\alpha = \frac{\rho_1 d_b |\mathbf{u}_b - \mathbf{u}_1|(1-\alpha_g)}{\mu_1} \tag{4.38}$$

$$f(\alpha_g) = \left\{ \frac{1+17.67(1-\alpha_g)^{9/7}}{18.67(1-\alpha_g)^{3/2}} \right\}^2 \tag{4.39}$$

Antal et al. (1991) proposed a drag coefficient considering the influence of the neighboring bubbles in laminar bubbly flow. The equations are written as

$$C_D = \frac{24}{Re_\alpha}(1+0.1Re_\alpha^{0.75}) \tag{4.40}$$

$$Re_\alpha = \frac{\rho_1 d_b |\mathbf{u}_b - \mathbf{u}_1|}{\mu_m} \tag{4.41}$$

$$\mu_m = \frac{\mu_1}{1-\alpha_g} \tag{4.42}$$

Jia et al. (2007) utilized the drag coefficient model:

$$C_D = \max\left[\frac{24}{Re_m}(1+0.15Re_m^{0.687}), \min(\tfrac{2}{3}Eo^{0.5}E(\alpha_g), \tfrac{8}{3}(1-\alpha_g)^2) \right] \tag{4.43}$$

$$Re_m = \frac{\rho_1 d_b |\mathbf{u}_b - \mathbf{u}_1|}{\mu_m} \tag{4.44}$$

$$\mu_m = \mu_1 (1-\alpha_g)^{-2.5\mu^*} \tag{4.45}$$

$$\mu^* = \frac{\mu_g + 0.4\mu_1}{\mu_g + \mu_1} \tag{4.46}$$

$$E(\alpha_g) = \frac{1+17.67 f_1(\alpha_g)^{6/7}}{18.67 f_1(\alpha_g)} \tag{4.47}$$

Ishii and Zuber (1979) and Ishii and Mishima (1984) proposed a model of drag coefficient that is appropriate for a large span in gas fractions from 0 to 1.0 and dependent on local gas holdups. This model is applicable for a wide range of regimes, i.e., from the Stokes regime through the distorted particle regime, and up to the churn turbulent flow. The drag coefficient is corrected according to the flow regime and has been validated against a large experimental database. The correlations of drag coefficient for multi-bubble cases are listed in Table 4.4.

Table 4.4   Correlations of the Drag Coefficient for Multi-Bubbles (Ishii and Zuber, 1979; Ishii and Mishima, 1984)

Regime	Equation	Number
Stokes regime	$C_D = \dfrac{24}{Re_m}$	(4.48)
Viscous regime	$C_D = \dfrac{24(1+0.1Re_m^{0.75})}{Re_m}$	(4.49)
Newton's regime	$C_D = 0.45\left\{\dfrac{1+17.67[f_1(\alpha_g)]^{6/7}}{18.67f_1(\alpha_g)}\right\}^2$	(4.50)
Distorted fluid particle regime	$C_D = \dfrac{2}{3}\sqrt{Eo}\left(\dfrac{1+17.67[f(\alpha_g)]^{6/7}}{18.67f(\alpha_g)}\right)^2$	(4.51)
Churn turbulent flow regime	$C_D = \frac{8}{3}(1-\alpha_g)^2$	(4.52)
Slug flow regime	$C_D = 9.8(1-\alpha_g)^3$	(4.53)

In Table 4.4, the parameters used are defined as follows (μ_m and μ^* are the same as the definitions of Eqs. (4.45) and (4.46)):

$$Re_m = \frac{\rho_1 d_b |\mathbf{u}_b - \mathbf{u}_1|}{\mu_m} \tag{4.54}$$

$$f(\alpha_g) = (1.0 - \alpha_g)^{1.5} \tag{4.55}$$

$$f_1(\alpha_g) = \frac{\mu_1}{\mu_m}\sqrt{1.0 - \alpha_g} \tag{4.56}$$

Jakobsen et al. (1997) argued that at present this approach based on experimental data was apparently the best justified way to adjust the generalized drag coefficient correlation for high gas concentrations. However, Jakobsen et al. (1997) and Ishii and Zuber (1979) found that the relative velocity between the gas and the liquid decreased with decreasing gas fraction, though this correlation was still valid in a churn turbulent regime in a turbulent bubble column. The fact of smaller relative velocities predicted using this model with increasing gas contents is in direct contradiction to the experimental data of Yao et al. (1991) and Grienberger and Hofmann (1992).

Morud and Hjertager (1996) developed a correlation that is dependent on flow regimes, to take into account bubble–bubble interaction. The flow is classified into three categories, i.e., homogeneous bubbly flow ($\alpha_g \leq 0.3$), heterogeneous bubbly flow ($0.3 < \alpha_g \leq 0.7$), and droplet flow ($\alpha_g > 0.7$). The coefficient for bubbly flow and droplet flow is given by

$$C_D = \tfrac{2}{3} d_b \sqrt{\frac{g\Delta\rho}{\sigma}} \left(\frac{1 + 17.67[f(\alpha_g)]^{6/7}}{18.67 f(\alpha_g)}\right)^2 \tag{4.57}$$

where

$$f(\alpha_g) = \begin{cases} (1.0 - \alpha_g)^{1.5}, & \alpha_g \leq 0.3 \\ \alpha_g^3, & \alpha_g > 0.7 \end{cases} \tag{4.58}$$

For churn turbulent flow, it is calculated from

$$C_D = \tfrac{8}{3}(1 - \alpha_g)^2, \qquad 0.3 < \alpha_g \leq 0.7 \tag{4.59}$$

In general, many correlations can be generalized as a function, which is dependent on gas holdup, multiplying by the drag coefficient of a single bubble. The following corrected correlation is preferred by many authors in practical applications:

$$C_D = C_d E(\alpha_g) \tag{4.60}$$

However, the function $E(\alpha_g)$ differs greatly in different cases and it is probably not only a function of the phase fraction, but also of some other numbers, for instance, the Archimedes and Reynolds numbers. Also, it is obvious that it should approach 1.0 when the gas fraction vanishes.

Azbel (1981) proposed a correction function dependent on gas holdup:

$$E(\alpha_g) = \frac{1 - \alpha_g^{5/3}}{(1 - \alpha_g)^2} \tag{4.61}$$

Rampure et al. (2007) put forward another correction function:

$$E(\alpha_g) = (1 - \alpha_g)^p \tag{4.62}$$

where the exponent p (in the range of $p = 1$–4) is dependent on the superficial gas velocity, and the higher the superficial gas velocity, the higher the value of exponent p. Behzadi et al. (2004) proposed a function that combined a power law with an exponential function to approximate the effects of neighbors, which is given by

$$E(\alpha_g) = \exp(3.64\alpha_g) + \alpha_g^{0.864} \tag{4.63}$$

In flow regions with high void fractions, Yeoh and Tu (2006) proposed a function given by

$$E(\alpha_g) = \left(\frac{1 + 17.67 \dfrac{\mu_l}{\mu_m}(1 - \alpha_g)^{3/7}}{18.67 \dfrac{\mu_l}{\mu_m}(1 - \alpha_g)^{1/2}}\right)^2 \tag{4.64}$$

León-Becerril et al. (2002) derived an expression that is even applicable to the deforming bubbles, as given by

$$C_D = C_d x^{2/3} E(\alpha_g, x) \tag{4.65}$$

with

$$E(\alpha_g) = [1 - p_2(x)\alpha_g]^{-2} \tag{4.66}$$

$$p_2(x) = \frac{1.43[2 + Z(x)]}{3} \tag{4.67}$$

$$Z(x) = 2\frac{(x^2 - 1)^{1/2} - \cos^{-1} x^{-1}}{\cos^{-1} x^{-1} - (x^2 - 1)^{1/2} / x^2} \tag{4.68}$$

$$x = \frac{a}{b} \tag{4.69}$$

$$Re_b = \frac{\rho_1 d_b |\mathbf{u}_b - \mathbf{u}_1|(1 - \alpha_g)}{\mu_1} \tag{4.70}$$

$$d_b = 2\sqrt[3]{a^2 b} \tag{4.71}$$

where a and b are the long axis and the small axis respectively. For a spherical bubble, $p_2 = 1.43$, and for an ellipticity $x = 1.8$ and $p_2 = 1.73$.

It is evident that different correction methods impact differently on the predicted results, and these corrections seem rather arbitrary. The influence of drag correction function on the predicted results should be examined further and it is very useful to the closure of interphase forces with bubbly flow for a special situation. Yang et al. (2011) investigated the correction by Eq. (4.62) with different exponent p and the predicted results of bubble columns are shown in Figure 4.10. The bigger the exponent p, the flatter the profile of gas holdup in the radial direction and hence the smaller the gas holdup predicted in the reactor. This is consistent with the investigation of Huang et al. (2008) considering the turbulence correction to the drag coefficient. Generally, the greater the value of drag coefficient used in the simulation, the more the gas holdup predicted in the reactor, due to the increased drag force. In addition, the bigger the drag coefficient employed in ALRs, the higher the circulating liquid velocity predicted.

4.3.2.3 Turbulent dispersion force

The turbulent dispersion force is used to describe the spread of bubbles due to the effect of flow turbulence, though arguments exist on its physical soundness. This force is partly physical in the sense that turbulent vortices exert an impact onto the bubbles and push them to move in the fluid. There are many expressions for this but they vary greatly, and two models that have been widely used in recent years will be addressed here.

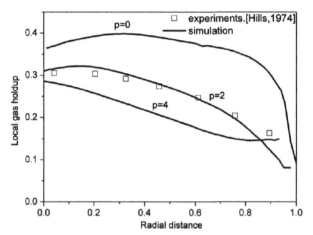

FIGURE 4.10 Effect of the correction factor on the predicted radial gas holdup profiles and comparison with experimental data in a bubble column (Yang et al., 2011).

The simplest and most common approach consists of modeling the effect of turbulence by analogy with the Einstein theory of Brownian motion. This approach was derived by Lahey et al. (1993) for the force per unit volume, and it reads

$$\mathbf{F}_{\text{TD,l}} = -\mathbf{F}_{\text{TD,g}} = C_{\text{TD}}\rho_1 k \nabla \alpha_g \tag{4.72}$$

where k is the turbulent kinetic energy. The key problem is to correlate the coefficient C_{TD} with flow conditions. As pointed out by Mao and Yang (2009), there is no general agreement so far on the physical basis and the estimation of the coefficient for the turbulent dispersion force. Therefore, many values for this coefficient have been adopted in the literature even for the bubbly flow in an air–water system to fit predictions to the experimental data. For the bubbly flow in air–water systems, the coefficient is in the range of 0.1–1.0. It is noteworthy that this model is an analogy with the thermal diffusion of air molecules in the atmosphere, and has no direct connection to the nonlinearity that it is meant to model.

Another typical expression for the turbulent dispersion force was developed in the frame of homogeneous turbulence. Talvy et al. (2007a) proposed that the turbulent dispersion force can be modeled as

$$\mathbf{F}_{\text{TD}} = -\rho_1 \overline{\mathbf{u}'_g \mathbf{u}'_1} \cdot \nabla \alpha_g \tag{4.73}$$

$$\left(\overline{u'_g u'_1}\right)_{ij} = 2k\frac{b+\eta_r}{1+\eta_r} \tag{4.74}$$

$$\eta_r = \frac{\tau^t_{12}}{\tau^F_{12}} \tag{4.75}$$

$$b = \frac{1+C_A}{\rho_g / \rho_1 + C_A} \tag{4.76}$$

where $C_A = 0.5$ denotes the added mass coefficient. Three time scales, i.e., the characteristic time of the particle entrainment by the continuous fluid motion τ_{12}^F, the characteristic time of energetic turbulent eddies τ_1^t and the characteristic time of the fluid turbulence viewed by bubbles τ_{12}^t, are given similar to Podila (2005) and Chen, X. H. (2004) by

$$\tau_{12}^F = \rho_1 \alpha_g K_{gl}^{-1} \left(\frac{\rho_g}{\rho_1} + C_A \right), \quad K_{gl} = \frac{3}{4} \frac{C_D}{d_b} \rho_1 \alpha_g \alpha_1 |\mathbf{u}_g - \mathbf{u}_1| \tag{4.77}$$

$$\tau_1^t = \tfrac{3}{2} C_\mu \frac{k}{\varepsilon} \tag{4.78}$$

$$\tau_{12}^t = \frac{\tau_1^t}{\sigma_k} (1 + C_\beta \zeta_r^2)^{-1/2} \tag{4.79}$$

$$\zeta_r = |\mathbf{u}_g - \mathbf{u}_1| / \sqrt{2k/3} \tag{4.80}$$

$$C_\beta = 1.8 - 1.35 \cos^2 \theta \tag{4.81}$$

where θ is the angle between the gas velocity and the slip velocity. The added mass coefficient is taken as 0.5, and the parameter C_β is set as 0.45 in the direction parallel to the mean relative velocity and as 1.80 in the normal direction.

In general, although great efforts have been made in developing a mathematical model to improve the prediction of bubbly flow in recent years, there is still much room for more accurate accounts of its mechanism. The turbulent dispersion force results from the turbulence of the continuous phase and the radial profiles of the gas void fraction, which leads to a more uniform distribution of gas in the radial direction.

4.3.2.4 Transversal or lift force

Lift force is referred to as that exerted on a bubble in the direction perpendicular to the bubble motion path, and the lateral force is known to be crucial in improving the predictions of radial profiles of velocities and void fractions. The physical phenomena giving rise to a transversal force on a single bubble in liquids can be roughly divided into two groups: the shear in the flow induces the Saffman force, and the rotation of a bubble leads to the Magnus force. This distinction seems to make theoretical sense, because a bubble will rotate in shear flow so these two forces appear together for freely moving bubbles.

It should be noted that, in the Eulerian–Eulerian framework, both Magnus and Suffman forces have an identical mathematical form. Therefore, a general expression for the lift force in an ideal fluid is written as

$$\mathbf{F}_1 = -C_1 V_b \rho_1 (\mathbf{u}_b - \mathbf{u}_1) \times (\nabla \times \mathbf{u}_1) \tag{4.82}$$

Table 4.5 Components of Lift Force in 3D Coordinate Systems Used in an Eulerian–Eulerian Two-Fluid Model

	Expression
Cylindrical coordinate system	$F_{l,r} = -C_l \alpha_g \rho_l \left[(v_g - v_l) \left(\frac{1}{r} \frac{\partial}{\partial r}(r v_l) - \frac{1}{r} \frac{\partial u_l}{\partial \theta} \right) - (w_g - w_l) \left(\frac{\partial u_l}{\partial z} - \frac{\partial w_l}{\partial r} \right) \right]$
	$F_{l,\theta} = -C_l \alpha_g \rho_l \left[(w_g - w_l) \left(\frac{1}{r} \frac{\partial w_l}{\partial \theta} - \frac{\partial v_l}{\partial z} \right) - (u_g - u_l) \left(\frac{1}{r} \frac{\partial}{\partial r}(r v_l) - \frac{1}{r} \frac{\partial u_l}{\partial \theta} \right) \right]$
	$F_{l,z} = -C_l \alpha_g \rho_l \left[(u_g - u_l) \left(\frac{\partial u_l}{\partial z} - \frac{\partial w_l}{\partial r} \right) - (v_g - v_l) \left(\frac{1}{r} \frac{\partial w_l}{\partial \theta} - \frac{\partial v_l}{\partial z} \right) \right]$
Cartesian coordinate system	$F_{l,x} = -C_l \alpha_g \rho_l \left[(v_g - v_l) \left(\frac{\partial v_l}{\partial x} - \frac{\partial u_l}{\partial y} \right) - (w_g - w_l) \left(\frac{\partial u_l}{\partial z} - \frac{\partial w_l}{\partial x} \right) \right]$
	$F_{l,y} = -C_l \alpha_g \rho_l \left[(w_g - w_l) \left(\frac{\partial w_l}{\partial y} - \frac{\partial v_l}{\partial z} \right) - (u_g - u_l) \left(\frac{\partial v_l}{\partial x} - \frac{\partial u_l}{\partial y} \right) \right]$
	$F_{l,z} = -C_l \alpha_g \rho_l \left[(u_g - u_l) \left(\frac{\partial u_l}{\partial z} - \frac{\partial w_l}{\partial x} \right) - (v_g - v_l) \left(\frac{\partial w_l}{\partial y} - \frac{\partial v_l}{\partial z} \right) \right]$

The final formulations for a 3D two-fluid model corresponding to cylindrical and Cartesian coordinate systems are listed in Table 4.5.

Many factors influence the lift coefficient C_l, and no general and reliable correlation for C_l is available. It is usually expressed as a function of Reynolds number, particle diameter, and local turbulent properties. For a solid sphere in linear shear flow, theoretically $C_l = 0.5$. However, there are many diversified forms of models to approximate the lateral coefficient with high gas void fraction, and the mechanism is not yet clear. Models of lift coefficient for engineering applications published in the open literature can be classified into four types, i.e., the lift coefficient depends on Reynolds number, gas holdup, bubble diameter, or local turbulence as discussed in the following.

Moraga et al. (1999) proposed a correlation of lift coefficient that is applicable to turbulent sheared multiphase flow. The model is correlated by the parameter $Re_b \cdot Re_{\nabla}$ and it is expressed as

$$C_l = \left[0.12 - 0.2 \exp\left(-\frac{Re_b \cdot Re_{\nabla}}{3.6 \times 10^5} \right) \exp\left(\frac{Re_b \cdot Re_{\nabla}}{3 \times 10^7} \right) \right]$$

$$6000 < Re_b \cdot Re_{\nabla} < 500,000,000 \qquad (4.83)$$

$$Re_b = \frac{\rho_l \left| \mathbf{u}_g - \mathbf{u}_l \right| d_b}{\mu_l}, \quad Re_{\nabla} = \frac{\rho_l \left| \nabla \times \mathbf{u}_l \right| d_b^2}{\mu_l}$$

Troshko et al. (2001) presented a model of the lateral coefficient using numerical simulation as follows:

$$C_1 = \begin{cases} 0.0767 & \text{for } Re_b \cdot Re_V \leq 6000 \\ -\left(0.12 - 0.2\exp\left(-\dfrac{Re_b \cdot Re_V}{36,000}\right)\right)\exp\left(\dfrac{-Re_b \cdot Re_V}{3 \times 10^7}\right) \\ & \text{for } 6000 < Re_b \cdot Re_V < 190,000 \\ -0.002 & \text{for } Re_b \cdot Re_V \geq 190,000 \end{cases} \quad (4.84)$$

Some researchers argued that the lift coefficient was gas voidage dependent. Behzadi et al. (2004) proposed a model as

$$C_1 = 0.000651\alpha_g^{-1.2} \quad (4.85)$$

This means the lift coefficient is always positive, but this is not true in some cases. Typically, it cannot get the result of gas holdup with "core peaking" under the condition of large bubbles.

Petersen (1992) presented a lift coefficient as follows:

$$C_1 = C_{La}[1.0 - 2.78\min(0.2, \alpha_g)] \qquad C_{La} = 0.01 - 0.5 \quad (4.86)$$

Ohnuki and Akimoto (2001) gave an expression for the lift coefficient that is dependent on the properties of fluids and can be written as

$$C_1 = C_{LF} + C_{WK}$$

$$C_{LF} = 0.288\tanh(0.121Re), \quad C_{WK} = \begin{cases} 0 & \text{for } Eo_d < 4 \\ -0.096Eo_d + 0.384 & \text{for } 4 \leq Eo_d \leq 10 \\ -0.576 & \text{for } Eo_d > 10 \end{cases} \quad (4.87)$$

$$\text{for} -5.5 \leq \log_{10} Mo \leq -2.8, \quad 1.39 \leq Eo \leq 5.74, \quad 0 \leq |du/dy| \leq 8.3$$

Tomiyama et al. (1995) reported a model calculated as

$$C_1 = -0.04Eo + 0.48 \quad (4.88)$$

Tomiyama et al. (2002) proposed a correlation for bubbles for a wide range of parameters ($1.39 < Eo < 5.74$, $-5.5 < \lg Mo < -2.8$, $0 < |\text{rot } \mathbf{u}_c| < 8.3 \text{ s}^{-1}$):

$$C_1 = \begin{cases} \min[0.288\tanh(0.121Re), f(Eo_d)] & \text{for } Eo_d < 4 \\ f(Eo_d) & \text{for } 4 \leq Eo_d \leq 10.7 \\ -0.29 & \text{for } Eo_d > 10.7 \end{cases}$$

$$f(Eo_d) = 0.00105Eo_d^3 - 0.0159Eo_d^2 - 0.0204Eo_d + 0.474$$

$$Eo = \frac{g(\rho_1 - \rho_g)d_b^2}{\sigma} \quad Eo_d = \frac{g(\rho_1 - \rho_g)d_h^2}{\sigma} \quad d_h = d_b(1 + 0.163Eo^{0.757})^{1/3}$$

$$(4.89)$$

In an air–water system at atmospheric pressure and room temperature, the proposed correlation of the lift coefficient yields the bubble diameter dependency and it is shown in Figure 4.11. It can be seen that the critical bubble diameter causing radial void profile transition from wall peaking to core peaking in the air–water bubbly flow is about 5.8 mm.

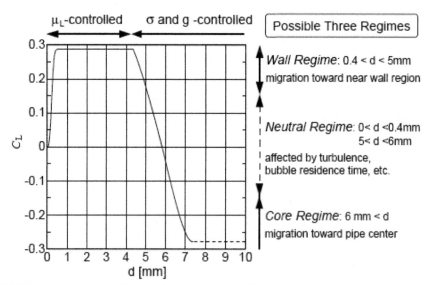

FIGURE 4.11 Relation of lift coefficients with bubble diameters proposed by Tomiyama et al. (2002).

However, the influence of the flow regimes, the properties of fluids, and gas contents on lift coefficient is not yet clear. No generally accepted correlation for the lift coefficient is available, and the models reported in the open literature give widely scattered values. The lift force is usually ignored by many researchers as compared to the drag force. Sometimes, the lift coefficient is even used as an adjustable parameter to match gas holdups with the experimental data.

4.3.2.5 Wall lubrication force

The wall lubrication force has been studied to improve the prediction of gaseous phase volume fraction profiles for a wide range of flow conditions, especially when the "wall peak" phenomenon occurs. Antal et al. (1991) proposed an additional geometry-independent wall lubrication force to model the repulsive force of a wall on a bubble, which is caused by the asymmetric fluid flow around bubbles in the vicinity of the wall due to the fluid boundary layer and can be determined as

$$\mathbf{F}_{w,r} = -C_w \frac{\alpha_g \rho_l |\mathbf{u}_g - \mathbf{u}_l|^2}{d_b} \mathbf{n}_w \tag{4.90}$$

with

$$C_w = \max\left[0, \left(C_1 + C_2 \frac{d_b}{y}\right)\right] \tag{4.91}$$

where y is the distance away from the wall. It should be noted that this force is in the normal direction away from a wall and is only valid in the near-wall region and should tend to zero when the distance from the wall increases. The wall lubrication force coefficients $C_1 = -0.01$ and $C_2 = 0.05$ have commonly been used (Chen, X. H. 2004; Frank et al., 2008), meaning that the wall force only acts within a distance of five times the diameter of bubbles. However, it is clearly shown that this formulation fails when the grid distribution is not fine enough.

It has been found from numerical simulations that the wall lubrication force predicted by the equation mentioned above is too small in amplitude to balance the strong lift forces, which results in overpredicted maxima of near-wall gas volume fraction with the highest gas void fraction reached in the grid element closest to the wall (Frank, 2005). Frank (2005) supposed a generalized formulation for the wall lubrication force as follows:

$$C_{\mathrm{W}} = -C_{\mathrm{W3}}(Eo) \cdot \max\left(0, \frac{1}{C_{\mathrm{WD}}} \frac{1 - y/C_{\mathrm{WC}}d_{\mathrm{b}}}{y(y/C_{\mathrm{WC}}d_{\mathrm{b}})^{p-1}}\right) \tag{4.92}$$

$$C_{\mathrm{W3}} = \begin{cases} \exp(-0.933Eo + 0.179) & 1 \le Eo \le 5 \\ 0.00599Eo - 0.0187 & 5 \le Eo \le 33 \\ 0.179 & 33 < Eo \end{cases} \tag{4.93}$$

This was verified with the experimental gas fraction profiles for vertical pipe flow and good agreement was obtained when the coefficients $C_{\mathrm{WC}} = 10.0$, $C_{\mathrm{WD}} = 6.8$, and $p = 1.7$ were adopted.

It is noteworthy that the radial profile of gas holdup is not determined only by one force, and it is controlled by a combination of three lateral forces, such as transverse lift force, turbulent dispersion force, and wall lubrication force.

4.3.2.6 Added mass force

When a bubble moves unsteadily in a liquid, it accelerates some of the liquid in its neighborhood. Due to the acceleration of the surrounding liquid induced by bubble motion, the bubbles experience a force greater than the drag for a bubble in steady motion, and the extra force is called virtual or added mass force. The acceleration induces an extra resisting force on the bubble equal to one-half the mass of the displaced fluid times the acceleration of the bubble by the theory of potential flow. If the dispersed density is far less than the density of the continuous phase, this force is significant.

The added mass force in bubbly flow is formulated as

$$\mathbf{F}_{\mathrm{A}} = -C_{\mathrm{A}}\alpha_{\mathrm{g}}\rho_{\mathrm{l}} \frac{\mathrm{D}(\mathbf{u}_{\mathrm{g}} - \mathbf{u}_{\mathrm{l}})}{\mathrm{D}t} \tag{4.94}$$

However, it is very difficult to calculate the relative acceleration and many researchers gave diversified forms of equations (Mao and Yang, 2009). One of the models proposed by Anderson and Jackson (1967) is

$$\frac{\mathrm{D}(\mathbf{u}_{\mathrm{c}} - \mathbf{u}_{\mathrm{p}})}{\mathrm{D}t} = \left(\frac{\partial \mathbf{u}_{\mathrm{c}}}{\partial t} + \mathbf{u}_{\mathrm{c}} \cdot \nabla \mathbf{u}_{\mathrm{c}}\right) - \left(\frac{\partial \mathbf{u}_{\mathrm{p}}}{\partial t} + \mathbf{u}_{\mathrm{p}} \cdot \nabla \mathbf{u}_{\mathrm{p}}\right) \tag{4.95}$$

The components of 3D steady models in cylindrical and Cartesian coordinate systems are outlined in Table 4.6.

A reliable correlation of the virtual mass force coefficient in viscous and turbulent flows is absent up to now. The virtual volume coefficient, C_A, for the potential flow around a sphere is 0.5. In reality, the bubbles are usually not spherical, and are often more or less oblate. The bubbles will follow a spiraling trajectory in such

Table 4.6   Components of Virtual Force in 3D Steady Coordinate Systems

	Expression
Cylindrical coordinate system	$F_{A,r,g} = -F_{A,r,l} = C_A \alpha_g \rho_l \left[\left(u_{l,r} \dfrac{\partial u_{l,r}}{\partial r} + \dfrac{u_{l,\theta}}{r} \dfrac{\partial u_{l,r}}{\partial \theta} - \dfrac{u_{l,\theta}^2}{r} + u_{l,z} \dfrac{\partial u_{l,r}}{\partial z} \right) - \left(u_{g,r} \dfrac{\partial u_{g,r}}{\partial r} + \dfrac{u_{g,\theta}}{r} \dfrac{\partial u_{g,r}}{\partial \theta} - \dfrac{u_{g,\theta}^2}{r} + u_{g,z} \dfrac{\partial u_{g,r}}{\partial z} \right) \right]$
	$F_{A,\theta,g} = -F_{A,\theta,l} = C_A \alpha_g \rho_l \left[\left(u_{l,r} \dfrac{\partial u_{l,\theta}}{\partial r} + \dfrac{u_{l,\theta}}{r} \dfrac{\partial u_{l,\theta}}{\partial \theta} + \dfrac{u_{l,\theta} u_{l,r}}{r} + u_{l,z} \dfrac{\partial u_{l,\theta}}{\partial z} \right) - \left(u_{g,r} \dfrac{\partial u_{g,\theta}}{\partial r} + \dfrac{u_{g,\theta}}{r} \dfrac{\partial u_{g,\theta}}{\partial \theta} + \dfrac{u_{g,\theta} u_{g,r}}{r} + u_{g,z} \dfrac{\partial u_{g,\theta}}{\partial z} \right) \right]$
	$F_{A,z,g} = -F_{A,z,l} = C_A \alpha_g \rho_l \left[\left(u_{l,r} \dfrac{\partial u_{l,z}}{\partial r} + \dfrac{u_{l,\theta}}{r} \dfrac{\partial u_{l,z}}{\partial \theta} + u_{l,z} \dfrac{\partial u_{l,z}}{\partial z} \right) - \left(u_{g,r} \dfrac{\partial u_{g,z}}{\partial r} + \dfrac{u_{g,\theta}}{r} \dfrac{\partial u_{g,z}}{\partial \theta} + u_{g,z} \dfrac{\partial u_{g,z}}{\partial z} \right) \right]$
Cartesian coordinate system	$F_{A,x,g} = -F_{A,x,l} = C_A \alpha_g \rho_l \left[\left(u_{l,x} \dfrac{\partial u_{l,x}}{\partial x} + u_{l,y} \dfrac{\partial u_{l,x}}{\partial y} + u_{l,z} \dfrac{\partial u_{l,x}}{\partial z} \right) - \left(u_{g,x} \dfrac{\partial u_{g,x}}{\partial x} + u_{g,y} \dfrac{\partial u_{g,x}}{\partial y} + u_{g,z} \dfrac{\partial u_{g,x}}{\partial z} \right) \right]$
	$F_{A,y,g} = -F_{A,y,l} = C_A \alpha_g \rho_l \left[\left(u_{l,x} \dfrac{\partial u_{l,y}}{\partial x} + u_{l,y} \dfrac{\partial u_{l,y}}{\partial y} + u_{l,z} \dfrac{\partial u_{l,y}}{\partial z} \right) - \left(u_{g,x} \dfrac{\partial u_{g,y}}{\partial x} + u_{g,y} \dfrac{\partial u_{g,y}}{\partial y} + u_{g,z} \dfrac{\partial u_{g,y}}{\partial z} \right) \right]$
	$F_{A,z,g} = -F_{A,z,l} = C_A \alpha_g \rho_l \left[\left(u_{l,x} \dfrac{\partial u_{l,z}}{\partial x} + u_{l,y} \dfrac{\partial u_{l,z}}{\partial y} + u_{l,z} \dfrac{\partial u_{l,z}}{\partial z} \right) - \left(u_{g,x} \dfrac{\partial u_{g,z}}{\partial x} + u_{g,y} \dfrac{\partial u_{g,z}}{\partial y} + u_{g,z} \dfrac{\partial u_{g,z}}{\partial z} \right) \right]$

cases. Some authors used $C_A = 2$ or even 3 to agree with the experimental data. For small void fractions, Pan et al. (1999) proposed the following correlation for bubble swarms:

$$C_A = 1 + 3.32\alpha_g \tag{4.96}$$

A similar correlation is given by Zuber (1964):

$$C_A = \frac{1}{2}\frac{1 + 2\alpha_g}{1 - \alpha_g} \tag{4.97}$$

In general, to date no reliable correlation for C_A is available, in particular for dense bubble swarms.

4.3.3 Closure of turbulence models

Another subject in dispute is the closure of turbulence models. All turbulence models of multiphase flows are developed from single-phase flows, and there is no consensus on how to model the turbulence in multiphase flows, especially for the closure considering the turbulence induced by the dispersed particles.

There are three types of turbulence models for multiphase flows to predict the turbulent stresses, namely the modeling of turbulence in the continuous phase, the model of homogeneous mixture, and the turbulence models separately for each phase. The commonly used two-equation turbulence models for a continuous phase and each phase are outlined in Table 4.7. However, in the majority of publications on simulation of turbulent bubbly flow, the standard k–ε model has been employed (Sokolichin et al., 2004). A no-slip boundary was usually used and the standard wall functions were applied for all the phases at the solid wall for practical engineering applications (Huang et al., 2007).

In addition, the mixture turbulence models including the standard k–ε model and the RNG k–ε model are also widely adopted in simulations of bubbly flow. In the mixture RNG k–ε model, the flow with swirl is considered by adding an additional sink source in the turbulence dissipation equation to account for non-equilibrium strain rates. The equations are expressed as (Yakhot and Orszag, 1986; Huang et al., 2007)

$$\nabla \cdot \rho_m \mathbf{u}_m k = \nabla \cdot \left(\frac{\mu_{eff}^t}{\sigma_k} \nabla k \right) + G_m - \rho_m \varepsilon \tag{4.98}$$

$$\nabla \cdot \rho_m \mathbf{u}_m \varepsilon = \nabla \cdot \left(\frac{\mu_{eff}^t}{\sigma_\varepsilon} \nabla \varepsilon \right) + \frac{\varepsilon}{k}(C_{1\varepsilon}G_m - C_{2\varepsilon}\rho_m \varepsilon) - R \tag{4.99}$$

The mixture properties in the above models are given by

$$\rho_m = \alpha_g \rho_g + \alpha_l \rho_l \tag{4.100}$$

$$\mathbf{u}_m = \frac{\alpha_g \rho_g \mathbf{u}_g + \alpha_l \rho_l \mathbf{u}_l}{\rho_m} \tag{4.101}$$

Table 4.7 Turbulence Models Used in Simulations for Practical Engineering Applications

k–ε model

$$\partial(\alpha\rho k)/\partial t + \nabla\cdot(\alpha\rho\mathbf{u}k) = \nabla\cdot(\alpha(\mu+\mu_t/\sigma_k)\nabla k) + \alpha G - \alpha\rho\varepsilon$$

$$\partial(\alpha\rho\varepsilon)/\partial t + \nabla\cdot(\alpha\rho\mathbf{u}\varepsilon) = \nabla\cdot(\alpha(\mu+\mu_t/\sigma_\varepsilon)\nabla\varepsilon) + \alpha\varepsilon(c_1 G - c_2\rho\varepsilon)/k$$

$$\mu_t = \rho_c C_\mu k^2/\varepsilon, \quad \text{with} \quad C_\mu = 0.09, \ c_1 = 1.44, \ c_2 = 1.92, \ \sigma_k = 1.0, \ \sigma_\varepsilon = 1.3$$

k–ε–A_p model

$$\partial(\alpha_c\rho_c k)/\partial t + \nabla\cdot(\alpha_c\rho_c\mathbf{u}_c k) = \nabla\cdot(\alpha_c(\mu_c+\mu_\varepsilon/\sigma_k)\nabla k) + \alpha_c G - \alpha_c\rho_c\varepsilon$$

$$\partial(\alpha_c\rho_c\varepsilon)/\partial t + \nabla\cdot(\alpha_c\rho_c\mathbf{u}_c\varepsilon) = \nabla\cdot(\alpha_c(\mu_c+\mu_{ct}/\sigma_\varepsilon)\nabla\varepsilon) + \alpha_c\varepsilon(c_1 G - c_2\rho_c\varepsilon)/k$$

$$\mu_{ct} = \rho_c C_\mu k^2/\varepsilon, \quad \text{with} \quad C_\mu = 0.09, \ c_1 = 1.44, \ c_2 = 1.92, \ \sigma_k = 1.0, \ \sigma_\varepsilon = 1.3$$

$$\mu_{dt} = \rho_d C_\mu k^2 R_p^2/\varepsilon, \quad R_p = 1 - \exp(-\tau_c/\tau_d), \quad \tau_c = 0.41k/\varepsilon, \quad \tau_d = \rho_d d_d^2/(18\mu_c)$$

k–ω model

$$\partial(\alpha\rho k)/\partial t + \nabla\cdot(\alpha\rho\mathbf{u}k) = \nabla\cdot(\alpha(\mu+\mu_t/\sigma_k)\nabla k) + \alpha G - \alpha\beta'\rho k\omega$$

$$\partial(\alpha\rho\omega)/\partial t + \nabla\cdot(\alpha\rho\mathbf{u}\omega) = \nabla\cdot(\alpha(\mu+\mu_t/\sigma_\omega)\nabla\omega) + \alpha\zeta\omega G/k - \alpha\beta\rho\omega^2$$

$$\mu_t = \rho_c k/\varepsilon, \quad \text{with} \quad \beta = 0.075, \ \beta' = 0.09, \ \zeta = 5/9, \ \sigma_k = 2.0, \ \sigma_\omega = 2.0$$

Low Reynolds number k–ε model

$$\partial(\alpha\rho k)/\partial t + \nabla\cdot(\alpha\rho\mathbf{u}k) = \nabla\cdot(\alpha(\mu+\mu_t/\sigma_k)\nabla k) + \alpha G - \alpha\rho\varepsilon$$

$$\partial(\alpha\rho\varepsilon)/\partial t + \nabla\cdot(\alpha\rho\mathbf{u}\varepsilon) = \nabla\cdot(\alpha(\mu+\mu_t/\sigma_\varepsilon)\nabla\varepsilon) + \alpha\varepsilon(c_1 f_1 G - c_2 f_2\rho\varepsilon)/k$$

$$\mu_t = \rho_c C_\mu f_\mu k^2/\varepsilon, \quad \text{with} \quad C_\mu = 0.09, \ c_1 = 1.35, \ c_2 = 1.8, \ \sigma_k = 1.0, \ \sigma_\varepsilon = 1.3$$

RNG–k–ε model

$$\partial(\alpha\rho k)/\partial t + \nabla\cdot(\alpha\rho\mathbf{u}k) = \nabla\cdot(\alpha(\mu+\mu_t/\sigma_{kRNG})\nabla k) + \alpha G - \alpha\rho\varepsilon$$

$$\partial(\alpha\rho\varepsilon)/\partial t + \nabla\cdot(\alpha\rho\mathbf{u}\varepsilon) = \nabla\cdot(\alpha(\mu+\mu_t/\sigma_{\varepsilon RNG})\nabla\varepsilon) + \alpha\varepsilon(c_{1RNG}G - c_{2RNG}\rho\varepsilon)/k$$

$$\mu_t = \rho_c C_{\mu RNG} k^2/\varepsilon, \quad \text{with} \quad C_{\mu RNG} = 0.085, \ c_{1RNG} = 1.42 - f_\eta, \ c_{2RNG} = 1.68,$$

$$\sigma_{kRNG} = 0.7179, \ \sigma_{\varepsilon RNG} = 0.7179, \ f_\eta = \eta(1-\eta/4.38)/(1+\beta_{RNG}\eta^3),$$

$$\eta = \sqrt{G/(\rho C_{\mu RNG}\varepsilon)}, \quad \beta_{RNG} = 0.012$$

The turbulent viscosity of the mixture and the production of the turbulence kinetic energy are computed from

$$\mu_m^t = \rho_m C_\mu \frac{k^2}{\varepsilon} \tag{4.102}$$

$$G_m = \mu_m^t (\nabla\mathbf{u}_m + (\nabla\mathbf{u}_m)^T) : \nabla\mathbf{u}_m \tag{4.103}$$

In Eq. (4.99), R is the extra strain rate term:

$$R = \frac{C_\mu \rho_m \eta^3 (1-\eta/\eta_0)}{(1+\beta\eta^3)} \frac{\varepsilon^2}{k} \tag{4.104}$$

where the constants in turbulence models are adopted as follows (Yakhot et al., 1992; Rahimi and Parvareh, 2005; Chow and Li, 2007):

$$C_\mu = 0.0845; \quad c_{\varepsilon 1} = 1.42; \quad c_{\varepsilon 2} = 1.68; \quad \sigma_k = 0.719; \quad \sigma_\varepsilon = 0.719$$

$$\beta = 0.012; \quad \eta_0 = 4.38; \quad \eta = E\frac{k}{\varepsilon}; \quad E^2 = 2E_{ij}E_{ij}; \quad E_{ij} = 0.5\left(\frac{\partial(u_m)_i}{\partial x_j} + \frac{\partial(u_m)_j}{\partial x_i}\right) \tag{4.105}$$

It is reported that the general applicability of the turbulence model to the dynamic prediction of bubbly flow is questionable since it neglects the turbulence induced by the dispersed gas phase. The three most popular models for bubble-induced turbulence are discussed here. The simplest model for consideration of the influence of bubbles on the continuous phase is the model given by Sato et al. (1981):

$$\mu_{\text{eff}}^{\text{t}} = \mu_{\text{m}}^{\text{t}} + \mu_{\text{BI}}^{\text{turb}}$$

(4.106)

$$\mu_{\text{BI}}^{\text{turb}} = 0.6 \rho_{\text{l}} \alpha_{\text{g}} d_{\text{b}} \left| \mathbf{u}_{\text{g}} - \mathbf{u}_{\text{l}} \right|$$

(4.107)

Another approach to modeling bubble-induced turbulence is proposed by Arnold et al. (1989), in which a linear superposition of the shear-induced single-phase turbulence and the bubble-induced pseudo-turbulence is assumed. The bubble-induced pseudo-stress tensor is represented by

$$\mathbf{T}_{\text{l,SI}}^{\text{turb}} = -\alpha_{\text{g}} \rho_{\text{l}} \left[\tfrac{1}{20}(\mathbf{u}_{\text{g}} - \mathbf{u}_{\text{l}})(\mathbf{u}_{\text{g}} - \mathbf{u}_{\text{l}}) + \tfrac{3}{20} \left| \mathbf{u}_{\text{g}} - \mathbf{u}_{\text{l}} \right|^2 \mathbf{I} \right]$$

(4.108)

where $\mathbf{I}$ denotes the identity tensor. However, Sokolichin et al. (2004) pointed out that both the above models may strongly underestimate bubble-induced turbulence.

The third approach incorporates the influence of gas bubbles on turbulence by means of additional source terms in the balance equations of turbulent kinetic energy and turbulent dissipation rate. There are many models using this approach in the literature. However, most additional terms can be represented as follows:

$$S_k = C_k f(\alpha_{\text{g}}, u_{\text{slip}}, \ldots)$$

(4.109)

$$S_\varepsilon = C_\varepsilon \cdot \frac{\varepsilon}{k} S_k$$

(4.110)

where the function $f(\alpha_g, u_{\text{slip}}, \ldots)$ has a diversified form based on different postulations. Ranade and Van den Akker (1994) proposed a model to account for the contribution of bubbles as

$$f(\alpha_{\text{g}}, u_{\text{slip}}, \ldots) = \left| \mathbf{F}_{\text{D}} \right| \cdot \left| \mathbf{u}_{\text{g}} - \mathbf{u}_{\text{l}} \right|$$

(4.111)

where $C_k = 0.02$, $C_\varepsilon = 1.44$, and $\mathbf{F}_{\text{D}}$ is the drag force. Unfortunately, the optimal parameter values of C_k and C_ε, determined by fitting experimental results, differ greatly for each particular case. It is not essential to illustrate with examples for selection of functions according to the nature of flow due to the lack of versatility. Therefore, this approach is not recommended here to predict the hydrodynamics in bubbly flow with practical applications. Compared to other two models, the model proposed by Sato et al. (1981) has been widely employed with great success in recent years.

4.3.4 **Numerical methods**

In the majority of numerical simulations for multiphase flows, there is strong coupling between the velocity, pressure, and volume fractions. The coupling of pressure–velocity–volume fraction is usually solved using traditional algorithms of the SIMPLE family, which is developed from single-phase flow where only pressure–velocity coupling is present. Two algorithms of mass conservation-based algorithms (MCBA; Moukalled and Darwish, 2004b) and geometric conservation-based algorithms (GCBA; Moukalled and Darwish, 2004a), which are extended from the SIMPLE algorithm for incompressible single flow to multi-fluid flow at all speeds, have been derived and verified. In this section, therefore, only some of the recent developments in numerical techniques for multiphase flow are introduced.

4.3.4.1 Unified high-order convection schemes
As pointed out by Li and Baldacchino (1995), the discretization schemes for the convective terms in Navier–Stokes equations and scalar transport equations are connected directly to solution accuracy, efficiency, and convergence. Low-order schemes, e.g., the first-order upwind scheme (FOU), exponential scheme, hybrid scheme, and power-law scheme, have been used by many researchers and the numerical diffusion caused by using these low-order schemes is significant. Therefore, high-order schemes, for instance the second-order scheme, third-order scheme or even higher ones, are increasingly adopted in computations. As an attempt to improve the numerical accuracy in a general finite-volume code, a generalized formulation for implementation of high-order advection schemes on a non-uniform grid is presented and examined by Li and Baldacchino (1995) and Li and Rudman (1995). The central difference (CD), quadratic upstream interpolation for convective kinematics (QUICK), second-order upwind (SOU), and second-order hybrid (SHYBRID) schemes are all generalized in this formulation.

The node notation in one direction (x direction here as an example, and similarly in other directions) is shown in Figure 4.12. It should be noted that the grid

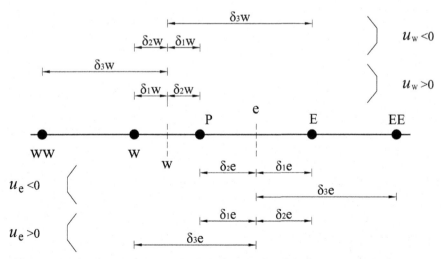

FIGURE 4.12 Grid-related parameters in the x direction for high-order schemes.

parameters δ_{ie} and δ_{iw} ($i = 1, 2$ and 3) have different meanings when the velocity at the local face changes direction.

It can be derived that the net convection flux at the west face can be summarized as

$$F_w \phi_w = (\alpha_{2w}\phi_W + \alpha_{1w}\phi_P - q_w(\phi_P - \beta_{2w}\phi_W + \beta_{1w}\phi_{WW}))F_w^+ + \\ (\alpha_{1w}\phi_W + \alpha_{2w}\phi_P - q_w(\phi_W - \beta_{2w}\phi_P + \beta_{1w}\phi_E))F_w^- \tag{4.112}$$

where $F_w = \rho u_w A_w$ is the mass flow rate across the west face of the control volume, and A_w and ρ are the area of the west face and density respectively. The upwind mass flow rate F_w^+ and F_w^- over the face w are defined as

$$F_w^+ = \frac{F_w + |F_w|}{2}, \quad F_w^- = \frac{F_w - |F_w|}{2} \tag{4.113}$$

The geometric parameters are defined as follows:

$$\alpha_{1w} = \frac{\delta_{1w}}{\delta_{1w} + \delta_{2w}} \qquad \beta_{1w} = \frac{\delta_{1w} + \delta_{2w}}{\delta_{3w} - \delta_{1w}} \\ \alpha_{2w} = \frac{\delta_{2w}}{\delta_{1w} + \delta_{2w}} \qquad \beta_{1w} = \frac{\delta_{2w} + \delta_{3w}}{\delta_{3w} - \delta_{1w}} \tag{4.114}$$

The scheme parameter, q_w, takes the appropriate form for different schemes. For the CD scheme, $q_w = 0$; for the SOU scheme, $q_w = \alpha_{1w}$; and for the QUICK scheme,

$$q_w = \frac{\delta_{2w}}{\delta_{2w} + \delta_{3w}}\alpha_{1w} \tag{4.115}$$

For the SHYBRID scheme, the scheme parameter can be expressed as a function of the local Peclet number ($Pe = F/D$):

$$q_w = 0 \qquad\qquad\qquad \text{if} \quad Pe_w = 0 \\ q_w = \max\left(0, \alpha_{1w} - \frac{1}{Pe_w}\right) \quad \text{if} \quad Pe_w \neq 0 \tag{4.116}$$

It is evident that the scheme parameter in the SHYBRID scheme is variable according to the magnitude of the local Peclet number, and the influence of downstream nodes diminishes as the value of the scheme parameter increases. When the transport is diffusion dominated, this value is small and the SHYBRID scheme is equivalent to the CD scheme; when the scheme parameter is very large or infinite, the SHYBRID scheme is the same as the SOU scheme; and when the parameter $Pe_f = (\delta_{2f}/\delta_{3f} + 1)/\alpha_{1f}$, the SHYBRID scheme is as good as the QUICK one.

Similarly, the net convection flux at the east face can be expressed as

$$F_e \phi_e = (\alpha_{1e}\phi_E + \alpha_{2e}\phi_P - q_e(\phi_E - \beta_{2e}\phi_P + \beta_{1e}\phi_W))F_e^+ \\ + (\alpha_{2e}\phi_E + \alpha_{1e}\phi_P - q_e(\phi_P - \beta_{2e}\phi_E + \beta_{1e}\phi_{EE}))F_e^- \tag{4.117}$$

The definitions of grid size parameters, geometrical parameters, and scheme parameters at the east face take a similar form as those at the west face.

However, the coefficient matrix obtained by employing high-order schemes may lose its diagonal dominance when highly convective flows occur at some locations. A robust and effective method, which was proposed by Khosla and Rubin (1974), is the deferred correction procedure. Since the matrix employing the first-order scheme is a diagonally dominant matrix, it can be adopted here to promote the convergence. The extra deferred correction term of high-order schemes can be treated explicitly as a source term. In this procedure, the net convective fluxes at the east and west faces can be expressed as the sum of the first-order upwind fluxes and the deferred correction fluxes:

$$F_e \phi_e = \phi_P F_e^+ + \phi_E F_e^- + (F\phi)_e^{dc} \qquad F_w \phi_w = \phi_W F_w^+ + \phi_P F_w^- + (F\phi)_w^{dc} \tag{4.118}$$

where

$$
\begin{aligned}
(F\phi)_e^{dc} &= (\alpha_{1e}\phi_E + (\alpha_{2e}-1)\phi_P - q_e(\phi_E - \beta_{2e}\phi_P + \beta_{1e}\phi_W))F_e^+ + \\
&\quad ((\alpha_{2e}-1)\phi_E + \alpha_{1e}\phi_P - q_e(\phi_P - \beta_{2e}\phi_E + \beta_{1e}\phi_{EE}))F_e^- \\
(F\phi)_w^{dc} &= ((\alpha_{2w}-1)\phi_W + \alpha_{1w}\phi_P - q_w(\phi_P - \beta_{2w}\phi_W + \beta_{1w}\phi_{WW}))F_w^+ + \\
&\quad (\alpha_{1w}\phi_W + (\alpha_{2w}-1)\phi_P - q_w(\phi_W - \beta_{2w}\phi_P + \beta_{1w}\phi_E))F_w^-
\end{aligned}
\tag{4.119}
$$

Accordingly, the resulting transport discretization equation can be written as

$$a_P \phi_P = \sum_{nb=e,w,n,s,t,b} a_{nb}\phi_{nb} + S \tag{4.120}$$

where e, w, n, s, t and b refer to the nodes of east, west, north, south, top, and bottom respectively. The coefficients are obtained from the FOU discretization. The source term of the discretization equation is

$$S = S_{FOU} + S^{dc} \tag{4.121}$$

where S_{FOU} is the source term in the transport equation using the FOU scheme. The extra deferred correction source S^{dc} can be computed as

$$S^{dc} = -(F\phi)_e^{dc} + (F\phi)_w^{dc} - (F\phi)_n^{dc} + (F\phi)_s^{dc} - (F\phi)_t^{dc} + (F\phi)_b^{dc} \tag{4.122}$$

As can be seen from the above derivation, the aforementioned method offers flexibility and ease of implementation for various schemes by using different specific scheme parameters.

4.3.4.2 Conservation of gas fraction with high-order schemes

Integration of the continuity equation of any phase over the control volume gives

$$f_{ke}\alpha_{ke} - f_{kw}\alpha_{kw} + f_{kn}\alpha_{kn} - f_{ks}\alpha_{ks} + f_{kt}\alpha_{kt} - f_{kb}\alpha_{kb} = 0 \tag{4.123}$$

where e, w, n, s, t, and b are denoted as the surfaces at east, west, north, south, top and bottom respectively, and k is the index of phase. However, the fluxes at each face can be obtained by the following relations:

$$
\begin{aligned}
f_{ke} &= (u_{kr})_e A_e, \quad f_{kw} = (u_{kr})_w A_w, \quad f_{kn} = (u_{k\theta})_n A_n, \\
f_{ks} &= (u_{k\theta})_s A_s, \quad f_{kt} = (u_{kz})_t A_t, \quad f_{kb} = (u_{kz})_b A_b
\end{aligned}
\tag{4.124}
$$

If the void fraction at the surface is expressed as the sum of the first-order upwind void fraction and the deferred correction void fraction, the final fluxes at the surfaces of the control volume can be written as follows:

$$f_{ke}\alpha_{ke} = [\alpha_{kP} + (\alpha_{ke}^{h+} - \alpha_{kP})]\max(f_{ke},0) - [\alpha_{kE} + (\alpha_{ke}^{h-} - \alpha_{kE})]\max(-f_{ke},0) \quad (4.125)$$

$$f_{kw}\alpha_{kw} = [\alpha_{kW} + (\alpha_{kw}^{h+} - \alpha_{kW})]\max(f_{kw},0) - [\alpha_{kP} + (\alpha_{kw}^{h-} - \alpha_{kP})]\max(-f_{kw},0) \quad (4.126)$$

$$f_{kn}\alpha_{kn} = [\alpha_{kP} + (\alpha_{kn}^{h+} - \alpha_{kP})]\max(f_{kn},0) - [\alpha_{kN} + (\alpha_{kn}^{h-} - \alpha_{kN})]\max(-f_{kn},0) \quad (4.127)$$

$$f_{ks}\alpha_{ks} = [\alpha_{kS} + (\alpha_{ks}^{h+} - \alpha_{kS})]\max(f_{ks},0) - [\alpha_{kP} + (\alpha_{ks}^{h-} - \alpha_{kP})]\max(-f_{ks},0) \quad (4.128)$$

$$f_{kt}\alpha_{kt} = [\alpha_{kP} + (\alpha_{kt}^{h+} - \alpha_{kP})]\max(f_{kt},0) - [\alpha_{kT} + (\alpha_{kt}^{h-} - \alpha_{kT})]\max(-f_{kt},0) \quad (4.129)$$

$$f_{kb}\alpha_{kb} = [\alpha_{kB} + (\alpha_{kb}^{h+} - \alpha_{kB})]\max(f_{kb},0) - [\alpha_{kP} + (\alpha_{kb}^{h-} - \alpha_{kP})]\max(-f_{kb},0) \quad (4.130)$$

The resulting discretization equation of void fraction for any phase is

$$a_{kP}\alpha_{kp} = a_{kE}\alpha_{kE} + a_{kW}\alpha_{kW} + a_{kN}\alpha_{kN} + a_{kS}\alpha_{kS} + a_{kT}\alpha_{kT} + a_{kB}\alpha_{kB} + b_k^{dc} \quad (4.131)$$

where

$$a_{kE} = \max(-f_{ke},0), \quad a_{kW} = \max(f_{kw},0) \quad (4.132)$$

$$a_{kN} = \max(-f_{kn},0), \quad a_{kS} = \max(f_{ks},0) \quad (4.133)$$

$$a_{kT} = \max(-f_{kt},0), \quad a_{kB} = \max(f_{kb},0) \quad (4.134)$$

$$\begin{aligned}
b_k^{dc} = &-[\max(f_{ke},0)(\alpha_{ke}^{h+} - \alpha_{kP}) - (\alpha_{ke}^{h-} - \alpha_{kE})\max(-f_{ke},0)] + \\
&[(\alpha_{kw}^{h+} - \alpha_{kW})\max(f_{kw},0) - (\alpha_{kw}^{h-} - \alpha_{kP})\max(-f_{kw},0)] - \\
&[(\alpha_{kn}^{h+} - \alpha_{kP})\max(f_{kn},0) - (\alpha_{kn}^{h-} - \alpha_{kN})\max(-f_{kn},0)] + \\
&[(\alpha_{ks}^{h+} - \alpha_{kS})\max(f_{ks},0) - (\alpha_{ks}^{h-} - \alpha_{kP})\max(-f_{ks},0)] - \\
&[(\alpha_{kt}^{h+} - \alpha_{kP})\max(f_{kt},0) - (\alpha_{kt}^{h-} - \alpha_{kT})\max(-f_{kt},0)] + \\
&[(\alpha_{kb}^{h+} - \alpha_{kB})\max(f_{kb},0) - (\alpha_{kb}^{h-} - \alpha_{kP})\max(-f_{kb},0)]
\end{aligned} \quad (4.135)$$

$$\begin{aligned}
a_{kP} = &\max(f_{ke},0) + \max(-f_{kw},0) + \max(f_{kn},0) + \\
&\max(-f_{ks},0) + \max(f_{kt},0) + \max(-f_{kb},0)
\end{aligned} \quad (4.136)$$

To meet the volumetric constraint in a two-fluid model:

$$\alpha_1 + \alpha_g = 1 \quad (4.137)$$

Bove (2005) proposed a method in which all the volume fractions are computed and normalized, to decouple the interactions between phases, and the volume fraction is normalized as follows:

$$\alpha_k^{\text{norm}} = \frac{\alpha_k}{\sum_{m=1}^{N} \alpha_m} \tag{4.138}$$

where N is the total number of phases.

Carver (1984) suggested that a new integrated equation of the dispersed phase in bubbly flow can be obtained by subtracting the integrated equation of the continuous phase after normalization by its respective density. The influence of volume fractions on each other is subtly considered in this method. If the high-order scheme is used, the void fraction at the surface can be expressed as the sum of the first-order upwind void fraction and the deferred correction void fraction as mentioned above. The discretization equations of gas and liquid can be expressed as

$$a_{gP}\alpha_{gP} = a_{gE}\alpha_{gE} + a_{gW}\alpha_{gW} + a_{gN}\alpha_{gN} + a_{gS}\alpha_{gS} + a_{gT}\alpha_{gT} + a_{gB}\alpha_{gB} + b_g^{dc} \tag{4.139}$$

$$a_{lP}\alpha_{lP} = a_{lE}\alpha_{lE} + a_{lW}\alpha_{lW} + a_{lN}\alpha_{lN} + a_{lS}\alpha_{lS} + a_{lT}\alpha_{lT} + a_{lB}\alpha_{lB} + b_l^{dc} \tag{4.140}$$

Recalling the relation of Eq. (4.137), one can get

$$a_{lP}(1-\alpha_{gP}) = a_{lE}(1-\alpha_{gE}) + a_{lW}(1-\alpha_{gW}) + a_{lN}(1-\alpha_{gN}) + \\ a_{lS}(1-\alpha_{gS}) + a_{lT}(1-\dot{\alpha}_{gT}) + a_{lB}(1-\alpha_{gB}) + b_l^{dc} \tag{4.141}$$

Subtracting the previous equation from Eq. (4.139) and rearranging yields

$$a_P\alpha_{gP} = a_E\alpha_{gE} + a_W\alpha_{gW} + a_N\alpha_{gN} + a_S\alpha_{gS} + a_T\alpha_{gT} + a_B\alpha_{gB} + S_u \tag{4.142}$$

where

$$a_E = \max(-f_{ge}, 0) + \max(-f_{le}, 0), \quad a_W = \max(f_{gw}, 0) + \max(f_{lw}, 0) \tag{4.143}$$

$$a_N = \max(-f_{gn}, 0) + \max(-f_{ln}, 0), \quad a_S = \max(f_{gs}, 0) + \max(f_{ls}, 0) \tag{4.144}$$

$$a_T = \max(-f_{gt}, 0) + \max(-f_{lt}, 0), \quad a_B = \max(f_{gb}, 0) + \max(f_{lb}, 0) \tag{4.145}$$

$$a_P = \max(f_{ge}, 0) + \max(f_{le}, 0) + \max(-f_{gw}, 0) + \max(-f_{lw}, 0) \\ + \max(f_{gn}, 0) + \max(f_{ln}, 0) + \max(-f_{gs}, 0) + \max(-f_{ls}, 0) \\ + \max(f_{gt}, 0) + \max(f_{lt}, 0) + \max(-f_{gb}, 0) + \max(-f_{lb}, 0) \tag{4.146}$$

$$S_u = (b_g^{dc} - b_l^{dc}) + \begin{bmatrix} \max(f_{le}, 0) + \max(-f_{lw}, 0) + \max(f_{ln}, 0) + \\ \max(-f_{ls}, 0) + \max(f_{lt}, 0) + \max(-f_{lb}, 0) \end{bmatrix} \\ - \begin{bmatrix} \max(-f_{le}, 0) + \max(f_{lw}, 0) + \max(-f_{ln}, 0) \\ + \max(f_{ls}, 0) + \max(-f_{lt}, 0) + \max(f_{lb}, 0) \end{bmatrix} \tag{4.147}$$

It should be noted that some of the source terms can be linearized to promote convergence. Although both methods satisfy the volume fractions constraint, they clearly do not satisfy the mass conservation for each phase unless the convergent solution is obtained. Carver's (1984) method involves less algebraic operations compared to the previous normalization method, and it is recommended here as a good method to decouple the interphase coupling.

4.3.4.3 Decoupling algorithms

The solution of the momentum balance equations in an Euler–Euler two-fluid model deserves special attention. The interphase decoupling algorithm is very important for numerical simulation, especially for steady-state multiphase flow. An adequate method could reduce computational expense and speed up a converged solution. In fact, for strong phase coupling or for relatively large drag coefficients, convergence becomes very slow. A method to promote the convergence rate of the momentum balance equation solution is to decouple the relationship between two phases. Two typical methods, i.e., the partial elimination algorithm (PEA) (Mudde and Van Den Akker, 2001; Bove, 2005) and the partial decoupling algorithm with SIMPLE (PAD-SIMPLE) (Huang et al., 2010), are described here.

The PEA is commonly used to weaken the strong coupling between the primary phase and the secondary phase by algebraic cross substitution of the velocity in the drag term to partially eliminate the effect of velocities. The central idea of the PEA is to render the discretized momentum equations more implicit by decoupling two sets of equations. To illustrate the method, the procedure of a two-phase flow model is presented as an example. If the sources of the momentum discretization equation include all other terms except the drag force term, the final equations of both phases for the u-component at the control volume of a grid point P can be written as

$$A_{P,u,l} u_{P,l}^{n+1} = \sum_{nbP} A_{nbP,u,l} u_{nbP,l}^{n+1} + S_{P,u,l}^{n+1} + F_l (u_{P,g}^{n+1} - u_{P,l}^{n+1}) \tag{4.148}$$

$$A_{P,u,g} u_{P,g}^{n+1} = \sum_{nbP} A_{nbP,u,g} u_{nbP,g}^{n+1} + S_{P,u,g}^{n+1} + F_g (u_{P,l}^{n+1} - u_{P,g}^{n+1}) \tag{4.149}$$

It is clear that each of the equations contains both velocity variables that need to be solved simultaneously. If we combine the similar terms, the above two equations become

$$u_{P,l}^{n+1} = \frac{1}{A_{P,u,l} + F_l} \left(\sum_{nbP} A_{nbP,u,l} u_{nbP,l}^{n+1} + S_{P,u,l}^{n+1} + F_l u_{P,g}^{n+1} \right) \tag{4.150}$$

$$u_{P,g}^{n+1} = \frac{1}{A_{P,u,g} + F_g} \left(\sum_{nbP} A_{nbP,u,g} u_{nbP,g}^{n+1} + S_{P,u,g}^{n+1} + F_g u_{P,l}^{n+1} \right) \tag{4.151}$$

Substituting Eq. (4.151) into Eq. (4.148) and rearranging, the following equation is obtained:

$$\left(A_{P,u,l} + F_l - \frac{F_l F_g}{A_{P,u,g} + F_g}\right) u_{P,l}^{n+1} = \sum_{nbP} A_{nbP,u,l} u_{nbP,l}^{n+1} + S_{P,u,l}^{n+1}$$
$$+ \frac{F_l}{A_{P,u,g} + F_g}\left(\sum_{nbP} A_{nbP,u,g} u_{nbP,g}^{n+1} + S_{P,u,g}^{n+1}\right) \qquad (4.152)$$

Analogously, substituting Eq. (4.150) into Eq. (4.149), we obtain

$$\left(A_{P,u,g} + F_g - \frac{F_l F_g}{A_{P,u,l} + F_l}\right) u_{P,g}^{n+1} = \sum_{nbP} A_{nbP,u,g} u_{nbP,g}^{n+1} + S_{P,u,g}^{n+1}$$
$$+ \frac{F_g}{A_{P,u,l} + F_l}\left(\sum_{nbP} A_{nbP,u,l} u_{nbP,l}^{n+1} + S_{P,u,l}^{n+1}\right) \qquad (4.153)$$

The same procedure can be used to obtain the partially decoupled equations for other velocity components. This treatment renders the equations more implicit and partly reduces the interlinkage of the two equations, which results in an enhancement of the convergence rate. It has been verified that there is no difficulty in convergence when the drag force or the velocities of both phases are very small, and the PEA method is not essential in these situations (Huang, 2008).

In the PAD-SIMPLE, the influence of interphase drag force is removed by using two revised coefficients in the pressure correction equations. For two-phase flows, the discretized momentum equations can be written in the following form:

$$A_{P,ke} u_{ke}^* = \sum_f a_{k,f} u_{k,f}^* + b + (P_P^* - P_E^*) A_{ew} \alpha_{ke} + (F_e)_k u_{je,j\neq k} \qquad (4.154)$$

$$A_{P,kn} u_{kn}^* = \sum_f a_{k,f} u_{k,f}^* + b + (P_P^* - P_N^*) A_{ns} \alpha_{kn} + (F_n)_k u_{jn,j\neq k} \qquad (4.155)$$

$$A_{P,kt} u_{kt}^* = \sum_f a_{k,f} u_{k,f}^* + b + (P_P^* - P_T^*) A_{tb} \alpha_{kt} + (F_t)_k u_{jt,j\neq k} \qquad (4.156)$$

In the above equations, $A_{P,k}$ and $a_{k,f}$ denote the coefficients for the center and neighboring nodes respectively, b is the source term, $(F_n)_k$ is the interphase exchange of drag force, and subscript f stands for t, b, n, s, e and w, referring respectively to the top, bottom, north, south, east and west surfaces of the center control volume. The resulting correction equation for pressure and velocity using the SIMPLE algorithm (Patankar, 1980) is

$$A_{P,ke} u_{ke}' = \left(\sum_f a_f u_f'\right)_k + (P_P' - P_E')\left(A_{ew} \alpha_{ke} + (F_e)_k \, du_j\big|_{j\neq k}\right) \qquad (4.157)$$

By ignoring the corrections due to the neighboring nodes, the final pressure correction equation becomes

$$u'_{ke} = \frac{\left(A_{ew}\alpha_{ke} + (F_e)_k \, du_j\big|_{j\neq k}\right)}{A_{P,ke}}(P'_P - P'_E) = d_{ke}(P'_P - P'_E) \tag{4.158}$$

For gas–liquid two-phase flows, the above correction equation is expanded into

$$du_g = \frac{A_{ew}\alpha_g + (F_e)_g \, du_l}{A_{P,ge}} \tag{4.159}$$

$$du_l = \frac{A_{ew}\alpha_g + (F_e)_l \, du_g}{A_{P,le}} \tag{4.160}$$

It is clear that the above correction equations of gas and liquid contain the mutual correction coefficients. The above two equations are solved simultaneously to result in the direct estimation of the corrections for the gas and liquid phases:

$$du_g = \frac{\dfrac{A_{ew}\alpha_g}{A_{P,ge}} + \dfrac{(F_e)_g A_{ew}\alpha_l}{A_{P,ge} A_{P,le}}}{1.0 - \dfrac{(F_e)_g (F_e)_l}{A_{P,ge} A_{P,le}}} \tag{4.161}$$

$$du_l = \frac{\dfrac{A_{ew}\alpha_l}{A_{P,le}} + \dfrac{(F_e)_l A_{ew}\alpha_g}{A_{P,ge} A_{P,le}}}{1.0 - \dfrac{(F_e)_g (F_e)_l}{A_{P,ge} A_{P,le}}} \tag{4.162}$$

Similarly, the other two corrections for the remaining velocity components in the other two directions in bubbly flow are

$$dv_g = \frac{\dfrac{A_{ns}\alpha_g}{A_{P,gn}} + \dfrac{(F_n)_g A_{ns}\alpha_l}{A_{P,gn} A_{P,ln}}}{1.0 - \dfrac{(F_n)_g (F_n)_l}{A_{P,gn} A_{P,ln}}} \tag{4.163}$$

$$dv_l = \frac{\dfrac{A_{ns}\alpha_l}{A_{P,ln}} + \dfrac{(F_n)_l A_{ns}\alpha_g}{A_{P,gn} A_{P,ln}}}{1.0 - \dfrac{(F_n)_g (F_n)_l}{A_{P,gn} A_{P,ln}}} \tag{4.164}$$

$$dw_g = \frac{\dfrac{A_{tb}\alpha_g}{A_{P,gt}} + \dfrac{(F_t)_g A_{tb}\alpha_l}{A_{P,gt} A_{P,lt}}}{1.0 - \dfrac{(F_t)_g (F_t)_l}{A_{P,gt} A_{P,lt}}} \tag{4.165}$$

$$dw_1 = \frac{\dfrac{A_{tb}\alpha_1}{A_{P,lt}} + \dfrac{(F_t)_l A_{tb}\alpha_g}{A_{P,gt} A_{P,lt}}}{1.0 - \dfrac{(F_t)_g (F_t)_l}{A_{P,gt} A_{P,lt}}} \tag{4.166}$$

This treatment renders the correction equations more implicit and enhances the rate of convergence. This algorithm is similar to the decoupling method of the PEA (e.g., Darwish and Moukalled, 2001; Bove, 2005), but with some further improvements. The PEA solves the momentum equations of gas and liquid in sequence after eliminating the velocity of one phase from the other. However, the decoupling algorithm of PAD-SIMPLE solves the momentum equations separately without considering the change of velocity for the other phase in the momentum equation but taking into account the change of the other phase in the pressure correction equation. Compared to the PEA, PAD-SIMPLE has the advantages of less memory storage and less computational expense. The improved procedure proves to be efficient in eliminating the interphase coupling of drag in bubbly flow. However, it should be noted that this decoupling algorithm is only applicable for two-phase flow where drag force is dominant, and the PEA can be extended for more phases.

4.3.4.4 Improved boundary conditions for steady simulation

In the majority of simulations with steady bubbly flow, the unsteady two-phase model is usually employed to avoid numerical divergence. The physical time of calculation should be long enough to get a steady solution. However, it is very troublesome to get a steady solution by using this method, especially when the target problem is totally time-independent. Although dynamic models have been widely adopted to investigate these complex phenomena, steady-state simulations have emerged in recent years. Although only a few researchers (e.g., Lin et al., 1997a, b; Mudde and Van Den Akker, 2001; Huang et al., 2007, 2008, 2010; Huang, 2008) employed the steady method to predict the bubbly flow, good results have been obtained in recent years and the feasibility of implementation is adequate if the boundary condition is correctly applied.

The top outflow boundary of an airlift loop reactor can be seen as a free surface. In some publications, a simple hypothesis to get a physically realistic solution was made. Mudde and Van Den Akker (2001) regarded the free surface boundary as a shear free surface for the water phase with zero normal velocity, while for the gas phase it acted as an outlet with a fixed vertical velocity of 0.2 m/s, which is approximately the bubble slip velocity in water for air bubbles of 2–5 mm sizes. Also, the free surface is assumed flat and all other variables are subject to

$$\frac{\partial \phi}{\partial z} = 0 \tag{4.167}$$

As we know, there are a wide range of bubbles in industrial reactors. However, the terminal slip velocity of the bubbles can be calculated by assuming the balance of pressure and drag for the bubbles in a stationary liquid:

$$F_D = F_P \tag{4.168}$$

$$F_D = \tfrac{3}{4} \rho_l \alpha_g \frac{C_D}{d_b} u_t^2 \approx (\rho_l - \rho_g) \alpha_g g \tag{4.169}$$

So the slip velocity of bubbles can be written as

$$u_t \approx \sqrt{\frac{4(\rho_l - \rho_g) g d_b}{3 \rho_l C_D}} \tag{4.170}$$

The above formula can be computed by combining Eq. (4.25) and the definition of the bubble Reynolds number:

$$Re \approx \frac{d_b u_t \rho_l}{\mu_l} \tag{4.171}$$

When the airlift loop reactor is operated with a continuous mode with both phases, special treatments for the upper boundary are required to promote the rate of convergence for steady simulations (Huang et al., 2008). It is very complicated to approximate this because of the coupling between the velocity and the fraction for each phase in bubbly flows. A boundary condition developed by Huang et al. (2008) is taken here: the free surface is thought to be flat and the relative velocity of gas and liquid is defined as the terminal slip velocity of bubbles responding to the working system. The revised boundary condition is as follows regarding the gas and the slurry entering into the reactor continuously:

$$\alpha_g = \alpha_{g,c0} \tag{4.172}$$

$$u_l = \max(0, u_{l,c0}) \tag{4.173}$$

$$u_g = u_l + u_{slip} \tag{4.174}$$

where the slip velocity of bubbles equals the corresponding terminal slip velocity.

For three-dimensional flows, some fluids may flow across the axis of the reactor, and the axis shows no restriction to fluid flow. Therefore, it is essential to maintain the continuity of physical quantities at the axis during simulation for a steady solution. Serre and Pulicani (2001) proposed a simple technique to specify the boundary value of velocity components at the axis, while Zhang et al. (2006) provided a more accurate description of the boundary conditions. For the radial and angular velocity components in a staggered grid (as indicated in Figures 3.5 and 3.6 in Chapter 3), the velocity vector in the horizontal plane at the axis is actually the average of all u vectors of all neighboring nodes at the axis in the r–θ plane. The flux across the axis can be treated the same as in the method of the radial velocity components, just ignoring the influence of angular velocity. For other variables,

$$\phi(0, j, k) = \frac{1}{N_\theta} \sum \phi(2, j, k) \tag{4.175}$$

where N_θ is the total grid number in the θ direction. Please refer to Section 3.2.6.2 in Chapter 3 for more details about the symmetry condition.

Another important issue is to determine the location of gas–liquid separation, which is commonly encountered in practical bubbly flows, when the dispersed phase is dominant at the top of the reactor and a head space of gas is present above the free liquid surface. However, it is very complicated to track the liquid-free interface between two phases. Zhang et al. (2009) argued that this problem can be simply resolved by using a special approximation of interphase forces above the free surface. The method has the following specification in the region with $\alpha_l < 0.55$:

$$C_d = 0.05, \quad C_l = 0, \quad C_A = 0 \tag{4.176}$$

Zhang et al. (2009) believed that the continuity and momentum equations reduced to those for single-phase flow. A small finite value for C_d was taken to guarantee the proper coupling of the two phases. The same method was used by Yang et al. (2011) to predict the hydrodynamics in a bubble column and achieved great success.

4.4 HYDRODYNAMICS AND TRANSPORT IN AIRLIFT LOOP REACTORS

4.4.1 Hydrodynamic behavior

ALRs are pneumatically agitated reactors characterized by buoyancy-driven flow with an upflow channel (i.e., riser) and a downflow channel (i.e., downcomer). Because of the presence of directional flow in the reactor, ALRs offer advantages such as scalability and operational flexibility over traditional reactors, for instance bubble columns.

The effects of the superficial gas velocity and the top/bottom clearances on the hydrodynamics in ALRs have been extensively examined. The experimental results suggest that although the flow structure in the riser and the downcomer are close to the plug flow, bypassing and stagnancy exist in the top and the bottom regions respectively. Compared to the low top clearances, the large top clearances provide a larger space for gas–liquid separation, facilitate the liquid phase changing its flow direction, and reduce the overall macromixing time in the reactor. A small bottom clearance tends to reduce the stagnancy zone and increase the hydraulic resistance, and thus slows down the liquid flow velocity and entrains fewer bubbles into the downcomer (Luo and Al-Dahhan, 2008). A larger gas input results in a larger gas holdup in the whole column and a faster liquid flow, and hence enhances the turbulence and the macromixing.

CFD simulations on the hydrodynamics of bubbly flow in an ALR using a two-fluid model have become popular and great success has been achieved (Roy et al., 2006; Talvy et al., 2007a; Huang et al., 2007, 2008). The results will not be given here for the sake of brevity, but the important information in IALRs will be addressed. The bubbles cannot be dragged into the downcomer by the circulating liquid and the bubbles are almost stationary in the top of the downcomer when the flow is in regime II (gas entrained but not recirculated) according to the classification of Heijnen et al. (1997). Heijnen et al. (1997), van Benthum et al. (1999), and Blažej et al. (2004b) argued that bubbles were balanced with drag and buoyancy in this regime and the liquid velocity in the downcomer was equal to the slip velocity of air bubbles. However, both the experimental data (van Baten and Krishna, 2003) and CFD results (Huang et al., 2010) show that the liquid velocity in the downcomer may be much greater than the slip velocity of bubbles. Huang et al. (2010) proposed an explanation of the balance of forces exerted on bubbles: the bubbles in the downcomer in these cases are balanced not only by the drag and buoyancy, but also the difference of static pressures. The force exerted on bubbles between the front and the rear resulting from the difference of static pressures has the same direction as the flow of liquid in the downcomer, but its direction is opposite in the riser. Therefore, a higher average liquid velocity than the slip velocity usually occurs in this regime, which results in a larger drag to compensate the change of static pressure difference.

Most research focused on the study of the overall gas phase and hydrodynamics in IALRs with large height-to-diameter ratios, while less work was done on those with low height-to-diameter ratios. Zhang et al. (2011) took the measurements of bubble behaviors in the riser such as local gas holdup, bubble size, and bubble rise velocity distribution by using a double-sensor conductivity probe. The effects of operation and structure parameters of the IALR with a low aspect ratio ($H/D \le 5$) on gas–liquid two-phase flow characteristics were investigated. It was found that bubbles are distributed equally across the whole riser volume for superficial gas velocity less than 0.06 m/s. The gas holdup in the riser increases linearly with superficial gas velocity. This corresponds to the homogeneous regime in Figure 4.13. However, as the superficial gas velocity increases above 0.06 m/s, large bubbles form due to

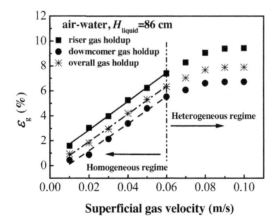

FIGURE 4.13 Gas holdup as a function of superficial gas velocity in an airlift loop reactor ($T_c = 0.10$ m, $B_c = 0.06$ m, D_r: $\phi 200 \times 7$ mm, perforated plate distributor) measured by Zhang et al. (2011).

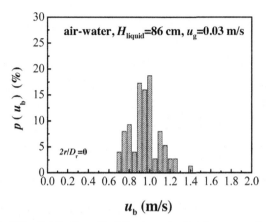

FIGURE 4.14 Probability distribution of bubble rise velocity at cross-section ($h/D = 4.62$, $T_c = 0.10$ m, $B_c = 0.06$ m, D_r: $\phi 200 \times 7$ mm, perforated plate distributor) measured by Zhang et al. (2011).

more frequent coalescence and aggregation. The bubbles move upward to the upper riser with higher rise velocities, and the gas holdup increases slowly. Usually, the IALR with a high aspect ratio has a low critical velocity (0.03 m/s) for flow regime transition. For the IALR with a low aspect ratio, a wider range of homogeneous regime is advantageous for higher gas holdups and lower liquid circulation velocities, and makes the bubble distribution more uniform in the axial and radial directions up to a gas superficial velocity of 0.06 m/s. The typical probability $p(u_b)$ histogram of the bubble rise velocity at $h/D = 4.62$ in Figure 4.14 illustrates that the radial profile of bubble velocities is rather uniform, as the rise velocity of a bubble depends mainly on its size.

4.4.2 Interphase transport phenomena

Besides the momentum transfer in bubbly flow, mass transfer and heat transfer in gas–liquid reactors are also important. Mass is usually transferred from the dispersed gas phase (air bubble) to the continuous liquid phase (water), and reactions take place in the continuous phase and at the same time the heat of reactions is released in the continuous phase. Examples of application are biochemical reactions in bioreactors, and hydrogenation and oxidation reactions in slurry reactors.

In chemical engineering, the global mass transfer efficiency of a reactor is usually evaluated using a global volumetric mass transfer coefficient $k_L a$. However, different researchers have used different models to estimate the mass transfer coefficients. Generally, there are four types of model for mass transfer, i.e., (1) phenomenological correlations and/or models derived from dimensional analysis and experimental data; (2) spatial models, of which the simplest version is the film model; (3) time models, the simplest version of which is Higbie's penetration model; and

(4) combined film-penetration models that have not been used in CFD up to now (e.g., Toor and Marchello, 1958). There are many different models of mass transfer coefficients corresponding to different theories and hypotheses. However, which of these is best for bubbly flow in an ALR is not yet clear. These expressions should be examined against experimental data and validated for further applications. A few models commonly used in recent years are examined and discussed here.

For air–water dispersions and for suspensions in which the suspending fluid is water-like, Acién Fernández et al. (2001) and Chisti (1989) calculated the mass transfer coefficient from the following correlation:

$$k_L a = 3.378 \times 10^{-4} \left(\frac{g D_L \rho_l^2 \sigma}{\mu_l^3} \right)^{0.5} \alpha_g e^{-0.131 c_s^2} \tag{4.177}$$

where c_s is the concentration of solids in a suspension (wt/vol%) and $c_s = 0$ in gas–liquid systems free of solids.

Talvy et al. (2007a) and Cockx et al. (1999, 2001) proposed a time model based on the penetration theory of Higbie (1935) to estimate the local mass transfer as follows:

$$k_L a = \frac{12 \alpha_g}{d_b} \sqrt{\frac{D_L U_{slip}}{\pi d_b}} \tag{4.178}$$

where D_L is the molecular diffusivity of gas in liquid.

Bird et al. (1960) introduced an expression for the mass transfer coefficient:

$$k_L a = \frac{12 \alpha_g}{d_b} \sqrt{\frac{D_L U_{slip}}{3 \pi d_b}} \tag{4.179}$$

However, Talvy et al. (2007a) argued that this equation only fitted for the mass transfer in creeping flow around a gas bubble, and the value of mass transfer coefficient was roughly half of that at relatively high Reynolds numbers.

Xue and Yin (2006) derived an expression for the mass transfer coefficient according to the Boussinesq hypothesis and the classical penetration theory of Higbie (1935). The final form of the time model is

$$k_L a = \sqrt{\frac{D_L}{\pi}} \left(\frac{\rho_l \varepsilon}{\mu_l} \right)^{1/4} \frac{12 \alpha_g}{d_b} \tag{4.180}$$

Tobajas et al. (1999) and Wen et al. (2005) established a correlation for the mass transfer coefficient based on the theory of Higbie (1935) and Kolmogoroff's theory of isotropic turbulence as follows:

$$k_L a = \sqrt{\frac{D_L}{\pi}} \left(\frac{U_{slip} \rho_l g \alpha_g}{\mu_l} \right)^{1/4} \frac{12 \alpha_g}{d_b} \tag{4.181}$$

For a mobile interface, the contact time with the liquid is short, Higbie's theory is valid, and accordingly Vasconcelos et al. (2003) found the following expression for the mass transfer coefficient:

$$k_{L}a = 6.78\frac{\alpha_{g}}{d_{b}}\sqrt{\frac{D_{L}U_{slip}}{d_{b}}}$$

(4.182)

This formula can also be used for deforming bubbles. For a rigid interface, the bubble behaves like a solid sphere. The mass transfer coefficient is then obtained theoretically from laminar boundary layer theory and a spatial model is found:

$$k_{L}a = c\frac{6\alpha_{g}}{d_{b}}\sqrt{\frac{U_{slip}}{d_{b}}}D_{L}^{2/3}v_{1}^{-1/6}$$

(4.183)

with $c \approx 0.6$. Experimental values of c have been found to vary from 0.4 to 0.95 (Griffith, 1960; Lochiel and Calderbank, 1964).

Huang et al. (2010) investigated the mass transfer of oxygen in air dissolved in water and found that the discrepancies between these models were very large. A comparison of the predicted mass transfer coefficients using the above-mentioned models with the experimental data of Juraščík et al. (2006) is shown in Figure 4.15, and the predicted mass transfer coefficients using Eq. (4.181) are not presented due to its largely overpredicted values. It is clear that the predicted results using Eqs. (4.177), (4.178), and (4.182) agree well with the experimental data. It is surprising that although Eq. (4.177) is a phenomenological correlation, it predicts the mass transfer coefficient well in all the test cases.

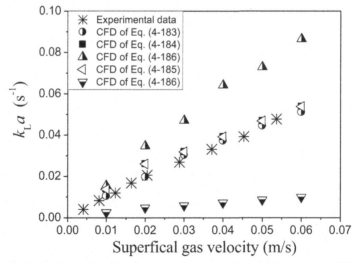

FIGURE 4.15 Comparison of predicted average mass transfer coefficients using different mass transfer models with experimental data (Huang et al., 2010).

Among these models, Eq. (4.178) has no adjustable regression coefficient and is a mechanism-based model, so it is recommended for modeling the mass transfer of bubbly flow in ALRs. Note that the predicted values of Eq. (4.182) are almost equal to those of Eq. (4.178), because these two equations have coefficients with negligible difference. Equation (4.180) behaves poorly, overpredicting the mass transfer coefficient by about 50% when the superficial gas velocity is above 0.03 m/s. The spatial model Eq. (4.183) underestimates the mass transfer coefficients substantially, and it is also not appropriate for estimation of mass transfer in current bubbly flow. Among all the models, the spatial model behaves poorly and the phenomenological relation of Eq. (4.177) may be risky for more complex bubbly flow. The time model has been widely used and has shown much better performance than the others, so it can be further developed and used for a wide range of bubbly flows.

The steady gas–liquid bubbly flow in an airlift loop reactor is considered here, so only the transport equations of concentration and heat will be briefly reviewed. Accordingly, chemical reactions and microbial growth can be incorporated into these transport models. In bubbly flow, gas is often used to provide the power of agitation and sometimes the reactants. Taking the direct liquefaction of coal, for example, the governing equations of concentrations of the hydrogen in gas, the dissolved hydrogen in liquid, the coal in liquid, and the product of the hydrogenation reaction in liquid are given by

$$\frac{\partial(\alpha_g \rho_g c_{Ag})}{\partial t} + \nabla \cdot \alpha_g \rho_g \mathbf{u}_g c_{Ag} = \nabla \cdot D_{tg} \alpha_g \rho_g \nabla c_{Ag} - \rho_g F_L \tag{4.184}$$

$$\frac{\partial(\alpha_l \rho_l c_{Al})}{\partial t} + \nabla \cdot \alpha_l \rho_l \mathbf{u}_l c_{Al} = \nabla \cdot \alpha_l \rho_l D_{tl} \nabla c_{Al} + \rho_l F_L - \rho_l \alpha_l r_A \tag{4.185}$$

$$\frac{\partial(\alpha_l \rho_l c_{Pl})}{\partial t} + \nabla \cdot \alpha_l \rho_l \mathbf{u}_l c_{Pl} = \nabla \cdot \alpha_l \rho_l D_{tl} \nabla c_{Pl} + \rho_l \alpha_l r_A \tag{4.186}$$

where r_A and F_L are the reaction rate and the interphase transfer of concentration of A respectively. The turbulent diffusion coefficients of hydrogen in the liquid and gas phases are expressed as

$$D_{tl} = \frac{v_t}{\sigma_t} + D_L = C \frac{k^2}{\varepsilon} + D_L \tag{4.187}$$

$$D_{tg} = \frac{v_t}{\sigma_t} + D_G = C \frac{k^2}{\varepsilon} + D_G \tag{4.188}$$

where D_G and D_L are the molecular diffusivities of hydrogen in gas and liquid respectively. The turbulent diffusion coefficient of coal in liquid is approximated by

$$D_{tcl} = \frac{v_t}{\sigma_t} = C_\mu \frac{k^2}{\varepsilon} \tag{4.189}$$

The interfacial transfer of concentration between the two phases can be expressed in terms of mass transfer coefficient and the driving force of concentration, and can be written as

$$F_L = k_L a(c_L^* - c_L) \tag{4.190}$$

where c_L^* and a are the saturation concentration of hydrogen and the interfacial area respectively. In the case of spherical bubbles, the interfacial area is expressed as

$$a = \frac{\alpha_g S_b}{V_b} = \frac{6\alpha_g}{d_b} \tag{4.191}$$

c_L^* is the saturation concentration of hydrogen in water and is generally given by Henry's law (Talvy et al., 2007a):

$$c_L^* = H_e P_{H_2} \tag{4.192}$$

where H_e is the Henry constant representing hydrogen solubility in water. The local saturation concentration c_L^* can be estimated from Henry's law and the local concentration of oxygen in gas:

$$c_L^* = m c_g \tag{4.193}$$

where m is defined as

$$m = H_e R T \tag{4.194}$$

with R and T the ideal gas constant and absolute temperature respectively.

In the multiphase model, there are respective enthalpies and temperature fields for each phase, and the heat is exchanged at the interface. The multiphase thermal energy equation for enthalpy is

$$\frac{\partial(\rho \alpha_k H_k)}{\partial t} + \nabla \cdot \left(\alpha_k \left(\rho_k \mathbf{u}_k H_k - \left(\frac{v_t}{\sigma_t} + \lambda_k \right) \nabla T_k \right) \right) = \sum_{\beta=1}^{n} (\Gamma_{k\beta}^+ h_{\beta s} - \Gamma_{\beta k}^+ h_{ks}) + Q_k + S_k \tag{4.195}$$

where $h_{ks}, T_k, \lambda_k, S_k, Q_k$, and $(\Gamma_{k\beta}^+ h_{\beta s} - \Gamma_{k\beta}^+ h_{ks})$ denote the enthalpy, temperature, thermal conductivity of phase k, external heat sources, interphase heat transfer to phase k across interfaces with other phases, and heat transfer induced by interphase mass transfer respectively. The total heat transfer per unit volume transferred to phase k due to thermal non-equilibrium across the phase interface is given by

$$Q_k = \sum_{\beta \neq k} Q_{k\beta} \tag{4.196}$$

where

$$Q_{k\beta} = -Q_{\beta k} \quad \Rightarrow \quad \sum_k Q_k = 0 \tag{4.197}$$

The source term due to chemical reactions can be expressed as

$$S_k = \alpha_k r_A \Delta H \tag{4.198}$$

The rate of heat transfer $Q_{k\beta}$ per unit time across a phase boundary of interfacial area per unit volume $A_{k\beta}$, from phase β to phase k, is

$$Q_{k\beta} = h_{k\beta} A_{k\beta} (T_\beta - T_k) \tag{4.199}$$

There are three categories of models to predict the heat transfer coefficient, namely the particle model correlation, mixture model correlation, and two resistance models. Due to its simplicity, the particle model correlation is recommended here, and the heat transfer coefficient can be expressed in terms of a dimensionless Nusselt number:

$$h_{k\beta} = \frac{\lambda_k Nu_{k\beta}}{d_\beta} \tag{4.200}$$

where λ_k, d_β, and $Nu_{k\beta}$ denote the thermal conductivity of the continuous phase, the mean diameter of the dispersed phase, and the dimensionless Nusselt number. For laminar forced convection around a spherical particle, theoretical analysis shows that $Nu = 2$. However, for a particle in a moving incompressible Newtonian fluid, the Nusselt number can be expressed in terms of the particle Reynolds number and the surrounding fluid Prandtl number ($Pr = \mu_c C_{pc}/\lambda_c$, which is based on the properties of the continuous phase) as

$$Nu = \begin{cases} 2 + 0.6 Re^{0.5} Pr^{0.33} & 0 \le Re < 776.06, 0 \le Pr < 250 \\ 2 + 0.27 Re^{0.62} Pr^{0.33} & 776.06 \le Re, 0 \le Pr < 250 \end{cases} \tag{4.201}$$

As we know, the enthalpy is related to temperature, namely $H = C_p T$. If the gas and liquid phases are thought to be incompressible, the energy conservation equation, which is dependent on temperature, can be written as

$$\frac{\partial(\alpha_k T_k)}{\partial t} + \nabla \cdot \left(\alpha_k \left(\mathbf{u}_k T_k - \left(\frac{v_t}{\sigma_t} + \frac{\lambda_k}{\rho_k C_{p,k}} \right) \nabla T_k \right) \right) = \sum_{\beta=1}^{n} \left(\Gamma_{k\beta}^+ h_{\beta s} - \Gamma_{\beta k}^+ h_{ks} \right) / \rho_k C_{p,k} \\ + Q_k / \rho_k C_{p,k} + \alpha_k r_A \Delta H / \rho_k C_{p,k} \tag{4.202}$$

Only the governing equations, appropriate closure terms and boundary conditions, and the transport phenomena including the flow, mass and heat transfer between two phases in gas–liquid bubbly flow can be well predicted. How to integrate these models together is not a trivial matter. Huang (2008) predicted the mass and heat transfer for direct coal liquefaction (DCL) in an IALR under elevated pressure and high temperature. The distributions of hydrogen in gas, dissolved hydrogen in liquid, mass transfer coefficient, coal in the slurry, product of hydrogenation, and temperatures of gas and liquid are illustrated in Figures 4.16–4.19 respectively. It is found that the IALR has good capability of mixing; the distributions of mass transfer coefficients, concentrations, and temperatures of both phases are very uniform except for a small region near the inlet and the IALR may be used as an ideal reactor for the process of direct coal liquefaction if it is well designed.

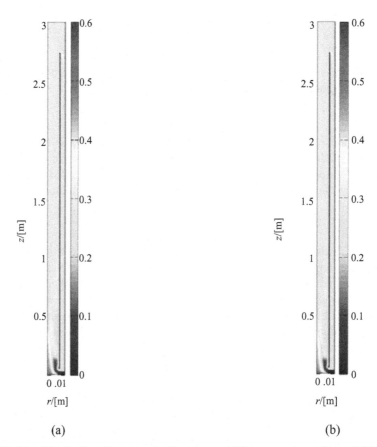

FIGURE 4.16 Distribution of total gas holdups in an airlift loop reactor at $U_{g,R}$ = 0.03 m/s and different $U_{l,R}$ (Huang, 2008).

(a) $U_{l,R}$ = 0 m/s. (b) $U_{l,R}$ = 0.002128 m/s.

4.5 MACROMIXING AND MICROMIXING

Mixing as a discipline has evolved from the foundations that were laid in the 1950s. The major techniques of mixing in industry are either mechanical or pneumatic agitation. Turbulent transport can be viewed as macroscopic convection plus a microscopic vortex-like velocity field, which is dominated by diffusion. Mixing is a multi-scale process involving large-scale blending (macromixing) and fine-scale homogenization (micromixing) reflecting the scale of interest in applied operations. Macromixing occurs due to the convective transport of large fluid elements. However, micromixing at the molecular scale occurs due to diffusion to decrease the intensity of segregation. Chemical reactions depend on reagents interacting at a molecular level and are therefore directly affected by the rate of micromixing.

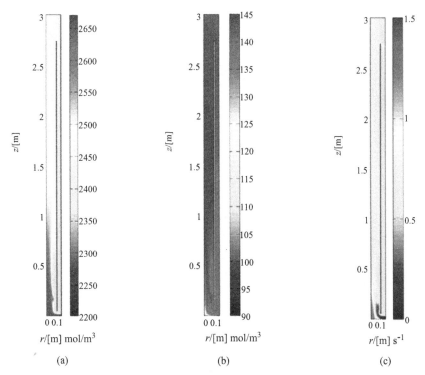

FIGURE 4.17 Distributions of hydrogen and mass transfer coefficient in an airlift loop reactor at the condition of $u_{g,R}$ = 0.03 m/s and $u_{l,R}$ = 0.002128 m/s (Huang, 2008).

(a) Gas hydrogen. (b) Liquid hydrogen. (c) Mass transfer coefficient.

As mixing performance is one of the most important factors in the design and scale-up of reactors, many engineering design principles have been developed. With the rapid development of modern technology, rational design of ALRs emphasizing the factors that affect mixing, such as the geometrical parameters of the reactors, gas input, the properties of working media, sparger modification, and measurement techniques for a desired process objective, has become possible. Macromixing and micromixing are both important: (a) the large-scale motion associated with flow structures leads to more uniform concentration distributions that differ markedly from those entering the streams, and (b) the overall mixing rate is influenced by the values of molecular diffusivities even at what are considered to be high Reynolds numbers (Broadwell and Godfrey Mungal, 1990).

4.5.1 Macromixing in airlift loop reactors

ALRs have been widely used in industries, particularly in bioprocessing and hydrometallurgy. Extensive studies about the flow dynamics in ALRs exist, but most of these studies are focused on global hydrodynamic parameters using conventional techniques.

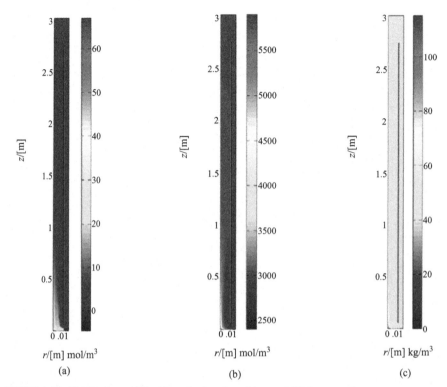

FIGURE 4.18 Distributions of liquid product and coal in an airlift loop reactor (Huang, 2008).

(a) Difference of concentrations between saturated hydrogen and dissolved hydrogen.
(b) Product. (c) Coal.

The local flow characteristics, such as macromixing and turbulence intensity, are crucial for reliable design and scale-up, and remain unclear. As we know, knowledge of liquid mixing time, circulation time, liquid circulation velocities, and axial mixing (characterized by axial dispersion coefficients and Bodenstein numbers) is important in the design and operation of reactors (Sánchez Mirón et al., 2004). Definitions of mixing time, axial dispersion coefficients, and Bodenstein numbers will be reviewed here.

Mixing time, t_m, is the time required for the injected tracer to attain a given uniformity quite close to the fully mixed state. The mixing time is a global index of mixing, and it is affected by axial and radial mixing and the effects of bulk flow. The mixing in airlift loop reactors is sometimes characterized by dimensionless mixing time, θ_m, defined as

$$\theta_m = \frac{t_m}{t_c} \tag{4.203}$$

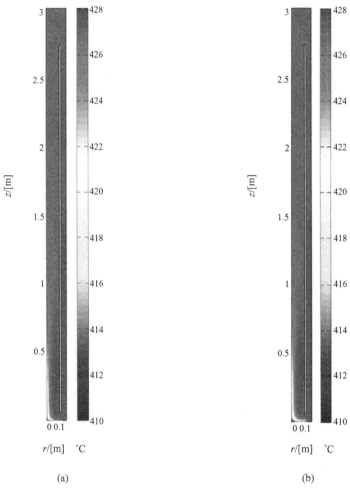

(a)   (b)

FIGURE 4.19 Distributions of liquid and gas temperatures in an airlift loop reactor (Huang, 2008).

(a) Liquid. (b) Gas.

where t_c is the circulation time, i.e., the time required for one passage through the circulation loop. The dimensionless mixing time is reported to depend only on the reactor geometry and not on the gas superficial velocity. The acid tracer method (Sánchez Mirón et al., 2004) is usually used to measure the mixing time, which is required to attain a ±5% deviation from complete homogeneity from the time of tracer addition. It has been widely reported that bubble columns have shorter mixing time at any aeration rate and the mixing time within is less sensitive to the aeration rate compared with ALRs due to the cyclic motion in bulk flow (Gumery et al., 2009). The values of the various mixing parameters in bubbly flow are quite comparable at identical aeration rates for any specific geometry of reactors.

Mixing in the axial direction of a reactor is described in terms of an axial dispersion coefficient, E_z, which is influenced mainly by the aeration rate, the geometry, and the properties of the fluid. In an ALR, the riser and the downcomer can have different values of E_z. In addition, an overall value of axial dispersion coefficient E_z can be identified for the entire circulation pathway. The axial dispersion coefficient (ADM) can be computed using

$$\frac{\partial c_r}{\partial \tau} = \frac{1}{Bo}\frac{\partial^2 c_r}{\partial Z^2} - \frac{\partial c_r}{\partial Z} \tag{4.204}$$

$$E_z = \frac{u_l L}{Bo} \tag{4.205}$$

where u_l is the average velocity of the liquid, L is the distance between the point of tracer injection and those of detection, c_r is the dimensionless concentration c/c_∞, τ is the dimensionless time t/t_c, and Bo is the Bodenstein number (or Peclet number). The Bodenstein number is the ratio between the mixing effects of the axial dispersion coefficient and the bulk movement of the fluid. Bodenstein number Bo can be written as

$$Bo = k\frac{t_m}{t_c} \tag{4.206}$$

where k is a constant that depends on the geometry of the reactor and the physical properties of the fluid. A reactor can be thought to be a continuous stirred tank reactor (CSTR) when the Bodenstein number is below 0.1, and plug flow is viewed as acceptable if the Bodenstein number is above 20. The computational details of the values of the axial dispersion coefficient and the Bodenstein number in ALRs can be found in Sánchez Mirón et al. (2004).

Macromixing is the mixing process driven by the largest scales of convective transport motions in the fluid medium. During transport through the reactor, the fluid elements stretch, fold or their thickness decreases to achieve an average homogenization throughout the working volume. Macromixing is always characterized by the mixing time in a system.

4.5.1.1 Experimental investigation of macromixing

Mixing time is defined as the time required to achieve a specified value of homogeneity Y after addition of some tracer agents:

$$Y = \left|\frac{c(t) - c_0}{c_\infty - c_0}\right| \tag{4.207}$$

where $c(t)$ is the instant conductivity at time t, c_∞ is the final average conductivity of the well-mixed liquid phase, and c_0 is the initial average conductivity of the liquid. If Y henceforth remains in the range between 0.95 and 1.05, the mixing is considered complete, and the time taken for this is the mixing time t_m.

Along with the increasing attention paid to mixing processes, several analytical techniques concerning mixing characteristics have been developed in the past decades. The methods of data acquisition mainly include contact measurement and noncontact measurement, such as the tracer response technique, flow follower method, thermo-anemometry, dye or coloring method, and ultrasonic Doppler velocimetry (Gumery et al., 2009).

Factors influencing macromixing time are often quite complex. There are many phenomenological models to relate the relationship between the operation variables and the observable hydrodynamic variables in ALRs. For example, Sánchez Mirón et al. (2004) proposed a correlation for predicting the mixing time in IALRs as follows:

$$t_{\mathrm{m}} = cU_{\mathrm{g}}^{-0.5} d^{1.4} \left(\frac{h_{\mathrm{D}}}{d}\right)^{1.2} \left(\frac{d_{\mathrm{d}}}{d}\right)^{-1.4} \left(1 - \frac{d_{\mathrm{d}}}{d}\right)^{-1.1}$$

(4.208)

where the constant c has a value of 2.2 (draft-tube sparged) or 2.6 (annulus sparged), h_{D} is the height of gas–liquid dispersion, d is the diameter of the reactor vessel, and d_{d} is the diameter of the draft tube. Equation (4.208) works when $0.11\,\mathrm{m} < d < 0.50\,\mathrm{m}$ and $5 < h_{\mathrm{D}}/d < 40$.

For the purpose of optimum performance of reactors, considerable research has been carried out to investigate the macromixing performance of ALRs. It has always been based on knowledge relating to hydrodynamics and design parameters such as flow pattern (Vial et al., 2001), gas input (Fadavi and Chisti, 2007), liquid circulation (Merchuk et al., 1996), downcomer-to-riser cross-sectional area ratio (Weiland, 1984), design of gas distributor (Merchuk et al., 1998) and gas–liquid separator (Choi et al., 1995), geometrical modifications (Fu et al., 2004), volumetric mass transfer coefficient (Huang et al., 2010), residence time distribution (Gavrilescu and Tudose, 1999), axial diffusion coefficient (Gavrilescu and Tudose, 1997), turbulent diffusion coefficient (Vial et al., 2005), Bodenstein number (Sánchez Mirón et al., 2004), etc. Gondo et al. (1973) investigated liquid mixing by large bubbles in a bubble column and found that the longitudinal dispersion coefficient of the liquid phase depended on the column diameter, rather than the column height and the mode of co- or counter-current contacts. Table 4.8 summarizes some of the empirical correlations of mixing time versus superficial gas velocity.

Table 4.8   Correlation Between Mixing Time and Gas Velocity

Reference	System	t_{m}
Weiland (1984)	External loop, two phases	$t_{\mathrm{m}} \propto u_{\mathrm{g}}^{-0.41}$
Kawase et al. (1994)	External loop, two phases	$t_{\mathrm{m}} \propto u_{\mathrm{g}}^{-1/3}$
Kennard and Janekeh (1991)	Three phases	$t_{\mathrm{m}} \propto u_{\mathrm{g}}^{-0.4}$
Lin et al. (1976)	Two phases, tower	$t_{\mathrm{m}} \propto u_{\mathrm{g}}^{-0.592}$
	Two phases, tower	$t_{\mathrm{m}} \propto u_{\mathrm{g}}^{-0.511}$
Cong et al. (2000)	Three phases, particle size: 180–315 μm	$t_{\mathrm{m}} = 20.82u_{\mathrm{g}}^{-0.371}$
	Three phases, particle size: 315–450 μm	$t_{\mathrm{m}} = 17.54u_{\mathrm{g}}^{-0.288}$
	Three phases, particle size: 450–600 μm	$t_{\mathrm{m}} = 17.86u_{\mathrm{g}}^{-0.251}$

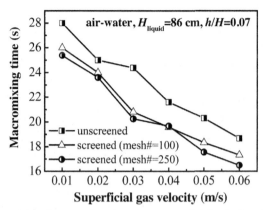

FIGURE 4.20 Macromixing time as a function of gas superficial velocity with different perforated plate distributor apertures ($D = 0.3$ m, $L = 0.70$ m, $T_c = 0.10$ m, $B_c = 0.06$ m, D_t: $\phi200 \times 7$ mm) by Zhang et al. (2012).

Gas distributors are usually used for flow distribution to homogenize energy input and to strengthen mass transfer by bubble dispersion. Many researchers compared different kinds of spargers by their power consumption and dispersion characteristics. The macromixing time is related to liquid circulation velocities determined by the energy balance in the reactor. The energy input into the riser occurs mainly due to isothermal expansion of gas as it rises up along the riser. Energy is dissipated by wall friction in the reactor, turbulent energy dissipation due to internal turbulence, and friction between the gas–liquid interface (Abashar et al., 1998). The feasibility of bubble redistribution by arranging a screen mesh above the perforated plate distributor was investigated with respect to macromixing time by Zhang et al. (2012) and the results are illustrated in Figure 4.20. It can be seen that, compared with the unscreened plate, the screened plate has a slightly beneficial effect on macromixing since it plays a role in the redistribution of bubbles. The kinetic energy of the gas jet from the sparger is usually small and can be neglected. Thus, there is no significant difference in the macromixing time with these two screened plates.

4.5.1.2 Mathematical models of macromixing

Understanding of the mixing parameters is very important to improve the productivity and product selectivity in reactors. In continuous reactors, mixing time is commonly related to the residence time distribution (RTD) and both are used extensively as key parameters in ALR analysis. The RTD, closely related with liquid recirculation in an ALR, can provide information on the level of mixing homogeneity.

Since Danckwerts (1953) introduced the concept of RTD, it has been used as an important tool in the analysis of chemical reactor performance. The tracer conductivity technique is commonly used to get the RTD by measuring the tracer concentration leaving a vessel $C(t)$ after a tracer pulse is injected into the entering stream of fluid at time $t = 0$. In theory, the mean residence time τ and its probability density function $E(t)$ can be calculated as follows:

$$\tau = \frac{\int_0^\infty tc(t)\,dt}{\int_0^\infty c(t)\,dt} \tag{4.209}$$

$$E(t) = \frac{c(t)}{\int_0^\infty c(t)\,dt} \tag{4.210}$$

An example of the E-curves measured and predicted with the model of Eq. (4.204) in the downcomer section of an ALR using the pulse tracer technique is illustrated in Figure 4.21, and a typical response curve is shown in Figure 4.22. These figures suggest that the circulating flow in ALR may be modeled as plug flow with axial dispersion.

Actually, the full transport equation for tracer concentrations in an axisymmetric cylindrical coordinate system is given by the following 3D equation:

$$\frac{\partial(\rho c)}{\partial t} + \frac{1}{r}\frac{\partial}{\partial r}(\rho\alpha_1 u_r rc) + \frac{1}{r}\frac{\partial}{\partial\theta}(\rho\alpha_1 u_\theta c) + \frac{\partial}{\partial z}(\rho\alpha_1 u_z c) = \frac{1}{r}\frac{\partial}{\partial r}\left(r\alpha_1 D_{eff}\frac{\partial c}{\partial r}\right) + \frac{1}{r}\frac{\partial}{\partial\theta}\left(\frac{D_{eff}}{r}\alpha_1\frac{\partial c}{\partial\theta}\right) + \frac{\partial}{\partial z}\left(D_{eff}\alpha_1\frac{\partial c}{\partial z}\right) \tag{4.211}$$

where D_{eff} is the effective diffusion coefficient and can be expressed as (Roy et al., 2006)

$$D_{eff} = \frac{\mu_{eff}}{\sigma_t} + D_m \tag{4.212}$$

in which μ_{eff} is the turbulent diffusion coefficient that can be obtained from the result of the flow field, $\sigma_t = 0.75$ is the turbulent Schmidt number, and D_m is the molecular

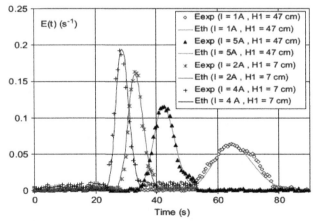

FIGURE 4.21 *E*-curves in the downcomer section for different current intensities and electrode positions (Essadki et al., 2011).

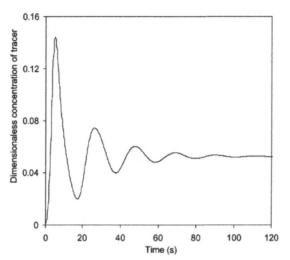

FIGURE 4.22 A typical response curve at the upper end of the riser for mixing time estimation (Essadki et al., 2011).

diffusion coefficient. This equation is sufficient to predict the macromixing behavior by solving the tracer concentration at the liquid outlet to get the desired $E(t)$ and to quantify the macromixing performance. With the tracer injection described correctly by the inlet boundary condition or equivalently the initial condition, numerical simulation of tracer concentration homogenization can be conducted based on the two-phase flow field resolved by numerical simulation of gas–liquid flow in the ALR.

Equation (4.204) is actually a simplified 1D form of Eq. (4.211) based on the assumption of axial plug flow and transverse uniformity. Solving Eq. (4.204), its analytical solution for a straight flow channel taking into account the boundary conditions, i.e., tracer injection at $z = 0$ and the detection point at $z = L$, can be expressed as

$$E(\theta) = \sqrt{\frac{Pe}{4\pi\theta}} \exp\left(-\frac{Pe(1-\theta)^2}{4\theta}\right), \quad \theta = \frac{t}{\tau} \tag{4.213}$$

in which $Pe = U_L(L/D)$ is the Peclet number, D is the axial dispersion coefficient, U_L is the liquid velocity, L is the length of the working column, and τ is the mean residence time.

Taking the tank-in-series model with N perfect reactors, the mean residence time of each real reactor is

$$\tau_1 = \tau_2 = \cdots = \tau_k = \cdots = \tau_N = \frac{\tau}{N} \tag{4.214}$$

$$\tau = \frac{V_R}{Q} \tag{4.215}$$

where V_R is the volume of the reactor, Q is the volumetric flow, and τ is the residence time of the real reactor. For the kth reactor, the mass balance equation for the injected tracer is

$$Qc_{k-1} = Qc_k + \frac{V_R}{N} \frac{dc_k}{dt} \tag{4.216}$$

The solution giving the $E(t)$ curve is

$$E(t) = \left(\frac{N}{\tau}\right)^N \frac{t^{N-1}}{(N-1)!} \exp\left(-\frac{Nt}{\tau}\right) \tag{4.217}$$

The tank-in-series model includes both a piston reactor ($N = \infty$) and a perfect mixed reactor ($N = 1$). If $E(t)$ has been determined either numerically or experimentally, the tank number N can be obtained by fitting $E(t)$ to Eq. (4.217). The value of N (which may be a real number) indexes the extent of macromixing in an ALR.

4.5.2 Micromixing in airlift loop reactors

Mixing is the process referring to the reduction of inhomogeneity concerning the intensity of species segregation. The smallest scale of macromixing is generated by the balance of turbulent stretching and diffusion, and is related to the Kolmogorv scale $\eta(x,t) = (v^3/\varepsilon(x,t))^{1/4}$ through the local Batchelor scale $\eta_B(x,t) = \eta(x,t)/Sc^{1/2}$ (Schumacher et al., 2005). Micromixing, on the other hand, is the process of interdiffusion of substances across the same phase. It is of interest in many chemical engineering applications, and is often the limiting step in fast reactions. In many processes the reactions are fast and the time required to bring homogeneity to the molecular level is greater than the time needed to complete the reactions. When more than one possible reaction route is involved, the yields of desired products will depend on the rate of micromixing. By optimizing the design and operation of industrial reactors to achieve more efficient micromixing, the processes can be upgraded to improve efficiency, increase product yield, and lower the burden in subsequent product purification (Rajab, 2005).

4.5.2.1 Experimental investigation of micromixing

During the last decades, much research has been performed on developing experimental methods to characterize and intensify the micromixing in ALRs (Assirelli et al., 2005; Nouri et al., 2008), e.g., by consecutive competing reactions (Bourne et al., 1981), parallel competing reactions (Bourne and Yu, 1994; Villermaux and Fournier, 1994), and other test reactions (Zhao et al., 2002). The main stoichiometric types of reactions commonly used in micromixing can be found in the contribution of Fournier et al. (1996).

Zhang et al. (2012) carried out investigations on micromixing determined experimentally in ALRs. A popular parallel competing test system proposed by Bourne and Yu (1994) is the neutralization and alkaline hydrolysis of ethyl monochloroacetate:

$$\text{NaOH (A)} + \text{HCl (B)} \xrightarrow{K_1} \text{NaCl (R)} + \text{H}_2\text{O} \tag{4.218}$$

$$\text{NaOH (A)} + \text{CH}_2\text{ClCOOC}_2\text{H}_5 \text{ (C)} \xrightarrow{K_2} \text{CH}_2\text{ClCOONa (Q)} + \text{C}_2\text{H}_5\text{OH} \tag{4.219}$$

where K_1 and K_2 are the rate constants of respective reactions. The segregation index X_Q is defined as the yield of Q relative to the limiting reagent A, i.e.,

$$X_Q = c_Q / (c_Q + c_R) \tag{4.220}$$

where c_Q and c_R are the concentrations of Q and R respectively, and the lower value of X_Q indicates better micromixing performance.

The effects of several operating parameters and geometric variables are taken into consideration and the related experimental results are shown in Figure 4.23. A change of diameter ratio directly affects the liquid circulation velocity. For a given H/D, the most unfavorable conditions are obtained at a low A_r/A_d ratio, which offers more opportunities for bubbles to escape from the surface rather than being entrained by the downward liquid into the annual space. An appropriate increase in A_r/A_d abates the above disadvantages and reduces the flow resistance of the riser so that the volumetric rate of liquid circulation is increased. Figure 4.23 shows that rather good micromixing is achieved at $A_r/A_d \approx 1$, when coalescence of bubbles in two-phase flow is almost inhibited (Weiland, 1984). The total resistance along the flow path is the smallest, leading to the lowest X_Q. Further increase of A_r/A_d ratio reduces micromixing efficiency due to the decreased turbulence level and the increased resistance for liquid circulation.

4.5.2.2 Mathematical models for micromixing
Baldyga et al. (1997) insisted that the mixing in a turbulent flow consisted of the following processes: (a) convection with average velocity; (b) turbulent dispersion by

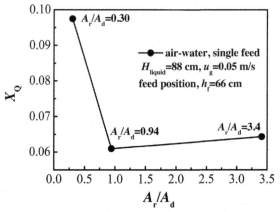

FIGURE 4.23 Segregation index as a function of area ratio of riser and downcomer ($T_c = 0.12$ m, $B_c = 0.06$ m, $t_f = 45$ min, feeding inside of the draft tube, unscreened perforated plate) (Zhang et al., 2012).

large-scale turbulent motions; (c) inertial convective disintegration of large eddies; (d) mixing on a molecular scale by engulfment, deformation, and diffusion inside small-scale turbulent motions. Determining the controlling mechanism of micromixing is important when formulating a micromixing model, as the full mechanism of micromixing is difficult to formulate mathematically and very difficult to solve. Large-scale mixing is best described with CFD in an Eulerian framework, whereas micromixing is better explained by taking a Lagrangian perspective as it can describe different subprocesses of micromixing separately with clarity. In view of this classification, several works have established some micromixing models, including the interaction by exchange with mean model (IEM; Tavare, 1995), the engulfment model (Eng; Baldyga, 1989; Baldyga and Bourne, 1989), engulfment deformation diffusion model (EDD; Baldyga and Bourne, 1984a–c), multiple-environment model (ME; Goto et al., 1975; Mehta and Tarbell, 1983), shrinking aggregate model (SA; Villermaux and David, 1983), stochastic coalescence–redispersion model (SCR; Pojman et al., 1991), and direct quadrature method of moments interaction by exchange with the mean model (DQMOM-IEM; Akroyd et al., 2010).

Lagrangian micromixing models can be roughly classified into two subgroups: empirical models based on RTD theory and the concept of fluid environments, and theoretical models based more on turbulence theory, such as the spectral analysis of turbulent kinetic energy and concentration fluctuation spectrum. Based on RTD theories, many empirical micromixing models have been proposed. Consistent with the given RTD, two extreme limits of micromixing have been defined: total segregation and maximum mixedness. The coalescence–redispersion (CD) model is another type of RTD-based model. Another class of empirical micromixing models is based on the concept of fluid environments. Adjacent fluid environments interact with each other due to micromixing. Many multi-environment models have been developed and applied so far, such as two-environment (2E) models, 3E models, and 4E models.

From the analysis of turbulent kinetic energy and scalar fluctuation spectra, the theoretical models attempt to distinguish different mechanisms of meso- and micromixing, and try to relate mathematically the micromixing time, which should be roughly equal to the reciprocal of the micromixing rate, to turbulence and fluid properties such as ε and μ. The IEM, as stated by Pohorecki and Baldyga (1983), seems to be the closest to reality because the micromixing is described as a continuous process of mass transfer between "points" and their environments. Thus, the IEM model has been most widely used compared with the two other types of diffusion models, and it still has some application now, mainly combined with CFD. The IEM model can be expressed as

$$\frac{dc_i}{dt} = \frac{1}{t_{micro}}(\langle c_i \rangle - c_i) + R_i$$

(4.221)

where c_i is the concentration of "points" and $\langle c_i \rangle$ is the average concentration of the environment. The IEM model assumes the reactor is well macromixed, neglecting the spatial inhomogeneities of flow and scalar fields. The determination of the micromixing time t_{micro} is a difficult and nontrivial matter due to the fact that t_{micro} is intimately related to the underlying flow field.

The engulfment–deformation–diffusion (EDD) model that describes the full mechanism of micromixing was developed in an outstanding and systematic work by Baldyga and Bourne (1984b, c). For $Sc < 4000$, engulfment becomes the dominant mechanism of micromixing and the deformation and diffusion steps can be neglected. Thus, for $Sc < 4000$, the full EDD model reduces to the well-known engulfment model (E-model), into which the chemical kinetics is directly incorporated (Baldyga and Bourne, 1984b). Taking into account the slowing down of the reaction zone growth by self-engulfment, the standard E-model (Baldyga and Bourne, 1984b) is given as

$$\frac{\mathrm{d}V_i}{\mathrm{d}t} = \langle E \rangle V_i (1 - X_i)$$

(4.222)

$$\frac{\mathrm{d}c_i}{\mathrm{d}t} = \langle E \rangle (1 - X_i)\left(\langle c_i \rangle - c_i\right) + R_i$$

(4.223)

where E is the engulfment rate, $E = 0.058(\varepsilon/\nu)^{1/2}$, and V_i and X_i denote the volume and the local volume fraction of an i-rich region, respectively.

Eulerian micromixing models are based on statistics. In an Eulerian framework, an averaging approach, i.e., Reynolds averaging, is generally employed to solve instantaneous scalar transport equations. However, information about the subgrid scale mixing and residence time is lost, and new problems will arise. Taking a typical non-premixed single second-order reaction, A + B → R, for example, the instantaneous scalar transport equation is expressed as

$$\frac{\partial c_\alpha}{\partial t} + \frac{\partial (u_i c_\alpha)}{\partial x_i} = \frac{\partial}{\partial x_i}\left(\Gamma_\alpha \frac{\partial c_\alpha}{\partial x_i}\right) + S_\alpha(\mathbf{C}), \quad \alpha = \mathrm{A}, \mathrm{B}$$

(4.224)

where Γ_α is the molecular diffusivity, $\mathbf{C}$ is the scalar vector, $\mathbf{C} = (c_\mathrm{A}, c_\mathrm{B})^\mathrm{T}$, and $S_\alpha(\mathbf{C})$ is the chemical source term, defined here by

$$S_\alpha(\mathbf{C}) = -kc_\mathrm{A}c_\mathrm{B}, \quad \alpha = \mathrm{A}, \mathrm{B}$$

(4.225)

Numerous investigations of micromixing in a stirred reactor have been carried out utilizing the aforementioned methods and models. The rotating packed bed, membrane reactor, tubular reactor, microreactor, impinging jet reactor, static mixer, and Couette flow reactor have also been studied from a micromixing point of view. To the authors' knowledge, few studies on the micromixing in bubble columns and ALRs have been reported to date.

4.6 GUIDELINES FOR DESIGN AND SCALE-UP OF AIRLIFT LOOP REACTORS

It is the difference in configuration that allows variation of operation of ALRs in terms of liquid circulation and gas disengagement. The main parameters for design and scale-up of ALRs include the operating parameters (i.e., flow regime, superficial gas and liquid velocities, temperature, pressure, solid loading, and catalyst size) and the structural parameters, namely the length of the draft tube, the ratio of height to diameter (H/D), the cross-section ratio of A_r/A_d, heights at the top clearance and the bottom clearance, structural shapes of the reactor, draft tube, gas degasification section, sparger, and internals. Some other factors are involved including total gas volume fraction, eliminating the dead zone in the reactor, mass transfer coefficient, liquid circulation velocity, mixing time and residence time, degree of degassing (or gaseous circulation or not), heat exchange, size and concentration of the particles in a slurry reactor, prohibition of particle sedimentation and choking in slurry reactors (this phenomenon can be prevented by designing the reactor with a high liquid velocity, which is greater than the ultimate settling velocity of the particles), and so on.

The shapes of ALRs can be cylindrical or rectangular tanks. However, the cylindrical reactor is favored due to its excellent pressure tolerance, which results in a good performance as regards safety and mechanical stability. Some researchers provided optimum structural parameters that were derived from the least energy consumption principle as follows (Li, 2004): $D_{riser}/D_{reactor} = 0.6–0.7$, length of the draft tube $= 7.5D_{reactor}$, ratio of the area in the top clearance for the fluid diverting to the cross-sectional area of the reactor $= 0.82$, while the ratio of the area in the bottom clearance for the fluid diverting to the cross-sectional area of the reactor $= 0.58$. It is noteworthy that there is only a small amount of gas in the downcomer when the principle of the geometric parameters mentioned above is adopted due to the large cross-sectional area in the downcomer. This may result in a large portion of useless volume with no mass transfer between gas and liquid, and it should be further optimized according to the specific application.

It is commonly accepted that the hydrodynamics is independent of the column size, if D, H, and H/D are larger than some minimum values. Different minimum values have been reported. For example, the diameter D should exceed 0.1–0.2 m, the height H should be larger than 0.3–0.5 m or even 1–3 m, and the aspect ratio H/D should be greater than 5. It should be pointed out that when using experimental results obtained in a laboratory reactor apparatus to design an industrial reactor, the laboratory apparatus must have a diameter larger than the critical value, say 0.15 m. The introduction of internals used to guide the flow of bubbles increases the resistance, and hence the liquid circulation is decreased. Under these circumstances, large bubbles are broken down into smaller ones, increasing the interfacial surface for higher mass transfer rates. However, the influence of internals becomes insignificant with the increase of distance downstream away from the internal, and the effective distance is 1.0–1.5 m (Wang et al., 2007).

For IALRs, Merchuck et al. (1994) pointed out that the bottom clearance had an important impact on the pressure drop, while the top clearance did not affect the pressure drop. When the free area for liquid flow between the bottom plate and the draft tube is very constrained, changes in the bottom clearance have a strong effect on the pressure drop. However, this is not so for clearances wider than 0.04 m for a reactor of 300 L with a draft tube diameter of 0.216 m. It was reported that the pressure drop data for a bottom clearance of 0.08 m were close to those corresponding to a bottom clearance of 0.04 m for a 300 L reactor. In addition, the influence of the gas superficial velocity on pressure drop is very small; the pressure drop increases at low gas superficial velocities and then tends to a plateau. Moreover, the top clearance has its maximal effect on gas holdup, and the bottom clearance is the leading parameter. The gas holdup increases in the riser as the bottom clearance decreases. The lower top clearance gives a shorter residence time in the separator and a larger bubble circulation, and therefore larger gas holdups.

The type of gas distributor has a significant impact on gas holdups in homogeneous bubbly flow, while this effect can be ignored when the pressure is greater than 5 MPa and the superficial gas velocity is high. In addition, the position of the sparger is also important in gas distribution and bubble coalescence. A gas distributor placed just inside the riser enhances gas distribution since the downcomer flow joins the stream of the riser, in contrast to a sparger placed at the entrance of the riser. In the latter scenario, the downcomer stream maldistributes and clusters the gas bubbles to the wall of the vessel, encouraging coalescence (Chisti, 1989). It was also reported that adding a small amount of a surfactant to water resulted in decreased bubble sizes and significantly higher gas holdups (Wang et al., 2007). The presence of electrolytes or impurities was also found to have the same effect. With an increase in temperature, the liquid viscosity and surface tension decrease and result in a smaller average bubble size and a narrower bubble size distribution. It should be noted that the influence of temperature on gas–liquid mass transfer is much more notable than gas holdups, due to the higher liquid diffusivity at high temperature. The superficial gas velocity at the flow regime transition point increases with system pressure. This can be explained by the fact that the increased gas density due to the elevated pressure reduces the interaction between neighboring bubbles (Krishna et al., 1994). In addition, the increased gas density affects the initial bubble diameter generated from the sparger. Generally, the higher the pressure, the smaller the initial bubble diameter produced, which results in a significant increase of gas voidage and volumetric mass transfer coefficient in the reactor. However, the increase of solid concentration may have a negative influence on the hydrodynamics. In a gas–liquid–solid slurry system, the particles are usually smaller than 100 μm. In such cases, the spatial profile of the solid concentration is almost uniform. The main influence of the solid concentration lies in an increase in the apparent viscosity, which can be well predicted using the Einstein equation (Muroyama et al., 2007; Viamajala et al., 2009). Increasing solid concentration generally increases the bubble size due to an increase in the apparent suspension viscosity. At high solid concentrations, the bubble coalescence tendency is enhanced so that the fraction of small bubbles becomes insignificant. Bubble breakup that occurs above the gas distributor is suppressed in the presence of fine particles in the suspension. Therefore, an increasing solid concentration generally decreases the gas holdup. Therefore, most results in the literature show that the volumetric mass transfer coefficient decreases with increasing solid concentration or liquid viscosity. However, it was also found that an increase or decrease of gas holdup in an air–water system with the addition of solids depends on the nature of the solids. Generally, addition of nonwettable solids reduces the gas holdup while addition of wettable solids increases the gas holdup (Shaikh and Al-Dahhan, 2007).

Although many ALRs are operated with entrained gas in the downcomer, others are designed so that the downcomer is gas free. The latter mode of operation should maximize the gas holdup difference between the riser and the downcomer to enhance the liquid circulation, and an EALR or an enlarged separator for an IALR is preferred to ensure near complete gas–liquid separation. On the contrary, the liquid circulation velocity will be too high in a tall industrial internal airlift reactor. In such a case, it is necessary to install some internals in the riser or design a multistage riser to decrease the liquid circulation velocity, increase the gas holdup, and enhance the gas–liquid mass transfer. Last but not least, if an IALR is used with a certain amount of gas existing in the downcomer, regime II of flow should be avoided because the key components to be transferred in gas bubbles in the downcomer will be depleted under these circumstances.

4.7 SUMMARY AND PERSPECTIVE

ALRs are considered to be ideal for many chemical processes, especially when a large amount of gas is involved as reactant or a cheap gas stream is available as agitating power source. The hydrodynamics, mass, and heat transfer rates therein can be well controlled if it is properly designed. Moreover, macromixing and micromixing in ALRs are generally satisfactory. The multi-scale phenomena of macro and micro scales characterized by the mixing in ALRs can be well captured and realized by experimental and numerical approaches.

Although great advances have been made up to now, new improvements are needed to improve the performance of ALRs for the sake of long-run safety, economy, and operation stability. Besides the transport process, process control and process intensification are two primary aspects that should be well understood. Efficient gas distributors should be designed to produce small bubbles. Additionally, catalyst separation and withdrawal is a challenging engineering problem, especially under conditions of elevated pressure and high temperature. A filtration and back-flushing system has been used in ALRs in latter years for separation of liquid from the slurry and has been successful, but it warrants further testing. It is noteworthy that separation of products from the liquid should be integrated with the reaction for a strong exothermic reaction system.

As a new method to predict the complex behavior of flow and transport in ALRs, the closure of CFD models should be verified with more experiments, and more robust algorithms for multiphase flow with decoupled computation of velocity, pressure, and fractions of voidage fields should be developed. Furthermore, some other numerical techniques to promote the balance of mass for individual phases are also required. In practice, most simulations are performed on laboratory or pilot scales; special algorithms with high-performance computing (HPC) should be developed to carry out simulations for large-scale industrial reactors.

The two-fluid model with the assumption of a constant bubble diameter can give reasonable predictions for the homogeneous regime because the bubble size distribution in such conditions is narrow and the bubble interaction is relatively weak. This method may not be valid in the transition or heterogeneous regimes. In the heterogeneous regime, it is known that the CFD–PBM coupled model, which combines the population balance model (PBM) with the CFD, is becoming an effective approach to describe the complex behavior of bubbles, including coalescence and breakup, and bubble–bubble and bubble–liquid interactions. However, most reported results using the CFD–PBM coupled model are either only for the homogeneous regime or for both the homogeneous and heterogeneous regimes, but the present simulation does not successfully predict a bimodal bubble size distribution in the heterogeneous regime. Additionally, CFD results show that the calculated bubble size distributions are quite different when different bubble coalescence and breakup models are used (Wang et al., 2005a). Therefore, it is an urgent task to set up a reliable PBM to account for the effects of coalescence and breakup in polydispersed systems.

NOMENCLATURE

A	area	m²
A, a	coefficient in the discretized equation	–
B_c	bottom clearance	m
C_A	added mass coefficient	–
C_D	drag coefficient of a swarm	–
C_d	drag coefficient of a single bubble	–
C_l	lift coefficient	–
C_{TD}	coefficient of the turbulent dispersion force	–
C_W	coefficient of the wall force	–
C_μ	model constant of k–ε model	–
c_0	nodes adjacent to the outlet boundary	
c_A	concentration of substance A	mol/m³ or kg/m³
c_p	concentration of product	mol/m³
D_L	molecular diffusivity of gas in liquid	m²/s
D, d	diameter	m
du, dv, dw	pressure correction coefficient	m/(Pa·s)
Eo	Eotvos number	–
Eo_d	modified Eotvos number	–
F	force	N
G	production of turbulent kinetic energy	kg/(m·s³)
g	acceleration due to gravity	m/s²
H	height of reactor	m
h	monitoring height	m
h_f	feed position	m
k	turbulent kinetic energy	m²/s²
$k_L a$	gas transfer coefficient	s⁻¹
L_p	height of packing	m

Mo	Morton number	–
N	number	
P	pressure	Pa
Re	Reynolds number	–
r	radial position	m
r_0	radial coordinate at the axis	m
S	source term in the discretized equation	
S_b	area of the bubble	m²
T_c	top clearance	m
T_{kij}	viscous stress tensor in phase k	kg/(m·s²)
t	time	s
t_c	circulation time	s
u, U	velocity	m/s
v	velocity	m/s
V_b	volume of the bubble	m³
w	velocity	m/s
We	Weber number	–
x, y, z	Cartesian coordinates	m

Greek letters

Γ	diffusion coefficient	
ΔH	heat of reaction	kJ/mol
α	voidage of phase	–
δ_{ij}	Kronecker delta	
ε	turbulent energy dissipation	m²/s³
ζ_r	ratio of characteristic time of turbulence in liquid and characteristic time scale of bubble necessary to cross the containing energy eddies	–
η_r	ratio of characteristic time scale of turbulence seen by gas phase and characteristic time scale of bubble entrainment by liquid motion	–
μ	viscosity	kg/(m·s)
ϕ	general variables	
v	kinematic viscosity	m²/s
σ	surface tension	N/m
σ_k	model constant of k–ε model	–
σ_t	turbulent Schmidt number	–
σ_ε	model constant of k–ε model	–

Subscripts

A	added force
ap	apparent
Bl	bubble induce
b	bubble
c	continuous phase; column
d	drag force; downcomer

eff	effective
FOU	first-order upwind
g	gas
H_2	hydrogen
i, j, k	directions of coordinate
k	phase
l	liquid or lift
m	mixture
p	dispersion phase
r	radial direction or riser
SI	shear induced
s	separator
slip	slip velocity
t	terminal velocity
θ	tangential direction
z	axial direction

Superscripts

dc	deferred correction
n	value at step n
norm	normalized quantity
T	matrix transposition
t, turb	turbulent
*	predicted value in last iteration
'	correction in this iteration

REFERENCES

Abashar, M. E., Narsingh, U., Rouillard, A. E., & Judd, R. (1998). Hydrodynamic flow regimes, gas holdup, and liquid circulation in airlift reactors. *Ind. Eng. Chem. Res.*, *37*(4), 1251–1259.

Acién Fernández, F. G., Fernández Sevilla, J. M., Sánchez Pérez, J. A., Molina Grima, E., & Chisti, Y. (2001). Airlift-driven external-loop tubular photobioreactors for outdoor production of microalgae: Assessment of design and performance. *Chem. Eng. Sci.*, *56*(8), 2721–2732.

Akroyd, J., Smith, A. J., McGlashan, L. R., & Kraft, M. (2010). Numerical investigation of DQMoM-IEM as a turbulent reaction closure. *Chem. Eng. Sci.*, *65*(6), 1915–1924.

Anderson, T. B., & Jackson, R. (1967). A fluid mechanical description of fluidized beds. *Ind. Eng. Chem. Fund.*, *6*(11), 527–539.

Antal, S. P., Lahey, R. T., & Flaherty, J. E. (1991). Analysis of phase distribution in fully developed laminar bubbly two-phase flow. *Int. J. Multiphase Flow*, *17*(5), 635–652.

Arnold, G. S., Drew, D. A., & Lahey, R. T. (1989). Derivation of constitutive equations for interfacial force and Reynolds stress for a suspension of spheres using ensemble cell averaging. *Chem. Eng. Comm.*, *86*(1), 43–54.

Assirelli, M., Bujalski, W., Eaglesham, A., & Nienow, A. W. (2005). Intensifying micromixing in a semi-batch reactor using a Rushton turbine. *Chem. Eng. Sci.*, *60*(8–9), 2333–2339.

Ayed, H., Chahed, J., & Roig, V. (2007). Hydrodynamics and mass transfer in a turbulent buoyant bubbly shear layer. *AIChE J.*, *53*(11), 2742–2753.

Azbel, D. (1981). *Two-phase flows in chemical engineering*. Cambridge: Cambridge University Press.

Azzaro, C., Duverneuil, P., & Couderc, J. P. (1992). Thermal and kinetic modelling of low-pressure chemical vapour deposition hot-wall tubular reactors. *Chem. Eng. Sci.*, *47*(15–16), 3827–3838.

Bakker, A., & Van den Akker, H. E. A. (1994). A computational model for the gas–liquid flow in stirred reactors. *Chem. Eng. Res. Des.*, *72*(A4), 573–582.

Bakshi, B. R., Zhong, H., Jiang, P., & Fan, L. -S. (1995). Analysis of flow in gas–liquid bubble columns using multi-resolution methods. *Chem. Eng. Res. Des.*, *73A*, 608–614.

Baldyga, J. (1989). Turbulent mixer model with application to homogeneous, instantaneous chemical reactions. *Chem. Eng. Sci.*, *44*, 1175–1182.

Baldyga, J., & Bourne, J. R. (1984a). A fluid mechanical approach to turbulent mixing and chemical reaction. Part I: Inadequacies of avilable methods. *Chem. Eng. Comm.*, *28*(4–6), 231–241.

Baldyga, J., & Bourne, J. R. (1984b). A fluid mechanical approach to turbulent mixing and chemical reaction. Part II: Micromixing in the light of turbulence theory. *Chem. Eng. Comm.*, *28*(4–6), 243–258.

Baldyga, J., & Bourne, J. R. (1984c). A fluid mechanical approach to turbulent mixing and chemical reaction. Part III: Computational and experimental results for the new micromixing model. Chem. Eng. Comm., *28*(4–6), 259–281.

Baldyga, J., & Bourne, J. R. (1989). Simplification of micromixing calculations. Part I: Derivation and application of new model. *Chem. Eng. J.*, *42*, 83–92.

Baldyga, J., Bourne, J. R., & Hearn, S. J. (1997). Interaction between chemical reactions and mixing on various scales. *Chem. Eng. Sci.*, *52*(4), 457–466.

Becker, S., Sokolichin, A., & Eigenberger, G. (1994). Gas–liquid flow in bubble columns and loop reactors: Part II. Comparison of detailed experiments and flow simulations. *Chem. Eng. Sci.*, *49*(24, Part 2), 5747–5762.

Behzadi, A., Issa, R. I., & Rusche, H. (2004). Modelling of dispersed bubble and droplet flow at high phase fractions. *Chem. Eng. Sci.*, *59*(4), 759–770.

Bird, R. B., Stewart, W. E., & Lightfoot, E. N. (1960). *Transport phenomena*. New York: John Wiley.

Blažej, M., Cartland Glover, G. M., Generalis, S. C., & Markoš, J. (2004a). Gas–liquid simulation of an airlift bubble column reactor. *Chem. Eng. Process.*, *43*(2), 137–144.

Blažej, M., Kiša, M., & Markoš, J. (2004b). Scale influence on the hydrodynamics of an internal loop airlift reactor. *Chem. Eng. Process.*, *43*(12), 1519–1527.

Boisson, N., & Malin, M. R. (1996). Numerical prediction of two-phase flow in bubble columns. *Int. J. Numer. Meth. Fluids*, *23*(12), 1289–1310.

Bourne, J. R., & Yu, S. (1994). Investigation of micromixing in stirred tank reactors using parallel reactions. *Ind. Eng. Chem. Res.*, *33*(1), 41–55.

Bourne, J. R., Kozicki, F., & Rys, P. (1981). Mixing and fast chemical reaction – I: Test reactions to determine segregation. *Chem. Eng. Sci.*, *36*(10), 1643–1648.

Bove, S. (2005). Computational fluid dynamics of gas-liquid flows including bubble population balances. *Ph. D. thesis*. Aalborg University: Denmark.

Broadwell, J. E., & Godfrey Mungal, M. (1990). Large-scale structures and molecular mixing. *Phys. Fluids*, *3*(5), 1193–1206.

Brucato, A., Grisafi, F., & Montante, G. (1998). Particle drag coefficients in turbulent fluids. *Chem. Eng. Sci.*, *53*(18), 3295–3314.

Carver, M. B. (1984). Numerical computation of phase separation in two-fluid flow. *J. Fluids Eng.*, *106*, 147–153.

Chandavimol, M. (2003). Experimental and simulation studies of two-phase flow in a stirred tank. *Ph. D. thesis*. University of Missouri-Rolla.

Chen, P. (2004). Modeling the fluid dynamics of bubble column flows. *Ph. D. thesis*. Saint Louis, MO: Washington University.

Chen, X. H. (2004). Application of computational fluid dynamics (CFD) to flow simulation and erosion prediction in single-phase and multiphase flow. *Ph. D. thesis*. University of Tulsa, OK.

Chisti, M. Y. (1989). *Airlift bioreactors*. London: Elsevier.

Chisti, Y. (1998). Pneumatically agitated bioreactors in industrial and environmental bioprocessing: Hydrodynamics, hydraulics, and transport phenomena. *Appl. Mech. Rev.*, *51*(1), 33–112.

Chisti, Y., & Moo-Young, M. (1993). Improve the performance of airlift reactors. *Chem. Eng. Progr.*, *89*(6), 38–45.

Choi, K. H., Chisti, Y., & Moo-Young, M. (1995). Influence of the gas–liquid separator design on hydrodynamic and mass transfer performance of split-channel airlift reactors. *J. Chem. Tech. Biotechnol.*, *62*(4), 327–332.

Chow, W. K., & Li, J. (2007). Numerical simulations on thermal plumes with k–ε types of turbulence models. *Build. Environ.*, *42*(8), 2819–2828.

Cockx, A., Do-Quang, Z., Liné, A., & Roustan, M. (1999). Use of computational fluid dynamics for simulating hydrodynamics and mass transfer in industrial ozonation towers. *Chem. Eng. Sci.*, *54*(21), 5085–5090.

Cockx, A., Do-Quang, Z., Audic, J. M., Liné, A., & Roustan, M. (2001). Global and local mass transfer coefficients in waste water treatment process by computational fluid dynamics. *Chem. Eng. Process.*, *40*(2), 187–194.

Cong, W., Liu, J., Ouyang, F., & Liao, Y. (2000). Mixing behaviour of three-phase internal-loop air-lift reactor. *Eng. Chem. Metall.*, *21*(1), 76–79 (in Chinese).

Danckwerts, P. V. (1953). Continuous flow systems: Distribution of residence times. *Chem. Eng. Sci.*, *2*(1), 1–13.

Darwish, M., & Moukalled, F. (2001). A unified formulation of the segregated class of algorithms for multifluid flow at all speeds. *Numer. Heat Tran.*, *40*, 99–137.

de Matos, A., Rosa, E. S., & Franca, F. A. (2004). The phase distribution of upward co-current bubbly flows in a vertical square channel. *J. Braz. Soc. Mech. Sci. Eng.*, *26*(3), 308–316.

Delnoij, E., Lammers, F. A., Kuipers, J. A. M., & van Swaaij, W. P. M. (1997). Dynamic simulation of dispersed gas–liquid two-phase flow using a discrete bubble model. *Chem. Eng. Sci.*, *52*(9), 1429–1458.

Deng, H., Mehta, R. K., & Warren, G. W. (1996). Numerical modeling of flows in flotation columns. *Int. J. Miner. Process.*, *48*(1–2), 61–72.

Drahoš, J., Zahradník, J., Punčochář, M., Fialová, M., & Bradka, F. (1991). Effect of operating conditions on the characteristics of pressure fluctuations in a bubble column. *Chem. Eng. Process.*, *29*(2), 107–115.

Drahoš, J., Bradka, F., & Punčochář, M., (1992). Fractal behaviour of pressure fluctuations in a bubble column. *Chem. Eng. Sci.*, *47*(15–16), 4069–4075.

Essadki, A. H., Gourich, B., Vial, C., & Delmas, H. (2011). Residence time distribution measurements in an external-loop airlift reactor: Study of the hydrodynamics of the liquid circulation induced by the hydrogen bubbles. *Chem. Eng. Sci.*, 66(14), 3125–3132.

Fadavi, A., & Chisti, Y. (2007). Gas holdup and mixing characteristics of a novel forced circulation loop reactor. *Chem. Eng. J.*, *131*(1–3), 105–111.

Fournier, M. C., Falk, L., & Villermaux, J. (1996). A new parallel competing reaction system for assessing micromixing efficiency – Experimental approach. *Chem. Eng. Sci.*, *51*(22), 5053–5064.

Frank, T. (2005). Advances in Computational Fluid Dynamics (CFD) of 3-dimensional gas–liquid multiphase flows. *NAFEMS seminar: Simulation of complex flows (CFD)*, (pp. 1–18), Wiesbaden, Germany.

Frank, T., Zwart, P. J., Krepper, E., Prasser, H. M., & Lucas, D. (2008). Validation of CFD models for mono- and polydisperse air–water two-phase flows in pipes. *Nucl. Eng. Des.*, *238*(3), 647–659.

Fu, C. -C., Lu, S. -Y., Hsu, Y. -J., Chen, G. -C., Lin, Y. -R., & Wu, W. -T. (2004). Superior mixing performance for airlift reactor with a net draft tube. *Chem. Eng. Sci.*, *59*(14), 3021–3028.

Gavrilescu, M., & Tudose, R. Z. (1997). Mixing studies in external-loop airlift reactors. *Chem. Eng. J.*, *66*(2), 97–104.

Gavrilescu, M., & Tudose, R. Z. (1999). Residence time distribution of the liquid phase in a concentric-tube airlift reactor. *Chem. Eng. Process.*, *38*(3), 225–238.

Gondo, S., Tanaka, S., Kazikuri, K., & Kusunoki, K. (1973). Liquid mixing by large gas bubbles in bubble columns. *Chem. Eng. Sci.*, *28*(7), 1437–1445.

Goto, H., Goto, S., & Matsubara, M. (1975). A generalized two-environment model for micromixing in a continuous flow reactor – II. *Identification of the model. Chem. Eng. Sci.*, *30*(1), 71–77.

Grienberger, J., & Hofmann, H. (1992). Investigations and modelling of bubble columns. *Chem. Eng. Sci.*, *47*(9–11), 2215–2220.

Griffith, R. M. (1960). Mass transfer from drops and bubbles. *Chem. Eng. Sci.*, *12*(3), 198–213.

Gumery, F., Farhad, E. -M., & Yaser, D. (2009). Characteristics of local flow dynamics and macro-mixing in airlift column reactors for reliable design and scale-up. *Int. J. Chem. Reactor Eng.*, *7*(1), 1–47.

Heijnen, J. J., Hols, J., van der Lans, R. G. J. M., van Leeuwen, H. L. J. M., Mulder, A., & Weltevrede, R. (1997). A simple hydrodynamic model for the liquid circulation velocity in a full-scale two- and three-phase internal airlift reactor operating in the gas recirculation regime. *Chem. Eng. Sci.*, *52*(15), 2527–2540.

Heijnen, S. J., Mulder, A., Weltevrede, R., Hols, P. H., & van Leeuwen, H. L. J. M. (1990). Large-scale anaerobic/aerobic treatment of complex industrial wastewater using immobilized biomass in fluidized bed and air-lift suspension reactors. *Chem. Eng. Technol.*, *13*(1), 202–208.

Higbie, R. (1935). The rate of absorption of a pure gas into a still liquid during short periods of exposure. *Trans. AIChE*, *35*, 365–389.

Huang, Q. (2008). Numerical simulation of multiphase transport and reaction in loop reactors. *Ph. D. thesis*. Beijing: Insititute of Process Engieering.

Huang, Q., Yang, C., Yu, G., & Mao, Z. -S. (2007). 3-D simulations of an internal airlift loop reactor using a steady two-fluid model. *Chem. Eng. Technol.*, *30*(7), 870–879.

Huang, Q., Yang, C., Yu, G., & Mao, Z. -S. (2008). Sensitivity study on modeling an internal airlift loop reactor using a steady 2D two-fluid model. *Chem. Eng. Technol.*, *31*, 1790–1798.

Huang, Q., Yang, C., Yu, G., & Mao, Z. -S. (2010). CFD simulation of hydrodynamics and mass transfer in an internal airlift loop reactor using a steady two-fluid model. *Chem. Eng. Sci.*, *65*(20), 5527–5536.

Ilegbusi, O., Iguchi, M., Nakajima, K., Sano, M., & Sakamoto, M. (1998). Modeling mean flow and turbulence characteristics in gas-agitated bath with top layer. *Metall. Mater. Trans. B*, *29*(1), 211–222.

Ishii, M., & Mishima, K. (1984). Two-fluid model and hydrodynamic constitutive relations. *Nucl. Eng. Des.*, *82*(2–3), 107–126.

Ishii, M., & Zuber, N. (1979). Drag coefficient and relative velocity in bubbly, droplet or particulate flows. *AIChE J.*, *25*(5), 843–855.

Jakobsen, H. A., Grevskott, S., & Svendsen, H. F. (1997). Modeling of vertical bubble-driven flows. *Ind. Eng. Chem. Res.*, *36*(10), 4052–4074.

Jakobsen, H. A., Lindborg, H., & Dorao, C. A. (2005). Modeling of bubble column reactors: Progress and limitations. *Ind. Eng. Chem. Res.*, *44*(14), 5107–5151.

Jia, X., Wen, J., Feng, W., & Yuan, Q. (2007). Local hydrodynamics modeling of a gas–liquid–solid three-phase airlift loop reactor. *Ind. Eng. Chem. Res.*, *46*(15), 5210–5220.

Juraščík, M., Blažej, M., Annus, J., & Markoš, J. (2006). Experimental measurements of volumetric mass transfer coefficient by the dynamic pressure-step method in internal loop airlift reactors of different scale. *Chem. Eng. J.*, *125*, 81–87.

Karamanev, D. G., & Nikolov, L. N. (1992). Free rising spheres do not obey Newton's law for free settling. *AIChE J.*, *38*(11), 1843–1846.

Kawase, Y., Omori, N., & Tsujimura, M. (1994). Liquid-phase mixing in external-loop airlift bioreactors. *J. Chem. Tech. Biotechnol.*, *61*(1), 49–55.

Kennard, M., & Janekeh, M. (1991). Two- and three-phase mixing in a concentric draft tube gas-lift fermentor. *Biotechnol. Bioeng.*, *38*(11), 1261–1270.

Kerdouss, F., Bannari, A., & Proulx, P. (2006). CFD modeling of gas dispersion and bubble size in a double turbine stirred tank. *Chem. Eng. Sci.*, *61*(10), 3313–3322.

Khopkar, A. R., & Ranade, V. V. (2006). CFD simulation of gas–liquid stirred vessel: VC, S33, and L33 flow regimes. *AIChE J.*, *52*(5), 1654–1672.

Khopkar, A. R., Kasat, G. R., Pandit, A. B., & Ranade, V. V. (2006). CFD simulation of mixing in tall gas–liquid stirred vessel: Role of local flow patterns. *Chem. Eng. Sci.*, *61*(9), 2921–2929.

Khosla, P. K., & Rubin, S. G. (1974). A diagonally dominant second-order accurate implicit scheme. *Computers Fluids*, *2*(2), 207–209.

Krishna, R., De Swart, J. W. A., Hennephof, D. E., Ellenberger, J., & Hoefsloot, H. C. J. (1994). Influence of increased gas density on hydrodynamics of bubble-column reactors. *AIChE J.*, *40*(1), 112–119.

Kuo, J. T., & Wallis, G. B. (1988). Flow of bubbles through nozzles. *Int. J. Multiphase Flow*, *14*(5), 547–564.

Lahey, R. T., Lopez de Bertodano, M., & Jones, O. C. (1993). Phase distribution in complex geometry conduits. *Nucl. Eng. Des.*, *141*(1–2), 177–201.

Lapin, A., Maul, C., Junghans, K., & Lübbert, A. (2001). Industrial-scale bubble column reactors: Gas–liquid flow and chemical reaction. *Chem. Eng. Sci.*, *56*(1), 239–246.

León-Becerril, E., Cockx, A., & Liné, A. (2002). Effect of bubble deformation on stability and mixing in bubble columns. *Chem. Eng. Sci.*, *57*(16), 3283–3297.

Li, F. (2004). *Study on the hydrodynamics of novel multistage airlift loop reactor. Ph. D. thesis*. Beijing: Tsinghua University.

Li, H., & Prakash, A. (2000). Influence of slurry concentrations on bubble population and their rise velocities in a three-phase slurry bubble column. *Powder Technol.*, *113*(1–2), 158–167.

Li, Y., & Baldacchino, L. (1995). Implementation of some higher-order convection schemes on non-uniform grids. *Int. J. Numer. Meth. Fluids*, *21*(12), 1201–1220.

Li, Y., & Rudman, M. (1995). Assessment of higher-order upwind schemes incorporating FCT for convection-dominated problems. *Numer. Heat Tran.*, *27*(1), 1–21.

Lin, C. H., Fang, B. S., Wu, C. S., Fang, H. Y., Kuo, T. F., & Hu, C. Y. (1976). Oxygen transfer and mixing in a tower cycling fermentor. *Biotechnol. Bioeng.*, *18*(11), 1557–1572.

Lin, W. C., Mao, Z. S., & Chen, J. Y. (1997a). Hydrodynamic studies on loop reactors (I): Liquid jet loop reactors. *Chinese J. Chem. Eng.*, *5*(1), 1–10.

Lin, W. C., Mao, Z. S., & Chen, J. Y. (1997b). Hydrodynamic studies on loop reactors (II): Airlift loop reactors. *Chinese J. Chem. Eng.*, *5*(1), 11–22.

Lochiel, A. C., & Calderbank, P. H. (1964). Mass transfer in the continuous phase around axisymmetric bodies of revolution. *Chem. Eng. Sci.*, *19*(7), 471–484.

Lucas, D., Krepper, E., & Prasser, H. M. (2005). Development of co-current air–water flow in a vertical pipe. *Int. J. Multiphase Flow*, *31*(12), 1304–1328.

Lucas, D., Krepper, E., & Prasser, H. M. (2007). Use of models for lift, wall and turbulent dispersion forces acting on bubbles for poly-disperse flows. *Chem. Eng. Sci.*, *62*(15), 4146–4157.

Luewisutthichat, W., Tsutsumi, A., & Yoshida, K. (1997). Chaotic hydrodynamics of continuous single-bubble flow systems. *Chem. Eng. Sci.*, *52*(21–22), 3685–3691.

Luo, H. -P., & Al-Dahhan, M. H. (2008). Macro-mixing in a draft-tube airlift bioreactor. *Chem. Eng. Sci.*, *63*(6), 1572–1585.

Mao, Z. -S. (2008). Knowledge on particle swarm: The important basis for multi-scale numerical simulation of multiphase flows. *Chin. J. Process. Eng.*, *8*(4), 645–659 (in Chinese).

Mao, Z. S., & Yang, C. (2009). Challenges in study of single particles and particle swarms. *Chinese J. Chem. Eng.*, *17*(4), 535–545.

Mehta, R. V., & Tarbell, J. M. (1983). Four environment model of mixing and chemical reaction. *Part I: Model development. AIChE J.*, *29*(2), 320–329.

Merchuck, J. C., Ladwa, N., Cameron, A., Bulmer, M., & Pickett, A. (1994). Concentric-tube airlift reactors: effects of geometrical design on performance. *AIChE J.*, *40*(7), 1105–1117.

Merchuk, J. C., Ladwa, N., Cameron, A., Bulmer, M., Berzin, I., & Pickett, A. M. (1996). Liquid flow and mixing in concentric tube air-lift reactors. *J. Chem. Tech. Biotechnol.*, *66*(2), 174–182.

Merchuk, J. C., Contreras, A., García, F., & Molina, E. (1998). Studies of mixing in a concentric tube airlift bioreactor with different spargers. *Chem. Eng. Sci.*, *53*(4), 709–719.

Moraga, F. J., Bonetto, F. J., & Lahey, R. T. (1999). Lateral forces on spheres in turbulent uniform shear flow. *Int. J. Multiphase Flow*, *25*(6–7), 1321–1372.

Morsi, S. A., & Alexander, A. J. (1972). An investigation of particle trajectories in two-phase flow systems. *J. Fluid Mech.*, *55*(2), 193–208.

Morud, K. E., & Hjertager, B. H. (1996). LDA measurements and CFD modelling of gas–liquid flow in a stirred vessel. *Chem. Eng. Sci.*, *51*(2), 233–249.

Moukalled, F., & Darwish, M. (2004a). Pressure-based algorithms for multifluid flow at all speeds – Part II: Geometric conservation formulation. *Numer. Heat Tran.*, *45*(6), 523–540.

Moukalled, F., & Darwish, M. (2004b). Pressure-based algorithms for multifluid flow at all speeds – Part I: Mass conservation formulation. *Numer. Heat Tran.*, *45*(6), 495–522.

Mudde, R. F., & Van Den Akker, H. E. A. (2001). 2D and 3D simulations of an internal airlift loop reactor on the basis of a two-fluid model. *Chem. Eng. Sci.*, *56*(21–22), 6351–6358.

Muroyama, K., Shimomichi, T., Masuda, T., & Kato, T. (2007). Heat and mass transfer characteristics in a gas–slurry–solid fluidized bed. *Chem. Eng. Sci.*, *62*(24), 7406–7413.

Nouri, L. H., Legrand, J., Benmalek, N., Imerzoukene, F., Yeddou, A. -R., & Halet, F. (2008). Characterisation and comparison of the micromixing efficiency in torus and batch stirred reactors. *Chem. Eng. J.*, *142*(1), 78–86.

Ohnuki, A., & Akimoto, H. (2001). Model development for bubble turbulent diffusion and bubble diameter in large vertical pipes. *J. Nucl. Sci. Technol.*, *38*(12), 1074–1080.

Pan, Y., Dudukovic, M. P., & Chang, M. (1999). Dynamic simulation of bubbly flow in bubble columns. *Chem. Eng. Sci.*, *54*(13–14), 2481–2489.

Pan, Y., Dudukovic, M. P., & Chang, M. (2000). Numerical investigation of gas-driven flow in 2-D bubble columns. *AIChE J.*, *46*(3), 434–449.

Patankar, S. V. (1980). *Numerical Heat Transfer and Fluid Flow*. New York: McGraw-Hill.

Petersen, E. E., & Margaritis, A. (2001). Hydrodynamic and mass transfer characteristics of three-phase gaslift bioreactor systems. *Crit. Rev. Biotechnol.*, *21*(4), 233–294.

Petersen, K. O. E. (1992). Experimentale numerique des ecoulements disphasiques dans les reacteurs chimiques. *Ph. D. Thesis*. Lyon: L'Universite Claude Bernald.

Podila, K. (2005). CFD modelling of turbulent bubbly flows in pipes. *M. Sc. thesis*. Canada: Dalhousie University.

Pohorecki, R., & Baldyga, J. (1983). New model of micromixing in chemical reactors. 1. General development and application to a tubular reactor. *Ind. Eng. Chem. Fund.*, *22*(4), 392–397.

Pojman, J. A., Epstein, I. R., Karni, Y., & Bar-Ziv, E. (1991). Stochastic coalescence-redispersion model for molecular diffusion and chemical reactions. 2. Chemical waves. *J. Phys. Chem.*, *95*(8), 3017–3021.

Pollard, D. J., Ison, A. P., Shamlou, P. A., & Lilly, M. D. (1998). Reactor heterogeneity with *Saccharopolyspora erythraea* airlift fermentations. *Biotechnol. Bioeng.*, *58*(5), 453–463.

Rahimi, M., & Parvareh, A. (2005). Experimental and CFD investigation on mixing by a jet in a semi-industrial stirred tank. *Chem. Eng. J.*, *115*(1–2), 85–92.

Rajab, A. (2005). *Influence of viscosity on the micromixing of non-Newtonian fluid in a stirred tank. Master thesis*. Beijing, China: Beijing University of Chemical Technology.

Rampure, M. R., Kulkarni, A. A., & Ranade, V. V. (2007). Hydrodynamics of bubble column reactors at high gas velocity: Experiments and computational fluid dynamics (CFD) simulations. *Ind. Eng. Chem. Res.*, *46*(25), 8431–8447.

Ranade, V. V., & Van den Akker, H. E. A. (1994). A computational snapshot of gas–liquid flow in baffled stirred reactors. *Chem. Eng. Sci.*, *49*(24 Part2), 5175–5192.

Richardson, J. F., & Zaki, W. N. (1954). Sedimentation and fluidisation. Part I. *Trans. Inst. Chem. Eng.*, *32*, 35–53.

Roy, S., Dhotre, M. T., & Joshi, J. B. (2006). CFD simulation of flow and axial dispersion in external loop airlift reactor. *Chem. Eng. Res. Des.*, *84*(A8), 677–690.

Sánchez Mirón, A., Cerón García, M. C., García Camacho, F., Molina Grima, E., & Chisti, Y. (2004). Mixing in bubble column and airlift reactors. *Chem. Eng. Res. Des.*, *82*(10), 1367–1374.

Sato, Y., Sadatomi, M., & Sekoguchi, K. (1981). Momentum and heat transfer in two-phase bubble flow – I. Theory. *Int. J. Multiphase Flow*, *7*, 167–177.

Schlueter, M., & Raebiger, N. (1998). Bubble swarm velocity in two phase flows. *Proceedings of ASME heat transfer division*. New York: ASME.

Schumacher, J., Sreenivasan, K. R., & Yeung, P. K. (2005). Very fine structures in scalar mixing. *J. Fluid Mech.*, *531*, 113–122.

Schwarz, M. P., & Turner, W. J. (1988). Applicability of the standard k–ε turbulence model to gas-stirred baths. *Appl. Math. Modelling*, *12*(3), 273–279.

Serre, E., & Pulicani, J. P. (2001). A three-dimensional pseudospectral method for rotating flows in a cylinder. *Computers Fluids*, *30*(4), 491–519.

Shaikh, A., & Al-Dahhan, M. H. (2007). A review on flow regime transition in bubble columns. *Int. J. Chem. Reactor Eng.*, *5*(R1), 1–68.

Sokolichin, A., Eigenberger, G., & Lapin, A. (2004). Simulation of buoyancy driven bubbly flow: Established simplifications and open questions. *AIChE J.*, *50*(1), 24–45.

Talvy, S., Cockx, A., & Liné, A. (2007a). Modeling hydrodynamics of gas–liquid airlift reactor. *AIChE J.*, *53*(2), 335–353.

Talvy, S., Cockx, A., & Liné, A. (2007b). Modeling of oxygen mass transfer in a gas–liquid airlift reactor. *AIChE J.*, *53*(2), 316–326.

Tavare, N. S. (1995). Mixng, reaction, and precipitation: Interaction by exchange with mean micromixing models. *AIChE J.*, *41*(12), 2537–2548.

Thorat, B. N., & Joshi, J. B. (2004). Regime transition in bubble columns: Experimental and predictions. *Exp. Therm. Fluid Sci.*, *28*(5), 423–430.

Tobajas, M., García-Calvo, E., Siegel, M. H., & Apitz, S. E. (1999). Hydrodynamics and mass transfer prediction in a three-phase airlift reactor for marine sediment biotreatment. *Chem. Eng. Sci.*, *54*(21), 5347–5354.

Tomiyama, A. (1998a). Struggle with computational bubble dynamics. *Multiphase Sci. Technol.*, *10*(4), 369–405.

Tomiyama, A. (1998b). Struggle with computational bubble dynamics. *Third international conference on multiphase flow*, Lyon, France.

Tomiyama, A. (2004). Drag lift and virtual mass forces acting on a single bubble. *Third international seminar two-phase flow modeling & experiments*, Pisa, Italy.

Tomiyama, A., Sou, A., Zun, I., Kanami, N., & Sakaguchi, T. (1995). Effects of Eötvös number and dimensionless liquid volumetric flux on lateral motion of a bubble in a laminar duct flow. *Proceeding of 2nd international conference on multiphase flow*, Kyoto, Japan.

Tomiyama, A., Tamai, H., Zun, I., & Hosokawa, S. (2002). Transverse migration of single bubbles in simple shear flows. *Chem. Eng. Sci.*, *57*, 1849–1858.

Toor, H. L., & Marchello, J. M. (1958). Film-penetration model for mass and heat transfer. *AIChE J.*, *4*(1), 97–101.

Troshko, A., Ivanov, N., & Vasques, S. (2001). Implementation of a general lift coefficient in the CFD model of turbulent bubbly flows. *Proceedings of 9th conference of the CFD Society of Canada*, Waterloo, Ontario Canada.

Turton, R., & Levenspiel, O. (1986). A short note on drag correlation for spheres. *Powder Technol.*, *47*(1), 83–86.

van Baten, J. M., & Krishna, R. (2003). Comparison of hydrodynamics and mass transfer in airlift and bubble column reactors using CFD. *Chem. Eng. Technol.*, *26*(10), 1074–1079.

van Baten, J. M., Ellenberger, J., & Krishna, R. (2003). Hydrodynamics of internal air-lift reactors: Experiments versus CFD simulations. *Chem. Eng. Process.*, *42*(10), 733–742.

van Benthum, W. A. J., van der Lans, R. G. J. M., van Loosdrecht, M. C. M., & Heijnen, J. J. (1999). Bubble recirculation regimes in an internal-loop airlift reactor. *Chem. Eng. Sci.*, *54*(18), 3995–4006.

Vasconcelos, J. M. T., Rodrigues, J. M. L., Orvalho, S. C. P., Alves, S. S., Mendes, R. L., & Reis, A. (2003). Effect of contaminants on mass transfer coefficients in bubble column and airlift contactors. *Chem. Eng. Sci.*, *58*(8), 1431–1440.

Vial, C., Poncin, S., Wild, G., & Midoux, N. (2001). A simple method for regime identification and flow characterisation in bubble columns and airlift reactors. *Chem. Eng. Process.*, *40*(2), 135.

Vial, C., Poncin, S., Wild, G., & Midoux, N. (2002). Experimental and theoretical analysis of the hydrodynamics in the riser of an external loop airlift reactor. *Chem. Eng. Sci.*, *57*(22–23), 4745–4762.

Vial, C., Poncin, S., Wild, G., & Midoux, N. (2005). Experimental and theoretical analysis of axial dispersion in the liquid phase in external-loop airlift reactors. *Chem. Eng. Sci.*, *60*(22), 5945–5954.

Viamajala, S., McMillan, J. D., Schell, D. J., & Elander, R. T. (2009). Rheology of corn stover slurries at high solids concentrations – Effects of saccharification and particle size. *Bioresour. Technol.*, *100*(2), 925–934.

Villermaux, J., & David, R. (1983). Recent advances in the understanding of micromixing phenomena in stirred reactors. *Chem. Eng. Comm.*, *21*(1–3), 105–122.

Villermaux, J., & Fournier, M. C. (1994). Investigation of micromixing in stirred tank reactors using parallel reactions. *AIChE Sympousium Series*, *299*, 50–54.

Wang, T., Wang, J., & Jin, Y. (2005a). Population balance model for gas–liquid flows: Influence of bubble coalescence and breakup models. *Ind. Eng. Chem. Res.*, *44*(19), 7540–7549.

Wang, T., Wang, J., & Jin, Y. (2005b). Theoretical prediction of flow regime transition in bubble columns by the population balance model. *Chem. Eng. Sci.*, *60*(22), 6199–6209.

Wang, T., Wang, J., & Jin, Y. (2007). Slurry reactors for gas-to-liquid processes. *Ind. Eng. Chem. Res.*, *46*(18), 5824–5847.

Weiland, P. (1984). Influence of draft tube diameter on operation behavior of airlift loop reactors. *German Chem. Eng.*, *7*, 374–385.

Wen, J. P., Jia, X. Q., & Feng, W. (2005). Hydrodynamic and mass transfer of gas–liquid–solid three-phase internal loop airlift reactors with nanometer solid particles. *Chem. Eng. Technol.*, *28*(1), 53–60.

Wu, X., & Merchuk, J. C. (2004). Simulation of algae growth in a bench scale internal loop airlift reactor. *Chem. Eng. Sci.*, *59*(14), 2899–2912.

Xue, S. W., & Yin, X. (2006). Numerical simulation of flow behavior and mass transfer in internal airlift-loop reactor. *Chem. Eng. (China)*, *34*(5), 23–27.

Yakhot, V., & Orszag, S. A. (1986). Renormalization group analysis of turbulence. I. Basic theory. *J. Sci. Comput.*, *1*(1), 3–51.

Yakhot, V., Orszag, S. A., Thangam, S., Gatski, T. B., & Speziale, C. G. (1992). Development turbulence models for shear flow by a double expansion technique. *Phys. Fluids*, *4*(7), 1510–1520.

Yang, N., Wu, Z., Chen, J., Wang, Y., & Li, J. (2011). Multi-scale analysis of gas–liquid interaction and CFD simulation of gas–liquid flow in bubble columns. *Chem. Eng. Sci.*, *66*(14), 3212–3222.

Yao, B. P., Zheng, C., Gasche, H. E., & Hofmann, H. (1991). Bubble behaviour and flow structure of bubble columns. *Chem. Eng. Process.*, *29*(2), 65–75.

Yeoh, G. H., & Tu, J. Y. (2006). Numerical modelling of bubbly flows with and without heat and mass transfer. *Appl. Math. Modell.*, *30*(10), 1067–1095.

Zhang, D., Deen, N. G., & Kuipers, J. A. M. (2009). Euler–Euler modeling of flow, mass transfer, and chemical reaction in a bubble column. *Ind. & Eng. Chem. Res.*, *48*(1), 47–57.

Zhang, J. P., Grace, J. R., Epstein, N., & Lim, K. S. (1997). Flow regime identification in gas–liquid flow and three-phase fluidized beds. *Chem. Eng. Sci.*, *52*(21–22), 3979–3992.

Zhang, W. P., Huang, Q., Yang, C., & Mao, Z. -S. (2011). Hydrodynamics, mixing and mass/heat transfer in an airlift internal loop reactor. *Ind. Eng. Chem. Res.*, 128, Beijing,China.

Zhang, W. P., Yong, Y. M., Zhang, G. J., Yang, C., & Mao, Z. -S. (2012). Micro-mixing characteristics and bubble behavior in an airlift internal loop reactor with low height-to-diameter. *Proceedings of 14th European conference on mixing*, 529, Warsaw, Poland.

Zhang, Y., Yang, C., & Mao, Z. S. (2006). Large eddy simulation of liquid flow in a stirred tank with improved inner–outer iterative algorithm. *Chinese J. Chem. Eng.*, *14*(3), 321–329.

Zhao, D., Muller-Steinhagen, H., & Smith, J. M. (2002). Micromixing in boiling and hot sparged systems – Development of a new reaction pair. *Chem. Eng. Res. Des.*, *80*(8), 880–886.

Zuber, N. (1964). On the dispersed two-phase flow in the laminar flow regime. *Chem. Eng. Sci.*, *19*(11), 897–917.

Zuber, N., & Findlay, J. A. (1965). Average volumetric concentration in two-phase flow systems. *J. Heat Transfer*, *87*, 453–468.

Preliminary investigation of two-phase microreactors

5.1 INTRODUCTION

Miniaturized chemical reactors, typically on-chip microchannel reactors, have become important in analytical and environmental monitoring (Reyes et al., 2002), as measuring devices for on-line process optimization (Ugi et al., 1996), as catalyst screening tools (Zech and Honicke, 2000), for production of micro fuel cells (Ehrfeld, 2000), and especially for microorganic synthesis/production in the pharmaceutical industry, where the test-rig stage in the development of a drug does not require the production of large quantities of the chemical (Fletcher et al., 2002). The most striking advantages of such microsystems are the high portability, reduced reagent consumption, minimization of waste production, remote (on-site) application, and efficient heat dissipation and mass transfer owing to high surface-area-to-volume ratios (Cooper et al., 2001). Quite a number of reactions executed in such systems are involved with immiscible reagent phases in aqueous–organic liquid (Doku et al., 2001), gas–liquid (Hetsroni et al., 2003; Koyama et al., 2004) and gas–liquid–solid (Losey et al., 2002; Mouza et al., 2002) systems, in which it is difficult to force a reactant in one phase to mix and react with another one in other phases. Therefore, techniques promoting phase contact and mixing become critical. The reaction rate in a reactor not only depends on the intrinsic kinetics of a reacting system, but also on the mass transfer within different phases. The design and operation of immiscible liquid–liquid and gas–liquid microreactor systems depend mainly on the method of dispersion/phase contact. For better design and utilization of microreactors, more and more attention has been paid to better understanding of the mechanisms of multiphase flow and interphase mass and heat transfer in microreactors. Much research and development effort needs to be repeated on microchemical engineering systems to gain the quantitative understanding needed for practical applications of these microreactors. Much progress has been made to this end.

The hydrodynamics of gas–liquid or liquid–liquid mixtures near the junctions of microchannels is influenced by interfacial tension, viscosity, inertia and channel walls, and has been a significant subject of research (Nisisako et al., 2002; Wang et al., 2009); this is the most important issue in a microfluidic system. Among two-phase microfluidic systems, liquid–liquid and gas–liquid systems behave differently, and the results for gas–liquid flows cannot be generalized to liquid–liquid flows (Dessimoz et al., 2008). Compared with the vast amount of work on gas–liquid flows in microchannels (Triplett et al., 1999b; Xu et al., 1999; Yu et al., 2007; Donata et al., 2008; Takamasa et al., 2008), there are fewer experimental and numerical studies on liquid–liquid flows in microchannels (Pandey et al., 2006). Several studies have indicated that the flow patterns of liquid–liquid systems in microchannels are controlled by the channel geometry, the interfacial properties of channel surface and fluids, as well as the flow rates (Xu et al., 2006; Kashid and Agar, 2007a), which are simultaneously considered to understand multiphase hydrodynamics in microchannels.

5.2 MATHEMATICAL MODELS AND NUMERICAL METHODS

While most of the studies on flow patterns (Zhao et al., 2006a, b, 2007), heat/mass transfer (Gong et al., 2007; Tan et al., 2008) and reactions in microfluidic liquid–liquid two-phase systems are experimental in nature, recent developments in computational fluid dynamics allow us to simulate multiphase flows in microfluidic devices, which offers a more cost-effective approach to characterize and optimize the performance of complex microfluidic multiphase flow systems (Liow, 2004; Shui et al., 2007a, b).

Traditional computational fluid dynamics (CFD) can be applicable for liquid, liquid–liquid, and gas–liquid systems (Yang and Mao, 2005; Wang et al., 2008, 2011; Lu et al., 2010; Raimondi and Prat, 2011) but is not good at incorporating microscopic interactions, which may become crucial in certain circumstances, for example the dynamics of wetting and solid–fluid interfacial slip. On the other hand, microscopic simulation approaches, including molecular dynamics (MD; Rapaport, 1995) and Monte Carlo (MC; Zhang, 2005) methods, study the fluid behavior by the evolution of individual molecules interacting with one another through intermolecular potentials, and the microscopic interactions can be well represented. However, the huge computational demand limits their applications to very small space and time scales (Házi et al., 2002).

Of all the numerical methods for microfluidic multiphase flows, the lattice Boltzmann (LB) method has emerged as a particularly useful numerical scheme. One of the challenges in simulating immiscible multiphase flows is to determine the location of the interface. The interface changes position constantly, and when two interfaces meet, breakup and merging need to be modeled (Junseok, 2005).

The early development of immiscible multiphase LB models (color LB model) can be traced back to Gunstensen et al. (1991) and Gunstensen and Rothman (1991a). The most simple and widespread among these approaches is the chromodynamic or color gradient method of Gunstensen and Rothman (1991b). It is formulated as a lattice gas approach containing two sets of LBM (lattice Boltzmann method) populations, one for each phase, and models phase separation and interface tension using a recoloring step. In each node, the algorithm attempts to separate two phases as much as possible by redistributing the two sets of populations. Interfaces are implicitly defined by the fluid fraction iso-surface, where the contents of two fluids are equal.

Because it is derived from the Boolean lattice gas method, the original form of the Rothman–Keller model suffers from the deficiencies associated with lattice artifacts and noise. In general the method is applicable only for small density and viscosity differences and in particular the recoloring step causes grid-dependent artifacts at the interface (Kehrwald, 2002).

Another type of immiscible multiphase LB model started with Shan and Chen (1993, 1994), who proposed a multi-component LB model based on the microscopic interaction between fluid particles and introduced the concept of interaction potentials. In this model, an additional momentum forcing term with a potential function and the strength of interparticle interaction is explicitly added into the velocity field after each time step. The corrected velocity is employed in the equilibrium distribution function. By introducing an additional forcing term, the model effectively mimics the intermolecular interactions (complex fluid behavior). Although it is possible to show that the total momentum in the whole computational domain is conserved, the momentum is not conserved locally. As a result, a spurious velocity always exists in the regions adjacent to the interface. The model has been quite successful in simulations of several fundamental interfacial phenomena. However, there are a few limitations to the Shan–Chen model, which make this method inferior in comparison to other methods for multiphase flows. The first serious problem is that one cannot introduce the temperature. The next problem relates to the way that this model represents capillary effects, which can not be quantified by the coefficient of surface tension. Finally, it is impossible to represent different viscosities in different phases.

Swift et al. (1995) developed another class of immiscible multiphase LB model using the free-energy approach. The general idea is to incorporate the phenomenological approaches of interface dynamics, using the concept of a free-energy function, and to utilize the discrete kinetic approach as a vehicle for coupling with complex-fluid hydrodynamics. The pressure tensor is defined using Cahn–Hilliard's approach for non-equilibrium thermodynamics. Strictly speaking, this model is phenomenological, in which the thermodynamic effects are introduced through a phenomenological equation of state. The term "free-energy-based" is attributed to the model chosen for pressure tensor. Multi-component versions of the free-energy-based model were developed in Swift et al. (1995) and Lamura et al. (1999).

Similar to the Shan–Chen model, the "free-energy-based" models do not utilize the "particle" nature of the discrete kinetic approach. The major drawback of this approach is that the model suffers from nonphysical Galilean invariance effects, coming from the non-Navier–Stokes terms, which appear at the level of the Chapman–Enskog analysis of the discrete Boltzmann equation. Efforts are being made to reduce this nonphysical effect (Holdych et al., 1998). The main advantage of this model over the Shan–Chen LBM model is that it is formulated to account for the equilibrium thermodynamics of non-ideal and multi-component fluids at a fixed temperature, thus allowing the introduction of well-defined temperature and thermodynamics. The model is consistent with Maxwell's equal-area reconstruction procedure. Furthermore, since the model allows local momentum conservation, the interfacial spurious velocity is nearly eliminated (Nourgaliev et al., 2002).

The above LB models have great advantages over conventional methods because they do not track interfaces explicitly. Instead, the interfaces arise naturally due to phase separation between fluid components. The drawbacks of these LB models are that the interfaces are diffuse over several lattice units, and the models usually become unstable when density and viscosity ratios are too high. Some recent developments, however, have alleviated these constraints (Sankananarayanan et al., 2002, 2003).

The LB method used in this chapter was developed by Santos et al. (2002), who extended the field mediator concept (Santos et al., 2005) for the lattice-gas models of immiscible fluids to LB models. Santos et al. (2002) applied the LB model to investigate the liquid junction potential at the interface between two electrolyte layers, and the LB solutions were validated against the results of analytical and finite difference methods for the evolution of concentration, net charge density, and electrostatic potential. Considering mutual cross collisions between lattice particles, the collision operator contains three independent relaxation times related to the species diffusivity and to the viscosity of each fluid, so the LB model maintains numerical stability at very high viscosity ratios of fluids. The interference between mediators and particles is modeled by considering the deviations in particle velocities, which are proportional to the mediator distribution at the sites. Interfacial tension is attained by modifying the collision term between two fluids, introducing long-range forces in the transition layer through field mediators. The mediator's action is restricted in the transition layer and the ideal gas state equation is recovered for each fluid far from the interface.

For the LB model based on field mediators for immiscible fluids, Santos et al. (2002) considered probability distribution functions $f_i^r(\mathbf{x},t)$ and $f_i^b(\mathbf{x},t)$ as the particle distributions of red (R) and blue (B) fluid particles in site $\mathbf{x}$ at time t, and similarly $M_i(\mathbf{x},t)$ as the particle distribution of mediators that are created just before the propagation step. $M_i(\mathbf{x},t)$ models the long-range interaction:

$$M_i(\mathbf{x}+\mathbf{e}_i,t+1) = c_1 M_i(\mathbf{x},t) + c_2 \frac{\sum f_i^r(\mathbf{x},t)}{\sum f_i^r(\mathbf{x},t) + \sum f_i^b(\mathbf{x},t)} \qquad (5.1)$$

where c_1 and c_2 are the weights used for setting the interaction length, and $c_1 + c_2 = 1$. In Eq. (5.1), the first term on the right-hand side is a recurrence relation, because $M_i(\mathbf{x},t)$ depends on $M_i(\mathbf{x}-\mathbf{e}_i,t-1)$, $f_i^r(\mathbf{x}-\mathbf{e}_i,t-1)$, and $f_i^b(\mathbf{x}-\mathbf{e}_i,t-1)$ for all neighboring sites around site $\mathbf{x} - \mathbf{e}_i$. When $c_1 = 0$ (or $c_2 = 1$), mediators are created at site $(\mathbf{x} - \mathbf{e}_i, t + 1)$ with the sole information of the concentration of R particles on the next neighboring site. In this case, the interaction length corresponds to one lattice unit. By increasing c_1, the interaction length can be arbitrarily increased.

The LB evolution equations for R and B fluid particles are

$$f_i^r(\mathbf{x}+\mathbf{e}_i,t+1) - f_i^r(\mathbf{x},t) = \Omega_i^r(\mathbf{x},t) \qquad (5.2)$$

$$f_i^b(\mathbf{x}+\mathbf{e}_i,t+1) - f_i^b(\mathbf{x},t) = \Omega_i^b(\mathbf{x},t) \qquad (5.3)$$

The collision operators for R and B are Ω_i^r and Ω_i^b, which model the particles' interaction and are required to satisfy mass and momentum conservation. Splitting the BGK collision operators, as proposed by Santos et al. (2002), a three-parameter BGK collision term satisfies the above restrictions:

$$\Omega_i^r = m^r \frac{f_i^{r\,eq}(\rho^r,\mathbf{u}^r) - f_i^r(\mathbf{x},t)}{\tau^r} + m^b \frac{f_i^{r\,eq}(\rho^r,\theta^b) - f_i^r(\mathbf{x},t)}{\tau^m} \tag{5.4}$$

$$\Omega_i^b = m^b \frac{f_i^{b\,eq}(\rho^b,\mathbf{u}^b) - f_i^b(\mathbf{x},t)}{\tau^b} + m^r \frac{f_i^{b\,eq}(\rho^b,\theta^r) - f_i^b(\mathbf{x},t)}{\tau^m} \tag{5.5}$$

where

$$\rho^r = \sum_{i=0}^{18} f_i^r(\mathbf{x},t) \tag{5.6}$$

$$\rho^b = \sum_{i=0}^{18} f_i^b(\mathbf{x},t) \tag{5.7}$$

$$\mathbf{u}^r = \frac{1}{\rho^r} \sum_{i=0}^{18} f_i^r(\mathbf{x},t)\mathbf{e}_i \tag{5.8}$$

$$\mathbf{u}^b = \frac{1}{\rho^b} \sum_{i=0}^{18} f_i^b(\mathbf{x},t)\mathbf{e}_i \tag{5.9}$$

where ρ^r, ρ^b, $\mathbf{u}^r$ and $\mathbf{u}^b$ are respectively the macroscopic densities and velocities of fluids R and B. τ^r, τ^b and τ^m are respectively the relaxation factors of fluids R, B, and field mediators, when their density distributions reach an equilibrium state.

The first term on the right-hand side of Eq. (5.4) is related to the relaxation of R particle distribution to an equilibrium state given by the R component density and velocity (R–R collisions). The second term considers R–B collisions and is related to the relaxation of R particles to the equilibrium state given by the density ρ^r and the B velocity (θ^b) modified by the action of mediators at the same site.

$$\theta^b = \mathbf{u}^b - A\hat{\mathbf{u}}^m \tag{5.10}$$

$$\theta^r = \mathbf{u}^r + A\hat{\mathbf{u}}^m \tag{5.11}$$

Here θ^r and θ^b are the local velocities modified by the action of mediators and the constant A is related to interfacial tension. We can obtain the interfacial tension between two fluids by adjusting the value of A. For ideal miscible fluids $A = 0$ and this collision term describes the relaxation of R particle distribution to an equilibrium state given by ρ^r and u^b, as a consequence of R–B cross collisions. For immiscible fluids, Eq. (5.10) means that R particles will be separated from B particles by long-range attractive forces from the R phase, and Eq. (5.11) is similar to Eq. (5.10). The mediator velocity at site $\mathbf{x}$ in Eqs. (5.10) and (5.11) is given by

$$\hat{\mathbf{u}}^{\mathrm{m}} = \sum_{i=1}^{18} M_i(\mathbf{x},t)\mathbf{e}_{\mathrm{l}} \bigg/ \bigg| \sum_{i=1}^{18} M_i(\mathbf{x},t)\mathbf{e}_{\mathrm{l}} \bigg| \qquad (5.12)$$

The boundary conditions for the mediator distribution are used for modulating the interaction process between the solid and the fluid. The wettability, or static contact angle, can be obtained by imposing, at each fluid site adjacent to a solid site, a constant value M_i^{solid} for the mediator distribution along direction i leading to the fluid phase in the propagation step. This quantity determines the static contact angle, since it determines the interaction between the solid and the fluid. As the mediator distribution carries the values of mass fraction, $M_i^{\mathrm{solid}} \in [-1, 1]$. When $M_i^{\mathrm{solid}} = 0$, the static contact angle will be $\alpha = 90°$. When $M_i^{\mathrm{solid}} > 0$, fluid R will behave as the wetting fluid, and when $M_i^{\mathrm{solid}} < 0$, fluid B will behave as the wetting fluid. Nevertheless, the precise value of the dynamic contact angle also depends on the relaxation parameters, τ^{r}, τ^{b} and τ^{m}.

5.3 SIMULATION USING LATTICE BOLTZMANN METHOD

5.3.1 Numerical simulation of two-phase flow in microchannels

The LBM is especially suitable for direct numerical simulation of multi-component and immiscible multiphase fluid flows. Since LB models belong to the class of diffuse-interface methods, they do not track interfaces and sharp interfaces are maintained via different mechanisms.

He et al. (1999) proposed a lattice Boltzmann scheme for simulation of multiphase flow in the nearly incompressible limit in which interfacial dynamics was modeled by incorporating molecular interactions and the interface between different phases was tracked using an index function. Two sets of distribution functions were used: one for tracking the pressure and velocity fields, and the other for the variable called the index function, which was used to obtain density and viscosity. Two-dimensional Rayleigh–Taylor instabilities were simulated and the results for initial linear growth rate and terminal bubble velocity were verified by comparison against previous theoretical and numerical results for single-mode instability. Figure 5.1 shows the evolution of the fluid interface from a 10% initial perturbation. During the early stages the heavy fluid falls as a spike and the light fluid rises to form bubbles. Starting from $t = 2.0$, the heavy fluid begins to form into two counter-rotating vortices. At a later tune ($t = 3.0$), these two vortices become unstable and a pair of secondary vortices appear at the tails of the roll-ups. With an increase in time ($t = 5.0$), the heavy fluid falling down gradually forms one central spike and two side spikes.

The process of slug formation in channels has been reported to be a strong function of the type of mixing zones and studied for different mixing zone shapes. The parameters of interest that have been studied by researchers are diameter of the

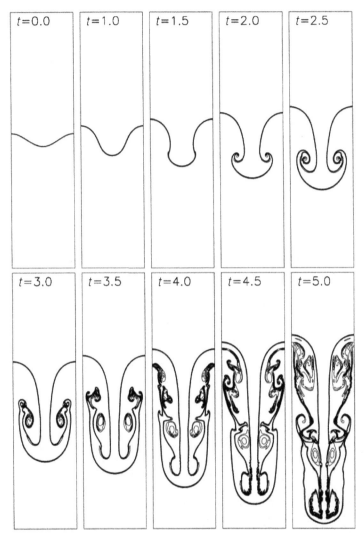

FIGURE 5.1 Evolution of fluid interface from a single-mode perturbation (He et al., 1999).

The Atwood number is 0.5; the Atwood number is equal to the ratio of the difference of the densities of heavy and light fluids to the sum of the densities of heavy and light fluids. The Reynolds number is 2048. A total of 19 density contours are plotted.

mixing zone, channel curvature (for meandered channels), superficial velocity, surface tension, viscous friction, and inertial and gravitational forces of the two phases. Yong et al. (2011) demonstrated that the developed LBM based on the field mediators simulator is a viable tool to study immiscible two-phase flows in a T-shaped microchannel. Figure 5.2 shows the process of formation of monodispersed droplets in a T-shaped microchannel, and the numerical results agree well qualitatively with

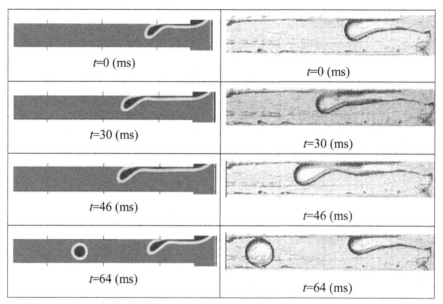

FIGURE 5.2 Comparison of predicted (left) and experimental (right) monodispersed kerosene droplet formation in a 300 μm × 600 μm microchannel ($q = 25$ where q denotes the volumetric flux ratio of water to kerosene, $We_{ks} = 1.483 \times 10^{-4}$, $We_{ws} = 0.12$) (Yong et al., 2011).

experimental data, showing the effectiveness of the numerical method. In this flow situation, the low value of the Weber number for kerosene, We_{ks}, assures that the interfacial tension overwhelms the inertial force for the kerosene phase. The value of the Weber number for water, We_{ws}, however, is 0.12, suggesting that the water phase inertial effects begin to play a role in kerosene droplet formation.

Two-phase flows present important challenges due to the nonlinear effect at the moving interface. The transport of liquid plugs in microchannels is very important and necessary for application in medical treatment of drug delivery in pulmonary airway trees (Jensen et al., 1994) and in extraction of oil from porous rocks (Lenormand et al., 1983). Many analytical, experimental, and computational studies considered the effects of mass transfer (Burns and Ramshaw, 2001), surfactant (Fujioka and Grotberg, 2005), inertia (Fujioka and Grotberg, 2004), gravity, plug volume (Kashid and Agar, 2007), and the stability of steady propagation (Hamida and Babadagli, 2008). Yong et al. (2013) studied a liquid plug flowing through a T-shaped microchannel for different driving pressures, contact angles, initial plug lengths, interfacial tensions, and viscosity ratios. The effect of gravity on the results is neglected because the T-shaped microchannel is placed horizontally. From Figure 5.3, it is clear that the contact angle also decides the flow patterns. When the blue fluid is nonwetting (Figure 5.3c), the plug remains at the junction of the T-shaped microchannel, which is to be avoided in applications of drug delivery and oil extraction. When the contact angle is 90°, the

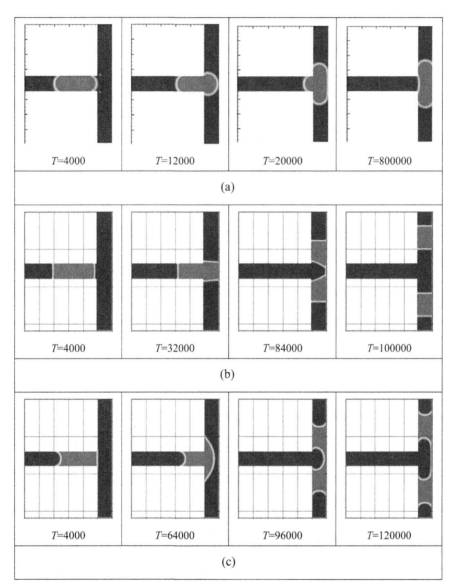

FIGURE 5.3 Transport flow of liquid plugs with different contact angles (T^r = 1.9, T^b = 1.0; ρ^r = 1.0, ρ^b = 1.0; $L_0 = 3l$, A = 0.7, T is nondimensional time) (Yong et al., 2013).

(a) M_i^{solid} = 1 (red fluid is nonwetting). (b) M_i^{solid} = 0.0. (c) M_i^{solid} = −1.0 (red fluid is wetting).

blocked plug is split into two daughter ones. When the blue fluid becomes wetting, the flow pattern is also split flow.

5.3.2 Numerical study of heat transfer in microchannels

Recognizing the significant potential of microchannel heat sinks in heat extraction, there has been a great deal of interest in the study of heat transfer in microchannels. The pioneering numerical work can be found in Tuckerman and Pease (1981) and many research works followed (Fedorov and Viskanta, 2004; Lee et al., 2005), as reviewed by Sobhan and Garimella (2001). Talimi et al. (2012) surveyed the numerical works on different areas of two-phase heat transfer. As can be seen in Table 5.1, significant research gaps exist in areas other than parallel plates and circulating flows. Nonsymmetrical thermal boundary conditions have not been considered. The prominent parameters that affect heat transfer in noncircular channels have been reported as *Pe*, slug length, channel geometry, interface shape (contact angle), and flow pattern (internal circulations).

The microstructures of microchannels can have a great effect on flow characteristics, and can significantly change heat exchange performance. Thus, to make an intensive study regarding the effect of microchannel structure on heat transfer performance is indeed very important. In Liu et al.'s (2011) paper, CFD and LBM were employed to simulate the liquid flow and heat transfer processes in microchannels, and both CFD and LBM can produce reasonably accurate predictions compared to experimental data. The heat exchange is very intensive at the inlet for each microchannel due to the inlet effect. It is the synergy level between velocity and temperature gradients that decides heat exchange performance. Various surface microstructures can all be treated, as the aim of these methods is to increase the synergy level. Liu et al. (2011) compared the predicted Nusselt numbers in the thermal developed

Table 5.1 Numerical Studies on Heat Transfer in Gas–Liquid Two-Phase Flow (Talimi et al., 2012)

Type	Constant wall temperature	Constant wall heat flux	Nonsymmetrical boundary conditions
Circulating flows	Lakehal et al. (2008), Narayanan and Lakehal (2008), Gupta et al. (2010)	Ua-arayaporn et al. (2005), He et al. (2007), Gupta et al. (2010), Mehdizadeh et al. (2011)	No study found
Parallel plates	Young and Mohseni (2008), Oprins et al. (2011), Talimi et al. (2011)	Young and Mohseni (2008)	No study found
Square, rectangle, triangle, ellipse	No study found	No study found	No study found

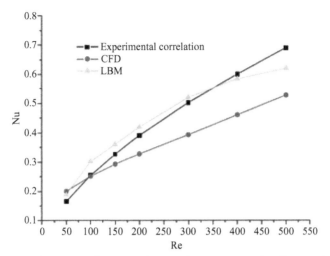

FIGURE 5.4 Comparison of predicted _Nu_ with experimental correlation (Peng and Peterson, 1996) for microstructure D (Liu et al., 2011).

region at different Reynolds numbers with a correlation to validate the numerical models, as shown in Figure 5.4.

Past numerical computations were limited to flow boiling in a simple geometry of microchannels. Lee et al. (2012) performed direct numerical simulation of boiling flow in a microchannel with transverse fins by employing a sharp-interface level set method to find the best conditions for heat transfer enhancement and investigate the effects of fin geometry on the bubble growth and heat transfer. The computational results show that the flow boiling in a microchannel was significantly enhanced when the liquid–vapor–solid interface contact region was increased with the addition of fins. The boiling heat transfer was pronounced when the fin height was 0.03 mm and the fin length was decreased. The fin segmentation near the sidewalls was effective in boiling heat transfer enhancement by enlarging the liquid layers between the bubble and the channel corners. The boiling heat transfer in the finned microchannel was enhanced up to 77% based on the bottom surface area and 33% based on the total fin surface area (Figures 5.5 and 5.6).

5.3.3 Numerical simulation of mass transfer in microchannels

When microchannel reactors are used for chemical kinetics experiments, it is imperative to ensure the absence of heat transfer and mass transfer influences to obtain the intrinsic rates. In microchannel reactors used for heterogeneous catalytic reactions, catalysts are generally deposited on to the walls of microchannels. Therefore, the reactants have to be transported, in the worst case, from the channel axis to the catalyst-coated walls. If mass transfer is slower than the reactions, the reactant concentrations at the catalyst surface will be lower than those at the axis, and the reaction processes become mass transfer controlled (Walter et al., 2005).

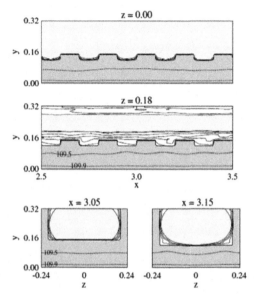

FIGURE 5.5 Effect of fin length on bubble shape and temperature field at $t = 3$ for $H_{fin} = 0.03$, $S_{fin} = 0.1$, and $L_{fin} = 0.1$ (Lee et al., 2012).

Yang et al. (2011) simulated a substrate solution with a specific inlet concentration flowing past a circular cylinder with biochemical reaction in an attached thin photosynthetic bacteria bioreactor for hydrogen production by applying the LB method. A non-equilibrium extrapolation method was employed to handle the velocity and concentration curved boundary and the model was validated by available theoretical and numerical results in terms of the drag and lift coefficients and concentration profiles. The velocity profile and concentration distributions of the substrate and hydrogen were determined, and the effect of Reynolds number on mass transfer characteristics was also discussed by introducing Sherwood number (in Figure 5.7). It can be seen from Figure 5.7a that the consumption efficiency η decreases with increasing Reynolds number. This can be understood since high Re results in high substrate loads in the channel and short HRT (hydraulic retention time) of the fluid, while the degradation capability of the biofilm with specific biomass cannot keep pace with the increasing load. It is also noted that the hydrogen yield γ is almost invariable with the increase of Re in Figure 5.7b. Since the biofilm simulated was considered to be in a steady and optimum state, and the inhibition effects from the high concentration of substrate and other subproducts were ignored, the hydrogen yield was maintained at a steady level. The substrate consumption efficiency is defined as

$$\eta = \left(\frac{C_{in} U_{in} - C_{out} U_{out}}{C_{in} U_{in}} \right)_{substrate} \times 100\% \tag{5.13}$$

and the hydrogen yield is given by

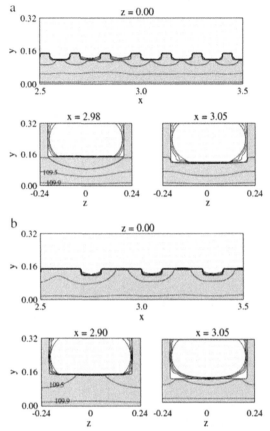

FIGURE 5.6 Effect of fin length on bubble shape and temperature field at $t = 3$ for $H_{fin} = 0.03$ and $S_{fin} = 0.1$ (Lee et al., 2012).

(a) $L_{fin} = 0.05$. (b) $L_{fin} = 0.2$.

$$\gamma = \frac{(C_{out}U_{out} - C_{in}U_{in})_{hydrogen}}{(C_{in}U_{in} - C_{out}U_{out})_{substrate}} \tag{5.14}$$

The impact of laminar flow is negative and its effects are still to be evaluated. Raimondi et al. (2008) investigated numerically liquid–liquid slug flow systems in a square channel of 50–960 μm depth and compared with three published reports, as shown in Figure 5.8. An interface-capturing technique without any interface reconstruction was used and the two-phase flow was described using the one-fluid approach, where a continuous function (the so-called volume fraction) allowed phase tracking because the volume fraction was zero in the continuous phase and one in the dispersed phase. Raimondi et al. (2008) compared the simulation results with the coefficients obtained from the models and correlations described in the literature.

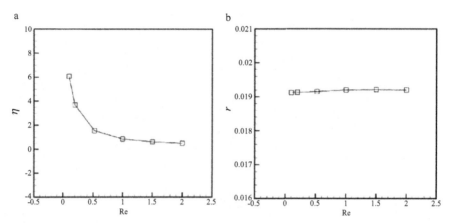

FIGURE 5.7 Variation of substrate consumption efficiency (a) and hydrogen yield (b) with *Re* (Yang et al., 2011).

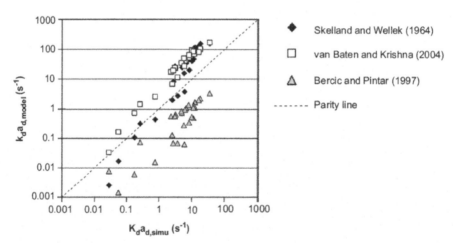

FIGURE 5.8 Comparison of mass transfer coefficients obtained by simulation and literature models (Skelland and Wellek, 1964; Berčič and Pintar, 1997; van Baten and Krishna, 2004)

The microreactors for development of liquid–liquid processes are promising since they are supposed to offer an enhancement of mass transfer compared to conventional devices due to the increase in surface/volume ratios. Traditional 2D models have been used to investigate the flow and mass transfer in a flat-plate microchannel bioreactor since the channel width is much larger than the channel height and the sidewall effect is negligible. However, the 3D configuration is more realistic in practical applications and a 2D model for mass transport simulation may have a limitation due to the difference of the velocity profiles through 2D and 3D channels. Therefore, to study a true 3D configuration makes the results more useful for practical applications. Zeng et al. (2007) used the commercial software FLUENT to solve the conservation equations for continuity, momentum, and species transport. The numerical result was verified by comparing against an analytical solution of oxygen concentration on the bottom in a 2D flat-bed microchannel bioreactor with a constant flow velocity in the flow direction and a constant reaction rate at the base. The effects of various mass transfer parameters on the species transport at the base were correlated using two dimensionless parameters; this was the first theoretical study on the dimensionless mass transfer coefficient in a 3D flat-plate rectangular microchannel bioreactor with Michaelis–Menten reaction kinetics at the base. Figure 5.9 shows the effect of dimensionless parameters on concentration distributions. In Figure 5.9a, at constant Damkohler number (Da, representing the ratio of

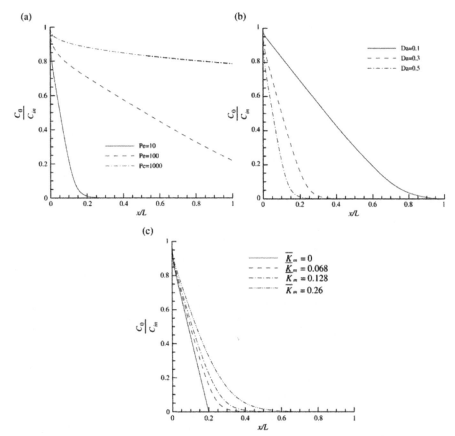

FIGURE 5.9 Effect of different mass transfer parameters on species concentration in the axial direction (Zeng et al., 2007) along the center line of the base plane ($y = 0$).

(a) Pe effect at $Da = 0.5$ and $\bar{K}_m = 0.068$. (b) Da effect at $Pe = 10$ and $\bar{K}_m = 0.068$.
(c) $\bar{K}_m$ effect at $Pe = 10$ and $Da = 0.3$.

reaction at the base to diffusion of species from the culture medium) and the dimensionless Michaelis–Menten constant, $\bar{K}_m$ (the ratio of Michaelis–Menten constant to the inlet concentration), the species concentration will be higher at higher Pe because of the increased convection of species by the flow. In Figure 5.9b, for constant Pe and $\bar{K}_m$, higher Da results in a lower concentration because of the higher reaction rate. In Figure 5.9c, compared with $\bar{K}_m = 0$, which means the maximum reaction rate at base, for constant Pe and Da, larger $\bar{K}_m$ results in a smaller absorption rate and thus a higher concentration.

5.4 EXPERIMENTAL

5.4.1 Flow pattern

Knowledge of flow patterns is fundamental for better prediction of complex two-phase flows as well as accompanying heat and mass transfer. Gas–liquid two-phase flows are widely encountered in the equipment of processing industries, such as stirred tanks, boilers, condensers, and distillers. However, there are significant differences between the gas–liquid flow characteristics in macro-scale tubes and micro-scale channels. Suo and Griffith (1964) proposed the transition boundaries among the elongated bubble, annular, and bubbly flows based on the velocity of the bubble and the volumetric flows of gas and liquid phases. Barnea et al. (1983) found that in small horizontal tubes with diameters ranging from 4 to 12 mm, the flow pattern transition among the dispersed flow, annular flow, and intermittent flow can be predicted by the Taitel–Dukler model, which has been widely used in large tubes. Triplett et al. (1999b) correlated the transition boundaries between the bubbly, churn, slug, slug–annular, and annular flows in small circular channels with diameters of 1.1 and 1.45 mm, and in semi-triangular channels with diameters of 1.09 and 1.49 mm, based on the superficial velocities of gas and liquid phases.

Over the past decades, many observations of flow patterns in microchannels were conducted (Zhao and Bi, 2001; Liu and Wang, 2008; Yue et al., 2008; Niu et al., 2009; Fu et al., 2010). Akbar et al. (2003) divided the flow map into surface tension-dominated, inertia-dominated and transition regions, and developed two-phase flow transition models based on Weber numbers. Waelchli and von Rohr (2006) incorporated the effect of fluid properties and channel geometry in a universal correlation for flow pattern transition. Choi et al. (2011) proposed a correlation based on gas and liquid superficial velocities to predict the flow pattern transition in microchannels with different aspect ratios. In addition, there have also been a few theoretical works to predict gas–liquid two-phase flow regimes (Taitel and Dukler, 1976; Dukler and Taitel, 1977a, b; Taitel et al., 1980; Mishima and Ishii, 1984; Brauner and Maron, 1992). These theoretical models were based on physical mechanisms and presented analytical predictions of flow pattern transitions for steady gas–liquid two-phase flows. These conventional flow regime maps were based primarily on gas and liquid superficial velocities or total mass flux and quality.

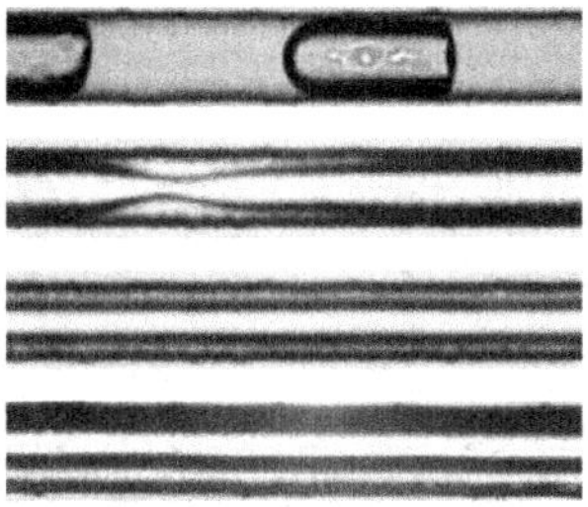

(a) slug flow (Q_L=3.6 mL/h, Q_G=10.8 mL/h)

(b) slug-annular flow (Q_L=14.4 mL/h, Q_G=108 mL/h)

(c) annular flow (Q_L=19.6 mL/h, Q_G =180 mL/h)

(d) parallel stratified flow (Q_L=28 mL/h, Q_G=360 mL/h)

FIGURE 5.10 Distinctive images of flow patterns by Wang et al. (2012).

In microfluidic systems, the Reynolds number is usually small, the flow is laminar, and the domination of surface force over volume forces prevails. In addition, the effects of geometry and the wetting property of the channel wall become important (Dongari et al., 2010; Jovanović et al., 2011). Wang et al. (2012) conducted experimental investigations of gas–liquid two-phase flows in microchannels with different surface wettabilities. Slug flow, annular flow, slug–annular flow, and stratified flow were observed, as shown in Figure 5.10. As the hydrophilicity of the channel surface is weakened, the transition boundaries for the slug flow to the slug–annular flow and the slug–annular flow to the annular flow are found to shift to lower Re_G/Re_L. The effect of wettability on microfluidics in the surface tension-dominated zone is more apparent than that in the inertia-dominated zone. The influence of the fluid's physical properties on the gas–liquid flow is evident. Both surface tension and liquid viscosity are factors in flow pattern transitions.

The heuristic universal flow regime transition models that involve fluid properties and the wetting condition of the channel wall are proposed as the following correlation of Ca $(= u_L \mu/\sigma)$ versus Reynolds number ratio (Wang et al., 2012).

The transition boundary between slug flow and slug–annular flow is given by

$$Ca = 4.0 \times 10^{-4} \left(\frac{\theta}{180°} \pi \right) \left(\frac{Re_G}{Re_L} \right)^{-0.76} \tag{5.15}$$

The transition boundary between slug–annular flow and annular flow is

$$Ca = 8.6 \times 10^{-4} \left(\frac{\theta}{180°} \pi \right) \left(\frac{Re_G}{Re_L} \right)^{-1.21} \tag{5.16}$$

The transition boundary between annular slow and parallel stratified flow is

$$Ca = 2.3 \times 10^{-2} \left(\frac{\theta}{180°} \pi \right) \left(\frac{Re_G}{Re_L} \right)^{-1.32} \tag{5.17}$$

A comparison of the presented models and experimental results is shown in Figure 5.11 (Wang et al., 2012).

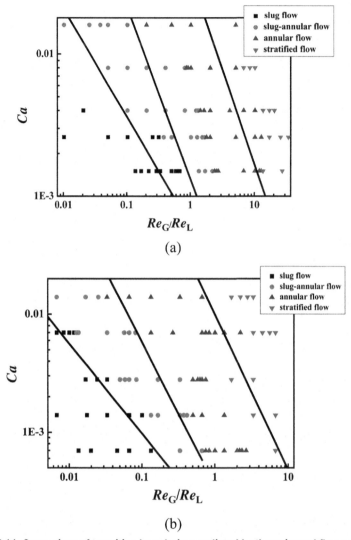

FIGURE 5.11 Comparison of transition boundaries predicted by the universal flow regime transition models (Wang et al., 2012) with experimental data in Choi et al. (2011).

(a) Nitrogen–kerosene in a glass channel. (b) Nitrogen–deionized water in a PDMS channel.

However, few models are known regarding the simultaneous flow of two immiscible liquids. There is no guarantee that the information available for gas–liquid cases can be extrapolated to liquid–liquid flows. There have been several studies on the generation of droplets in microfluidic devices and studies are focused on a specific type of flow pattern. Studies covering a wide range of flow rates, which might lead to other new flow patterns, are relatively rare.

5.4.2 **Pressure drop**

Pressure drop is an essential design parameter for micro-devices as it is closely related to energy consumption and process economy, and also a very crucial factor in the tradeoff between the enhancement of heat exchange and mass transfer at the expense of increased flow resistance (Pehlivan et al., 2006). A number of researchers have proposed different experiment-based models to reveal the characteristics of the two-phase pressure drop in microchannels. In general, these models can be classified into two groups: the flow pattern-independent models, which involve little information of the flow regime; and the flow pattern-dependent models, which are developed for certain flow regimes and involve flow pattern-specific parameters.

Up to now, the first group of pressure drop models has been widely used to predict two-phase pressure drops in microchannels, because they are simple and demand less detail of the two-phase flow. In this group, the homogeneous flow model (Ungar and Cornwell, 1992; Triplett et al., 1999a; Chen et al., 2002; Kawahara et al., 2002) and the Lockhart and Martinelli model (L-M model) (Lockhart and Martinelli, 1949) have been used most frequently. The homogeneous flow model treats the two-phase mixture as a pseudo-single-phase fluid with homogeneous fluid properties. The key parameter of the pseudo-phase is the viscosity of the mixture, and a few viscosity models have been proposed in previous work (Owen, 1961; Dukler et al., 1964; Beattie and Whalley, 1982; Lin et al., 1991). The L-M model suggests the friction pressure gradient of two-phase flow can be estimated by the friction pressure gradient of the liquid phase and a two-phase multiplier Φ_L^2, which is calculated from the Martinelli parameter X:

$$\left(-\frac{dP}{dL}\right)_S = \Phi_L^2 \left(-\frac{dP}{dL}\right)_L \tag{5.18}$$

$$\Phi_L^2 = 1 + \frac{C}{X} + \frac{1}{X^2} \tag{5.19}$$

$$\left(-\frac{dP}{dL}\right)_S = \left(-\frac{dP}{dL}\right)_G + \left(-\frac{dP}{dL}\right)_L + C\sqrt{\left(-\frac{dP}{dL}\right)_G \left(-\frac{dP}{dL}\right)_L} \tag{5.20}$$

Chisholm (1967) proposed a Chisholm factor C associated with the flow conditions of gas and liquid phases to relate Φ_L^2 and X of the L-M model. Over the past decades, modified L-M models with more parameters introduced or novel correlations of the Chisholm factor C have been proposed (Mishima and Hibiki, 1996; Yue et al., 2004, 2008; Lee and Lee, 2008; Niu et al., 2009; Ma et al., 2010; Su et al., 2010). However, the accuracy of both homogeneous and L-M models is limited due to neglect of specific hydrodynamic information.

Flow pattern-dependent models have been developed based on the characteristics of specific flow patterns. Bretherton (1961) proposed a correlation for the pressure drop over a Taylor bubble moving in circular tubes based on the study on liquid film thickness surrounding the bubble. Kreutzer et al. (2005) and Liu and Wang (2008) proposed semi-empirical pressure drop models for Taylor flow based on the Hagen–Poiseuille equation. Warnier et al. (2010) expanded the work of Kreutzer et al. (2005) and modified the pressure drop model by incorporating a correction in the liquid slug length. Although the characteristics of Taylor flow have attracted more attention, the investigation on the overall pressure drop for gas-phase continuous flow in microchannels is still needed.

5.4.3 **Mass transfer performance**

Microreactors are considered as efficient tools for intensification of mass transfer by taking advantage of the large surface-to-volume ratio and short transport path (Kashid et al., 2011; Roudet et al., 2011). As shown in Table 5.2, the mass transfer coefficient in a microreactor is much higher than that in a conventional reactor. Yue et al. (2007) investigated the absorption process of CO_2 into water and a buffer solution in a 667 μm Y-shaped microchannel, and Yue et al. (2010) extended their research on flow distribution and mass transfer performance to a parallel microchannel contactor integrated with constructal distributors. They found that the liquid side volumetric mass transfer coefficients (up to 4 s^{-1}) in a microchannel were at least one or two orders of magnitude higher than those in conventional contactors, and this conclusion was confirmed by Liu et al. (2011). The characterization of mass transfer in Taylor flow in microchannels and capillaries has attracted a great deal of attention (Pohorecki, 2007; Yue et al., 2009; Tan et al., 2012). Although previous work proposed correlations of overall mass transfer coefficient as functions of Reynolds number, Schmidt number, superficial velocity and even pressure drop, other hydrodynamic parameters such as surface effects have not been considered.

Table 5.2   Comparison of Mass Transfer Parameters in Different Gas–Liquid Contactors (Su et al., 2010)

Type	$k_L \times 10^5$ (m/s)	a (m^2/m^3)
Bubble columns	10–40	50–600
Couette–Taylor flow reactor	9–20	200–1200
Impinging jet absorbers	29–66	90–2050
Packed columns, co-current	4–60	10–1700
Packed columns, counter-current	4–20	10–350
Spray column	12–19	75–170
Static mixers	100–450	100–1000
Stirred tank	0.3–80	100–2000
Tube reactors, horizontal and coiled	10–100	50–700
Tube reactors, vertical	20–50	100–2000
Gas–liquid microchannel contactor	40–160	3400–9000

5.4.4 **Micromixing**

Micromixers can be broadly classified as active and passive mixers. For an active micromixer, external energy suppliers such as ultrasound, time-dependent electric and magnetic fields, etc. are used. In a passive micromixer, there is no external energy input apart from the pressure drop to drive the flow. Mixing occurs as a result of diffusion or chaotic advection (Tofteberg et al., 2010). Liu et al. (2012) studied the micromixing performance in micromixers with various configurations using both experimental methods and CFD simulation. They concluded that the segregation index of the Villermaux–Dushman method, which indicates the poorness in micromixing performance, had a nearly linear relationship with Re in the range from 2000 to 10,000, while it was independent of the width of the mixing channel. Under typical operating conditions, the flow in passive microstructured devices is laminar and molecular diffusion is the major mechanism. To make mixing more efficient, some measures have been taken to enhance the micromixing in passive micromixers. Chung and Shih (2008) presented a planar micromixer with rhombic microchannels and a converging–diverging element. Nimafar et al. (2012) compared the mixing process in T-shaped, O-shaped, and H-shaped microchannels, and found that the H-shaped micromixer had the best mixing efficiency. Ahn et al. (2008) quantified the secondary flow and pattern of two-liquid mixing inside a meandering square microchannel. The complex nature of the chip required a novel probe acting like a scanning electron microscope in the semiconductor industry and assured quality control by diagnosing structure and flow non-invasively in real time. They found that spectral-domain Doppler optical coherence tomography was a strong candidate for the novel probe apart from improvement of the configurations of micromixers, and the operating conditions were another important factor in micromixing. Su et al. (2012) reported that the micromixing performance of miscible liquid–liquid two phases can be intensified in Taylor flow regime by introducing gas phase. They also found that the micromixing efficiency in packed microchannels was better than that in empty ones. The micromixing performance was associated with the packing length and the appropriate packing position (Su et al., 2011).

5.5 **SUMMARY AND PERSPECTIVE**

Microreactor studies have made striking progress in the key technologies of design, manufacture, integration, and scale-up of microchannels. In this chapter, we sum up the recent progress in the mathematical modeling, numerical simulation, and experimental studies on microreactors:

1. The multiphase lattice Boltzmann models have considerable advantages over conventional methods because they do not track the interfaces explicitly, so that the lattice Boltzmann method has emerged as a particularly useful numerical scheme, which is suitable for multi-component and immiscible multiphase flow, mass and heat transfer processes in microdevices. A true 3D simulation makes the results more useful for practical applications.

2. The effects of flow patterns, surface and inertia forces, geometry, wetting property of channel walls, and pressure drop on mass transfer and mixing processes have been introduced for experimental work and numerical simulations.
3. There are other issues such as circulation patterns within the slug, pressure drop, type of mixing element to generate the slug flow, and stability of the flow, all of which should be considered while designing such microreactors.
4. The prominent parameters affecting heat transfer in noncircular channels have been reported as Pe, slug length, channel geometry, interface shape (contact angle), and flow pattern.
5. The mass transfer is related to the flow, reaction process and even heat transfer processes, and Re, Da and Sh numbers decide which process will control the mass transfer process.

The laminar flow under low Reynolds numbers in microfluidic devices implies that turbulence is not available to help enhance heat/mass transport and interphase chemical reactions. For intensification of passive mixing and transport rates, the geometry of a micromixer can be improved to generate secondary flow and chaotic advection. Developments in the near future should be devoted to establishing microstructured mixers as ordinary pilot-scale and production apparatus. It is very important to resolve reliable numbering up by maintaining the benefits of successful design of a single microreactor. The capability to integrate multiple functions such as heating and sensing in a single device will become an essential ingredient for commercial success, along with gradual progress via fundamental research.

Recently, microreactor studies have made significant progress in the key technologies of design, manufacture, integration, and scale-up of microchannels, especially those of micropumps, micromixers, microreactors, microheaters, microseparators, and so on. However, a series of problems need to be solved if the microreactors are really to replace traditional reactors, such as easy blockage of microchannels, design of catalyst, integration of sensors and controllers, and scale-up of microreactors.

The following popular issues of microreactor application are pending: (1) Research on the principle of microreactors to build up systematically more sophisticated models and simulation methods for design, scale-up, and optimization of microreactors. (2) Exploring new ways to make the microchemical production more environmentally friendly, economic, and energy saving. (3) Strengthening the basic study of microsystems to finally form a complete framework theory of microsystems.

As a new numerical method to simulate fluid flow, the lattice Boltzmann method is already used to probe many flow phenomena. At present, the popular issues of the LBM are to resolve the forces in the lattice Boltzmann equation, to use non-uniform meshes for domains with large gradients, to enforce second-order boundary conditions near the curved walls, to analyze the LB models with single and multi-relaxation time, and so on. Nevertheless, conventional CFD methods are still powerful tools in simulating low Reynolds number flows, in the range where the continuum assumption remains valid. It has been noticed that in submillimeter channels the intermolecular mechanisms play more prominent roles than in centimeter-scale devices. Therefore, conventional CFD models should be developed into multi-scale ones with more mechanisms on meso- and micro-scales incorporated.

The LB method is a mesoscopic and dynamic description of the physics of fluids, so it can model problems wherein both macroscopic hydrodynamics and microscopic statistics are important, which as yet have not been easily described using macroscopic equations. The LB method can describe the interface of two immiscible phases and is a useful numerical method to successfully predict the flow process of immiscible liquid–liquid two-phase microreactors. The LB model is still under development. The development of a reliable LB method for mass transfer and thermal systems will allow simulation of heat and mass transfer and surface phenomena simultaneously.

NOMENCLATURE

c	weight used for settling the interaction length	–
C	concentration	mol/m^3
Ca	capillary number	–
Da	Darcy number	–
e_i	site displacement per unit time	–
f	distribution function of fluid particles	–
H	fin height	m
K_d	mass transfer coefficient	m/s
K_m	dimensionless Michaelis–Menten constant	–
L	fin length	m
m	mass	kg
Nu	Nusselt number	–
P	pressure	Pa
Pe	Peclet number	–
Q	mass flux	mL/h
Re	Reynolds number	–
S	fin surface	m^2
t	time	s
u	velocity	m/s
We	Weber number	–
X	Martinelli parameter	–
x	site	–
x, y, z	Cartesian coordinates	m

Greek letters

ρ	density	kg/m^3
θ	modified velocity	m/s
α	angle	°
Φ_L	two-phase multiplier	–
Ω	collision operator	–

Subscripts

d	drag force or downcomer
i	direction of discrete velocity
ks	kerosene
L	liquid
G	gas

Superscripts

b	blue fluid particles
eq	equilibrium state
m	mediator
r	red fluid particles

REFERENCES

Ahn, Y. C., Jung, W., & Chen, Z. P. (2008). Optical sectioning for microfluidics: Secondary flow and mixing in a meandering microchannel. *Lab Chip*, *8*, 125–133.

Akbar, M. K., Plummer, D. A., & Ghiaasiaan, S. M. (2003). On gas–liquid two-phase flow regimes in microchannels. *Int. J. Multiphase Flow.*, *9*, 1163–1177.

Barnea, D., Luninski, Y., & Taitel, Y. (1983). Flow pattern in horizontal and vertical two phase flow in small diameter pipes. *Can. J. Chem. Eng.*, *61*, 617–620.

Beattie, D. R. H., & Whalley, P. B. (1982). A simple two-phase flow frictional pressure drop calculation method. *Int. J. Multiphase Flow.*, *8*, 83–87.

Berčič, G., & Pintar, A. (1997). The role of gas bubbles and liquid slug lengths on mass transport in the Taylor flow through capillaries. *Chem. Eng. Sci.*, *52*(21–22), 3709–3719.

Brauner, N., & Maron, D. M. (1992). Analysis of sratified/non-stratified transitional boundaries in inclined gas–liquid flows. *Int. J. Multiphase Flow.*, *18*, 541–557.

Bretherton, F. P. (1961). The motion of long bubbles in tubes. *J. Fluid Mech.*, *10*, 166–168.

Burns, J. R., & Ramshaw, C. (2001). The intensification of rapid reactions in multiphase systems using slug flow in capillaries. *Lab Chip*, *1*, 10–15.

Chen, I. Y., Yang, K. S., & Wang, C. C. (2002). An empirical correlation for two-phase frictional performance in small diameter tubes. *Int. J. Heat Mass Transfer*, *45*, 3667–3671.

Chisholm, D. (1967). A theoretical basis for the Lockhart–Martinelli correlation for two-phase flow. *Int. J. Heat Mass Transfer*, *10*, 1767–1778.

Choi, C. W., Yu, D. I., & Kim, M. H. (2011). Adiabatic two-phase flow in rectangular microchannels with different aspect ratios: Part I – Flow pattern, pressure drop and void fraction. *Int. J. Heat Fluid Flow.*, *54*, 616–624.

Chung, C. K., & Shih, T. R. (2008). Effect of geometry on fluid mixing of the rhombic micromixers. *Microfluid. Nanofluid.*, *4*, 419–425.

Cooper, J., Disley, D., & Cass, T. (2001). Microsystems special – lab-on-a-chip and microarrays. *Chem. & Ind.*, 653–655.

Dessimoz, A. L., Cavin, L., Renken, A., & Kiwi-Minsker, L. (2008). Liquid–liquid two-phase flow patterns and mass transfer characteristics in rectangular glass micro-reactors. *Chem. Eng. Sci.*, *63*(16), 4035–4044.

Doku, G. N., Haswell, S. J., McCreedy, T., & Greenway, G. M. (2001). Electric field-induced mobilization of multiphase solution systems based on the nitration of benzene in a micro reactor. *Analyst*, *126*(1), 14–20.

Donata, M. F., Franz, T., & Philipp, R. V. R. (2008). Segmented gas–liquid flow characterization in rectangular microchannels. *Int. J. Multiphase Flow.*, *34*(12), 1108–1118.

Dongari, N., Durst, F., & Chakraborty, S. (2010). Predicting microscale gas flows and rarefaction effects through extended Navier–Stokes–Fourier equations from phoretic transport considerations. *Microfluid. Nanofluid.*, *9*, 831–846.

Dukler, A. E., & Taitel, Y. (1977a). Flow regime transitions for vertical upward gas liguid flow: A preliminary approach through physical modeling. *Progress Report, No. 1*, NUREG-0162.

Dukler, A. E., & Taitel, Y. (1977b). Flow regime transitions for vertical upward gas liguid flow: A preliminary approach through physical modeling. *Progress Report No. 2*, NUREG-0163.

Dukler, A. E., Wicks, M., & Clevel, R. G. (1964). Frictional pressure drop in two-phase flow: B. An approach through similarity analysis. *AIChE J.*, *10*, 44–51.

Ehrfeld, W. (2000). Microreaction technology: Industrial prospects. *The 3rd international conference on microreaction technology*, Frankfurt.

Fedorov, A. G., & Viskanta, R. (2004). Three-dimensional analysis of heat transfer in a micro-heat sink with single phase flow. *Int. J. Heat Mass Transfer*, *47*(19–20), 4215–4231.

Fletcher, P. D. I., Haswell, S. J., Pombo-Villar, E., Warrington, B. H., Watts, P., Wong, S. Y. F., & Zhang, X. (2002). Mirco reactors: Principles and applications in organic synthesis. *Tetrahedron*, *58*, 4735–4757.

Fu, T. T., Ma, Y. G., Funfschilling, D., & Li, H. Z. (2010). Gas–liquid flow stability and bubble formation in non-Newtonian fluids in microfluidic flow-focusing devices. *Microfluid. Nanofluid*, *8*, 799–821.

Fujioka, H., & Grotberg, J. B. (2004). Steady propagation of a liquid plug in a two-dimensional channel. *ASME J. Biomech. Eng.*, *126*, 567–577.

Fujioka, H., & Grotberg, J. B. (2005). The steady propagation of a surfactant-laden liquid plug in a two-dimensional channel. *Phys. Fluids*, *17*(8), 082102.

Gong, X. C., Lu, Y. C., Xiang, Z. Y., Zhang, Y. N., & Luo, G. S. (2007). Preparation of uniform microcapsules with silicone oil as continuous phase in a micro-dispersion process. *J. Microencapsul.*, *24*(8), 767–776.

Gunstensen, A. K., & Rothman, D. H. (1991a). A lattice-gas model for three immiscible fluids. *Phys. D: Nonlinear Phenomena*, *47*(1–2), 47–52.

Gunstensen, A. K., & Rothman, D. H. (1991b). A Galilean-invariant immiscible lattice gas. *Phys. D: Nonlinear Phenomena*, *47*(1–2), 53–63.

Gunstensen, A. K., Rothman, D. H., Zaleski, S., & Zanetti, G. (1991). Lattice Boltzmann model of immiscible fluids. *Phys. Rev. A*, *43*(18), 4320–4327.

Gupta, R., Fletcher, D. F., & Haynes, B. S. (2010). CFD modeling of flow and heat transfer in the Taylor flow regime. *Chem. Eng. Sci.*, *65*, 2094–2107.

Hamida, T., & Babadagli, T. (2008). Displacement of oil by different interfacial tension fluids under ultrasonic waves. *Colloids Surfaces A: Physicochem. Eng. Aspects*, *316*, 176–189.

Házi, G., Imre, A. R., Mayer, G., & Farkasa, I. (2002). Lattice Boltzmann methods for two-phase flow modeling. *Annals of Nuclear Energy*, *29*, 1421–1453.

He, Q., Fukagata, K., & Kasagi, N. (2007). Numerical simulation of gas–liquid two-phase flow and heat transfer with dry-out in a micro tube. *6th international conference on multiphase flow*, Leipzig, Germany.

He, X. Y., Chen, S. Y., & Zhang, R. Y. (1999). A lattice Boltzmann scheme for incompressible multiphase flow and its application in simulation of Rayleigh–Taylor instability. *J. Comput. Phys.*, *152*, 642–663.

Hetsroni, G., Mosyak, A., Segal, Z., & Pogrebnyak, E. (2003). Two-phase flow patterns and heat transfer in parralel micro-channels. *Int. J. Heat Mass Transfer*, *29*(3), 341–360.

Holdych, D. J., Rovas, D., Geogiadis, J. G., & Buckius, R. O. (1998). An improved hydrodynamic formulation for multiphase flow lattice Boltzmann models. *Int. J. Mod. Phys. C*, *9*, 1393.

Jensen, O. E., Halpern, D., & Grotberg, J. B. (1994). Transport of a passive solute by surfactant-driven flows. *Chem. Eng. Sci.*, *49*(8), 1107–1117.

Jovanović, J., Zhou, W. Y., Rebrov, E. V., Nijhuis, T. A., Hessel, V., & Schouten, J. C. (2011). Liquid–liquid slug flow: Hydrodynamics and pressure drop. *Chem. Eng. Sci.*, *66*, 42–54.

Junseok, K. (2005). A diffuse-interface model for axisymmetric immiscible two-phase flow. *Appl. Math. Comput.*, *160*(2), 589–606.

Kashid, M. N., & Agar, D. W. (2007). Hydrodynamics of liquid–liquid slug flow capillary micro-reactor: Flow regimes, slug size and pressure drop. *Chem. Eng. J.*, *131*(1–3), 1–13.

Kashid, M. N., Renken, A., & Kiwi-Minsker, L. (2011). Gas–liquid and liquid–liquid mass transfer in microstructured reactors. *Chem. Eng. Sci.*, *66*, 3876–3897.

Kawahara, A., Chung, P. M. Y., & Kawaji, M. (2002). Investigation of two-phase flow pattern, void fraction and pressure drop in a microchannel. *Int. J. Multiphase Flow*, *28*, 1411–1435.

Kehrwald, D. (2002). *Numerical analysis of immiscible lattice BGK*. Dissertation, Fachbereich Mathematik, University at Kaiserslautern.

Koyama, S., Lee, J., & Yonemoto, R. (2004). An investigation on void fraction of vapor–liquid two-phase flow for smooth and microfin tubes with R134a at adiabatic condition. *Int. J. Multiphase Flow*, *30*, 291–310.

Kreutzer, M. T., Kapteijn, F., Moulijn, J. A., Kleijn, C. R., & Heiszwolf, J. J. (2005). Inertial and interfacial effects on pressure drop of Taylor flow in capillaries. *AIChE J.*, *51*, 2428–2440.

Lakehal, D., Larrignon, G., & Narayanan, C. (2008). Computational heat transfer and two-phase flow topology in miniature tubes. *Microfluid. Nanofluid*, *4*, 261–271.

Lamura, A., Gonnella, G., & Yeomans, J. M. (1999). A lattice Boltzmann model of ternary fluid mixtures. *Europhys. Lett.*, *99*, 314–320.

Lee, C. Y., & Lee, S. Y. (2008). Pressure drop of two-phase plug flow in round mini-channels: Influence of surface wettability. *Exp. Therm. Fluid Sci.*, *32*, 1716–1722.

Lee, P. S., Garimella, S. V., & Liu, D. (2005). Investigation of heat transfer in rectangular microchannels. *Int. J. Heat Mass Transfer*, *48*(9), 1688–1704.

Lee, W., Son, G., & Yoon, H. Y. (2012). Direct numerical simulation of flow boiling in a finned microchannel. *Int. Commun. Heat Mass Transfer*, *39*, 1460–1466.

Lenormand, R., Zarcone, C., & Sarr, A. (1983). Mechanisms of the displacement of one fluid by another in a network of capillary ducts. *J. Fluid Mech.*, *135*, 337–353.

Lin, S., Kwok, C. C. K., Li, R. Y., Chen, Z. H., & Chen, Z. Y. (1991). Local frictional pressure drop during vaporization for R-12 through capillary tubes. *Int. J. Multiphase Flow*, *17*, 95–102.

Liow, J. L. (2004). Numerical simulation of drop formation in a T-shaped microchannel. *15th Australasian fluid mechanics conference*, Sydney, Australia.

Liu, D. S., & Wang, S. D. (2008). Flow pattern and pressure drop of upward two-phase flow in vertical capillaries. *Ind. Eng. Chem. Res.*, *47*, 243–255.

Liu, Y., Cui, J., Jiang, Y. X., & Li, W. Z. (2011). A numerical study on heat transfer performance of microchannels with different surface microstructures. *Appl. Therm. Eng.*, *31*, 921–931.

Liu, Z. D., Lu, Y. C., Wang, J. W., & Luo, G. S. (2012). Mixing characterization and scaling-up analysis of asymmetrical T-shaped micromixer, Experiment and CFD simulation. *Chem. Eng. J.*, 181–182, 597–606.

Lockhart, R. W., & Martinelli, R. C. (1949). Proposed correlation of data for isothermal two-phase, two-component flow in pipes. *Chem. Eng. Prog.*, 45, 39–48.

Losey, M. W., Jackman, R. J., Firebaugh, S. L., Schmidt, M. A., & Jensen, K. F. J. (2002). Design and fabrication of microfluidic devices for multiphase mixing and reaction. *Microelectromech. Syst.*, 11, 709–715.

Lu, P., Wang, Z. H., Yang, C., & Mao, Z. -S. (2010). Experimental investigation and numerical simulation of mass transfer during drop formation. *Chem. Eng. Sci.*, 65(20), 5517–5526.

Ma, Y. G., Ji, X. Y., Wang, D. J., Fu, T. T., & Zhu, C. Y. (2010). Measurement and correlation of pressure drop for gas–liquid two-phase flow in rectangular microchannels. *Chinese J. Chem. Eng.*, 18, 940–947.

Mehdizadeh, A., Sherif, S. A., & Lear, W. E. (2011). Numerical simulation of thermofluid characteristics of two-phase slug flow in microchannels. *Int. J. Heat Mass Transfer*, 54, 3457–3465.

Mishima, K., & Hibiki, T. (1996). Some characteristics of air–water two-phase flow in small diameter vertical tubes. *Int. J. Multiphase Flow*, 22, 703–712.

Mishima, K., & Ishii, M. (1984). Flow regime transition criteria for upward two-phase flow in vertical tubes. *Int. J. Heat Mass Transfer*, 27, 723–737.

Mouza, A. A., Paras, S. V., & Karabelas, A. J. (2002). The influence of small tube diameter on falling film and flooding phenomena. *Int. J. Multiphase Flow*, 28(8), 1311–1331.

Narayanan, C., & Lakehal, D. (2008). Two-phase convective heat transfer in miniature pipes under normal and microgravity conditions. *J. Heat Transfer*, 130(7), 074502–074505.

Nimafar, M., Viktorov, V., & Martinelli, M. (2012). Experimental comparative mixing performance of passive micromixers with H-shaped sub-channels. *Chem. Eng. Sci.*, 76, 37–44.

Nisisako, T., Torii, T., & Higuchi, T. (2002). Droplet formation in a microchannel network. *Lab Chip*, 2, 24–26.

Niu, H. N., Pan, L. W., Su, H. J., & Wang, S. D. (2009). Flow pattern, pressure drop, and mass transfer in a gas–liquid concurrent two-phase flow microchannel reactor. *Ind. Eng. Chem. Res.*, 48, 1621–1628.

Nourgaliev, R. R., Dinh, T. N., & Sehgal, B. R. (2002). On lattice Boltzmann modeling of phase transitions in an isothermal non-ideal fluid. *Nucl. Eng. Des.*, 211, 153–171.

Oprins, H., Danneels, J., Van Ham, B., Vandevelde, B., & Baelmans, M. (2008). Convection heat transfer in electrostatic actuated liquid droplets for electronics cooling. *Microelectronics J.*, 39, 966–974.

Owen, W. L. (1961). Two-phase pressure gradient. *Int. dev. in heat transfer* (Pt II). New York: ASME.

Pandey, S., Gupta, A., Chakrabarti, D. P., Das, G., & Ray, S. (2006). Liquid–liquid two phase flow through a horizontal T-junction. *Chem. Eng. Res. Des.*, 84(10), 895–904.

Pehlivan, K., Hassan, I., & Vaillancourt, M. (2006). Experimental study on two-phase flow and pressure drop in millimeter-size channels. *Appl. Therm. Eng.*, 26, 1506–1514.

Peng, X. F., & Peterson, G. P. (1996). Convective heat transfer and flow friction for water flow in microchannel structures. *Int. J. Heat Mass Transfer*, 39(12), 2599–2608.

Pohorecki, R. (2007). Effectiveness of interfacial area for mass transfer in two-phase flow in microreactors. *Chem. Eng. Sci.*, 62, 6495–6498.

Raimondi, N. D. M., & Prat, L. (2011). Numerical study of the coupling between reaction and mass transfer for liquid–liquid slug flow in square microchannels. *AIChE J.*, *57*, 1719–1732.

Raimondi, N. D. M., Prat, L., Gourdon, C., & Cognet, P. (2008). Direct numerical simulations of mass transfer in square microchannels for liquid–liquid slug flow. *Chem. Eng. Sci.*, *63*, 5522–5530.

Rapaport, D. C. (1995). *The art of molecular dynamics simulation*. Cambridge: Cambridge University Press.

Reyes, D. R., Iossifidis, D., Auroux, P. A., & Manz, A. (2002). Micro total analysis systems. 1. Introduction, theory and technology. *Analyt. Chem.*, *74*(12), 2623–2636.

Roudet, M., Loubiere, K., Gourdon, C., & Cabassud, M. (2011). Hydrodynamic and mass transfer in inertial gas–liquid flow regimes through straight and meandering millimetric square channels. *Chem. Eng. Sci.*, *66*, 2974–2990.

Sankaranarayanan, K., Shan, X., Kevrekidis, I. G., & Sundaresan, S. (2002). Analysis of drag and virtual mass forces in bubbly suspensions using an implicit formulation of the lattice Boltzmann method. *J. Fluid Mech.*, *452*, 61–96.

Sankaranarayanan, K., Kevrekidis, I. G., Sundaresan, S., Lu, J., & Tryggvason, G. (2003). A comparative study of lattice Boltzmann and front-tracking finite-difference methods for bubble simulations. *Int. J. Multiphase Flow*, *29*(1), 109–116.

Santos, L. O. E., Facin, P. C., & Philippi, P. C. (2002). Lattice-Boltzmann model based on field mediators for immiscible fluids. *Phys. Rev. E*, *68*(5), 056302.

Santos, L. O. E., Wolf, F. G., & Philippi, P. C. (2005). Dynamics of interface displacement in capillary flow. *J. Stat. Phys.*, *121*(1–2), 197–207.

Shan, X., & Chen, H. (1993). Lattice Boltzmann model for simulating flows with multiple phases and components. *Phys. Rev. E*, *47*(3), 1815–1819.

Shan, X., & Chen, H. (1994). Simulation of non-ideal gases and liquid–gas phase transition by the Lattice Boltzmann equation. *Phys. Rev. E*, *49*(4), 2941.

Shui, L. L., Eijkel, J. C. T., & van den Berg, A. (2007a). Multiphase flow in microfluidic systems – Control and applications of droplets and interfaces. *Adv. Colloid Interface Sci.*, *133*(1), 35–49.

Shui, L. L., Eijkel, J. C. T., & van den Berg, A. (2007b). Multiphase flow in micro- and nano-channels. *Sensors Actuators B*, *121*(1), 263–276.

Skelland, A. H. P., & Wellek, R. M. (1964). Resistance to mass transfer inside droplets. *AIChE J.*, *10*(4), 491–496.

Sobhan, C. B., & Garimella, S. V. (2001). A comparative analysis of studies on heat transfer and fluid flow in microchannels. *Nanoscale Microscale Thermophys. Eng.*, *5*(4), 293–311.

Su, H. J., Niu, H. N., Pan, L. W., Wang, S. D., Wang, A. J., & Hu, Y. K. (2010). The characteristics of pressure drop in microchannels. *Ind. Eng. Chem. Res.*, *49*, 3830–3839.

Su, Y. H., Chen, G. W., & Yuan, Q. (2011). Ideal micromixing performance in packed microchannels. *Chem. Eng. Sci.*, *66*, 2912–2919.

Su, Y. H., Chen, G. W., & Yuan, Q. (2012). Influence of hydrodynamics on liquid mixing during Taylor flow in a microchannel. *AIChE J.*, *58*, 1660–1670.

Suo, M., & Griffith, P. (1964). Two phase flow in capillary tubes. *J. Basic Eng.*, *86*, 576–582.

Swift, M. R., Osborn, W. R., & Yeomans, J. M. (1995). Lattice Boltzmann simulation of nonideal fluids. *Phys. Rev. Lett.*, *75*(5), 830–833.

Taitel, Y., & Dukler, A. E. (1976). A model for predicting flow regime transitions in horizontal and near horizontal gas–liquid flow. *AIChE J.*, *22*, 47–55.

Taitel, Y., Bornea, D., & Dukler, A. E. (1980). Modeling flow patterns transitions for steady upward gas-liquid flow in vertical tubes. *AIChE J.*, *26*, 345–354.

Takamasa, T., Hazuku, T., & Hibiki, T. (2008). Experimental study of gas–liquid two-phase flow affected by wall surface wettability. *Int. J. Heat Fluid Flow*, *29*(6), 1593–1602.

Talimi, V., Muzychka, Y. S., & Kocabiyik, S. (2011). A numerical study on shear stress and heat transfer of segmented flow between parallel plates. *Proceedings of the 9th international conference on nanochannels, microchannels, and minichannels*, Edmonton, Canada.

Talimi, V., Muzychka, Y. S., & Kocabiyik, S. (2012). A review on numerical studies of slug flow hydrodynamics and heat transfer in microtubes and microchannels. *Int. J. Multiphase Flow*, *39*, 88–104.

Tan, J., Xu, J. H., Li, S. W., & Luo, G. S. (2008). Drop dispenser in a cross-junction microfluidic device: Scaling and mechanism of break-up. *Chem. Eng. J.*, *136*(2–3), 306–311.

Tan, J., Lu, Y. C., Xu, J. H., & Luo, G. S. (2012). Mass transfer performance of gas–liquid segmented flow in microchannels. *Chem. Eng. J.*, *181–182*, 229–235.

Tofteberg, T., Skolimowski, M., Andreassen, E., & Geschke, O. (2010). A novel passive micromixer: Lamination in a planar channel system. *Microfluid. Nanofluid*, *8*, 209–215.

Triplett, K. A., Ghiaasiaan, S. M., Abdel-Khalik, S. I., LeMouel, A., & McCord, B. N. (1999a). Gas–liquid two-phase flow in microchannels. Part II: Void fraction and pressure drop. *Int. J. Multiphase Flow*, *25*, 395–410.

Triplett, K. A., Ghiaasiaan, S. M., Abdel-Khalik, S. I., & Sadowski, D. L. (1999b). Gas–liquid two-phase flow in microchannels. Part I: Two-phase flow pattern. *Int. J. Multiphase Flow*, *25*(3), 377–394.

Tuckerman, D. B., & Pease, R. F. W. (1981). High-performance heat sinking for VLSI. *IEEE Electron Device Letters*, *2*(5), 126–129.

Ua-arayaporn, P., Fukagata, K., Kasagi, N., & Himeno, T. (2005). Numerical simulation of gas–liquid two-phase convective heat transfer in a micro tube. *ECI international conference on heat transfer and fluid flow in microscale*, (pp. 25–30), Castelvecchio Pascoli.

Ugi, I., Almstetter, M., Gruber, B., & Heilingbrunner, M. (1996). Molecular libraries in liquid phase via UGI-MCR. *Research on Chemical Intermediates*, *22*(7), 625–644.

Ungar, E.K., & Cornwell, J.D. (1992). Two-phase pressure drop of ammonia in small diameter horizontal tubes. *AIAA 17th aerospace ground testing conf.*, Nashville, TN, 6–8 July.

Van Baten, J. M., & Krishna, R. (2004). CFD simulations of mass transfer from Taylor bubbles rising in circular capillaries. *Chem. Eng. Sci.*, *59*, 2535–2545.

Waelchli, S., & Von Rohr, P. R. (2006). Two-phase flow characteristics in gas–liquid microreactors. *Int. J. Multiphase Flow*, *32*, 791–806.

Walter, S., Malmberg, S., Schmidt, B., & Liauw, M. A. (2005). Mass transfer limitations in microchannel reactors. *Catal. Today*, *110*, 15–25.

Wang, J. F., Lu, P., Wang, Z. H., Yang, C., & Mao, Z.-S. (2008). Numerical simulation of unsteady mass transfer by the level set method. *Chem. Eng. Sci.*, *63*(12), 3141–3151.

Wang, J. F., Wang, Z. H., Lu, P., Yang, C., & Mao, Z. S. (2011). Numerical simulation of the Marangoni effect on transient mass transfer from single moving deformable drops. *AIChE J.*, *57*(10), 2670–2683.

Wang, K., Lu, Y. C., Xu, J. H., & Luo, G. S. (2009). Determination of dynamic interfacial tension and its effect on droplet formation in the T-shaped microdispersion process. *Langmuir*, *25*(4), 2153–2158.

Wang, X., Yong, Y. M., Fan, P., Yu, G. Z., Yang, C., & Mao, Z. S. (2012). Flow regime transition for concurrent gas–liquid flow in micro-channels. *Chem. Eng. Sci.*, *69*, 578–586.

Warnier, M. J. F., de Croon, M. H. J. M., Rebrov, E. V., & Schouten, J. C. (2010). Pressure drop of gas–liquid Taylor flow in round micro-capillaries for low to intermediate Reynolds numbers. *Microfluid. Nanofluid*, *8*, 33–45.

Xu, J. H., Luo, G. S., Li, S. W., & Chen, G. G. (2006). Shear force induced monodisperse droplet formation in a microfluidic device by controlling wetting properties. *Lab Chip*, *6*(1), 131–136.

Xu, J. L., Cheng, P., & Zhao, T. S. (1999). Gas–liquid two-phase flow regimes in rectangular channels with mini/microgaps. *Int. J. Multiphase Flow*, *25*, 411–432.

Yang, C., & Mao, Z. S. (2005). Numerical simulation of interphase mass transfer with the level set approach. *Chem. Eng. Sci.*, *60*(10), 2643–2660.

Yang, Y., Liao, Q., Zhu, X., Wang, H., Wu, R., & Lee, D. J. (2011). Lattice Boltzmann simulation of substrate flow past a cylinder with PSB biofilm for bio-hydrogen production. *Int. J. Hydrogen Energy*, *36*(21), 14031–14040.

Yong, Y. M., Yang, C., Jiang, Y., Joshi, A., Shi, Y. C., & Yin, X. L. (2011). Numerical simulation of immiscible liquid–liquid flow in microchannels using lattice Boltzmann method. *Sci. China: Chem.*, *54*(1), 244–256.

Yong, Y. M., Li, S., Yang, C., & Yin, X. L. (2013). Transport of wetting and nonwetting liquid plugs in a T-shaped microchannel. *Chinese J. Chem. Eng.*, *21*(5), 1–10.

Young, P., & Mohseni, K. (2008). The effect of droplet length on Nusselt numbers in digitized heat transfer. *11th IEEE intersociety conference on thermal and thermomechanical phenomena in electronic systems* (pp. 352–359), Orlando, FL.

Yu, Z., Hemminger, O., & Fan, L. S. (2007). Experiment and lattice Boltzmann simulation of two-phase gas–liquid flows in microchannels. *Chem. Eng. Sci.*, *62*(24), 7172–7183.

Yue, J., Chen, G. W., & Yuan, Q. (2004). Pressure drops of single and two-phase flows through T-type microchannel mixers. *Chem. Eng. J.*, *102*, 11–24.

Yue, J., Chen, G. W., Yuan, Q., Luo, L. A., & Gonthier, Y. (2007). Hydrohynamics and mass transfer characteristics in gas–liquid flow through a rectangular microchannel. *Chem. Eng. Sci.*, *62*, 2096–2108.

Yue, J., Luo, L. A., Gonthie, Y., Chen, G. W., & Yuan, Q. (2008). An experimental investigation of gas–liquid two-phase flow in single microchannel contactors. *Chem. Eng. Sci.*, *63*, 4189–4202.

Yue, J., Luo, L. A., Gonthier, Y., Chen, G. W., & Yuan, Q. (2009). An experimental study of air–water Taylor flow and mass transfer inside square microchannels. *Chem. Eng. Sci.*, *64*, 3697–3708.

Yue, J., Boichot, R., Luo, L. A., Gonthier, Y., Chen, G. W., & Yuan, Q. (2010). Flow distribution and mass transfer in a parallel microchannel contactor integrated with constructal distributors. *AIChE J.*, *56*, 298–317.

Zech, T., & Honicke, D. (2000). Efficient and reliable screening of catalysts for microchannel reactors by combinatorial methods. *Proceedings of the 4th international conference on microreaction technology* (pp. 379–389), Atlanta, GA.

Zeng, Y., Lee, T. S., Yu, P., & Low, H. T. (2007). Numerical study of mass transfer coefficient in a 3D flat-plate rectangular microchannel bioreactor. *Int. Commun. Heat Mass Transfer*, *34*, 217–224.

Zhang, J. F. (2005). *Fluid–fluid and solid–fluid interfacial studies by means of the lattice Boltzmann method*. Ph. D. thesis, University of Alberta.

Zhao, T. S., & Bi, Q. C. (2001). Co-current air–water two-phase flow patterns in vertical triangular microchannels. *Int. J. Multiphase Flow*, *21*, 765–782.

Zhao, Y. C., Chen, G. W., & Yuan, Q. (2006a). Liquid–liquid two-phase flow patterns in a rectangular microchannel. *AIChE J.*, *52*(12), 4052–4060.

Zhao, Y. C., Ying, Y., Chen, G. W., & Yuan, Q. (2006b). Characterization of micro-mixing in T-shaped micromixer. *J. Chem. Ind. Eng. (China)*, *57*(8), 1184–1190.

Zhao, Y. C., Chen, G. W., & Yuan, Q. (2007). Liquid–liquid two-phase mass transfer in the T-junction microchannels. *AIChE J.*, *53*(12), 3042–3053.

Crystallizers: CFD–PBE modeling

6.1 INTRODUCTION

Crystallization may be the oldest unit operation in chemical engineering. It is one of the best and cheapest ways for production of one or several substances from an amorphous, liquid, or gaseous state to the crystalline state. Research on crystallization and crystallizers covers a vast range of aspects such as theory, experimentation, crystallization kinetics, industrial techniques, design of crystallizers, crystallization control, etc. In this chapter we focus particularly on the modeling of crystallizers, especially reactive crystallization processes. Techniques for solving the equation describing the crystallization process, i.e., the population balance equation (PBE), are discussed in detail. Idealized models like mixed-suspension mixed-product removal (MSMPR) have been used to model a specified crystallizer (Randolph and Larson, 1988). With the MSMPR model, the solution of the PBE is only necessary for a chemically homogeneous system, thus the model is simplified significantly and sometimes even an analytical solution may be available. However, well-mixed magma only exists in a laboratory crystallizer with strong stirring, and mixed-product removal also only exists in idealized cases. Moreover, for precipitators where the precipitation process occurs mostly around the feeding nozzles, the MSMPR assumption is obviously inaccurate. In actual crystallization systems, the spatial distributions of supersaturation, flow fields, and solid concentration fields are important and the PBE is commonly coupled with computational fluid dynamics (CFD), i.e., CFD–PBE modeling, to take into account these spatial heterogeneities. The precipitation of barium sulfate ($BaSO_4$) has been widely used as a model reaction and studied in the past decades. Thus, procedures for simulating the precipitation process of $BaSO_4$ in crystallizers such as stirred tanks will be addressed as a typical example in this chapter.

The PBE is a well-established mathematical framework for dealing with crystallization processes (Costa et al., 2007). Since the PBE is a hyperbolic integro-partial-differential equation, analytical solutions exist only in a few very simplified cases, and therefore numerical techniques or closure schemes often have to be employed. So far, many numerical methods have been proposed and applied, and generally they can be classified into four main categories: moment method, multi-class method (MCM), weighted residuals method, and stochastic method (Cheng et al., 2012a).

Numerical Simulation of Multiphase Reactors with Continuous Liquid Phase. DOI: 10.1016/B978-0-08-099919-7.00006-8

The moment method includes mainly the standard method of moments (Hulburt and Katz, 1964), the quadrature method of moments (QMOM; McGraw, 1997), and extensions of the QMOM such as the direct quadrature method of moments (DQMOM; Marchisio and Fox, 2005). For the multi-class method, the entire continuous size range is divided into a number of small contiguous subclasses (intervals, bins or sections), and then the PBE is converted into a set of discretized equations. Hounslow's method (Hounslow et al., 1988), the method of classes (Marchal et al., 1988), the fixed pivot technique (Kumar and Ramkrishna, 1996a), and the cell average technique (Kumar et al., 2006) all belong to the MCM category. These schemes or techniques differ in the type of size grid (geometric, uniform or irregular), the handling of birth and death terms, the way of conservation of properties during aggregation and breakage, and the computation of fluxes at the size bin boundaries. The weighted residuals methods include the weighted residual method with global functions and the finite element method (Nicmanis and Hounslow, 1996; Mahoney and Ramkrishna, 2002). When mentioning the stochastic method, we generally refer to the Monte Carlo simulation (Falope et al., 2001). Of all these methods, the moment methods and the MCM, especially the fixed pivot techniques, are the most widely employed in simulations of crystallization and other particulate processes. Thus, they will be discussed in detail in the following sections. The weighted residuals method and the stochastic method will not be discussed in this chapter.

6.2 MATHEMATICAL MODELS AND NUMERICAL METHODS

6.2.1 General population balance equation

The PBE is used to describe variations in the state of particles in space and time. A general form of the PBE for a spatially inhomogeneous system is given as (Randolph and Larson, 1988)

$$\frac{\partial[\rho n(\xi; \mathbf{x}, t)]}{\partial t} + \nabla \cdot [\rho \mathbf{u} n(\xi; \mathbf{x}, t)] = -\rho \frac{\partial}{\partial \xi_j}[\zeta_j n(\xi; \mathbf{x}, t)] + \rho h(\xi; \mathbf{x}, t) \qquad (6.1)$$

where $\mathbf{x}$ is the spatial coordinate vector or the external coordinates, $\mathbf{u}$ is the mean velocity vector, $n(\xi; \mathbf{x}, t)$ is the number density, $\xi \equiv (\xi_1, \ldots, \xi_n)$ is called the property vector or the internal coordinates describing the states of crystals such as volume, length, voidage, and surface area, $h(\xi; \mathbf{x}, t)$ represents the net birth rate due to aggregation and breakage, and ζ_j is defined as (for simplicity, independent variables $\mathbf{x}$ and t are omitted in following sections)

$$\zeta_j = \frac{d\xi_j}{dt}, \quad j \in 1, \ldots, n \qquad (6.2)$$

For turbulent flows, employing the Reynolds time-averaged method and omitting the fluctuation terms of nucleation, growth, aggregation and breakage, we have

$$\frac{\partial[\rho \bar{n}(\xi)]}{\partial t} + \nabla \cdot [\rho \bar{\mathbf{u}} \bar{n}(\xi)] - \nabla \cdot [-\overline{\mathbf{u}' n'(\xi)}] = -\rho \frac{\partial}{\partial \bar{\xi}_j}[\overline{\zeta_j} \bar{n}(\xi)] + \rho \bar{h}(\xi) \qquad (6.3)$$

The correlation of velocity fluctuation and number density fluctuation is modeled as

$$-\overline{\mathbf{u}'n'(\xi)} = \frac{\mu_{\text{eff}}}{Sc_t}\nabla\overline{n}(\xi) = \Gamma_{\text{eff}}\nabla\overline{n}(\xi) \tag{6.4}$$

where μ_{eff} is the effective viscosity and Sc_t is the turbulent Schmidt number. Omitting the overbar symbols for each variable, then the Reynolds-averaged PBE is

$$\frac{\partial[\rho n(\xi)]}{\partial t} + \nabla\cdot[\rho\mathbf{u}n(\xi)] - \nabla\cdot[\Gamma_{\text{eff}}\nabla n(\xi)] = -\rho\frac{\partial}{\partial\xi_j}[\zeta_j n(\xi)] + \rho h(\xi) \tag{6.5}$$

As for $\xi \equiv (\xi_1,\ldots,\xi_n)$, multi-dimensional internal property coordinates will give a more detailed description of the states of crystals. However, the resulting PBEs are very complicated. The multi-dimensional PBE could be a possible future research topic in PBE modeling, but in this chapter we will discuss only the one-dimensional PBE (generally volume v or characteristic length L as the internal coordinate), which is employed most often in CFD–PBE modeling. In the case of particle volume v as the internal coordinate, i.e., $\xi_1 \equiv v$, the PBE is described as

$$\frac{\partial[\rho n(v)]}{\partial t} + \nabla\cdot[\rho\mathbf{u}n(v)] - \nabla\cdot[\Gamma_{\text{eff}}\nabla n(v)] = -\rho\frac{\partial}{\partial v}\left[\frac{dv}{dt}n(v)\right] + \rho h(v) \tag{6.6}$$

and

$$h(v) = B^{\text{a}}(v) - D^{\text{a}}(v) + B^{\text{b}}(v) - D^{\text{b}}(v) \tag{6.7}$$

where $B^{\text{a}}(v)$, $D^{\text{a}}(v)$, $B^{\text{b}}(v)$, and $D^{\text{b}}(v)$ represent the birth and death rates due to aggregation and breakage and can be described as (Marchisio et al., 2003a)

$$B^{\text{a}}(v) = \tfrac{1}{2}\int_0^v \beta(v-\epsilon,\epsilon)n(v-\epsilon)n(\epsilon)d\epsilon \tag{6.8}$$

$$D^{\text{a}}(v) = n(v)\int_0^{+\infty}\beta(v,\epsilon)n(\epsilon)d\epsilon \tag{6.9}$$

$$B^{\text{b}}(v) = \int_v^{+\infty}\psi(\epsilon)b(v|\epsilon)n(\epsilon)d\epsilon \tag{6.10}$$

$$D^{\text{b}}(v) = \psi(v)n(v) \tag{6.11}$$

where $\beta(v,\epsilon)$ is the aggregation rate, $\psi(v)$ is the breakage rate, and $b(v|\epsilon)$ denotes the size probability distribution of the particle with volume ϵ after it breaks.

In the case of particle length L as the internal coordinate, i.e., $\xi_1 \equiv L$, the PBE is

$$\frac{\partial[\rho n(L)]}{\partial t} + \nabla\cdot[\rho\mathbf{u}n(L)] - \nabla\cdot[\Gamma_{\text{eff}}\nabla n(L)] = \frac{-\rho\partial[G(L)n(L)]}{\partial L}$$
$$+ \rho[B^{\text{a}}(L) - D^{\text{a}}(L) + B^{\text{b}}(L) - D^{\text{b}}(L)] \tag{6.12}$$

where $G(L) = dL/dt$ is the linear growth rate, and

$$B^a(L) = \frac{L^2}{2} \int_0^L \frac{\beta((L^3 - \lambda^3)^{1/3}, \lambda)}{(L^3 - \lambda^3)^{2/3}} n[(L^3 - \lambda^3)^{1/3}]n(\lambda)\,d\lambda \tag{6.13}$$

$$D^a(L) = n(L) \int_0^{+\infty} \beta(L, \lambda)n(\lambda)d\lambda \tag{6.14}$$

$$B^b(L) = \int_L^{+\infty} \psi(\lambda)b(L|\lambda)n(\lambda)d\lambda \tag{6.15}$$

$$D^b(L) = \psi(L)n(L) \tag{6.16}$$

The volume-based form of the PBE can be converted into the corresponding length-based form and vice versa. For pure aggregation and/or breakage events, the consistency with mass balance can be more easily achieved using the volume as an internal coordinate (Verkoeijen et al., 2002). However, as pointed out by Mahoney and Ramkrishna (2002), in crystallization and when the particle size L is very small (L almost 0), since the length-based growth rate is still finite, the corresponding volumetric growth rate approaches zero at this small L, and a singularity in the volume-based number density may occur in some cases. The singular behavior will create very steep gradients in the solution, which are difficult to handle. For this reason, the PBE is expressed using the length as the internal coordinate at the expense of complicating the aggregation calculation a little. Another reason is that the length-based PBE and the growth rate are commonly adopted in published works on precipitation process modeling (e.g., Marchisio et al., 2002; Cheng et al., 2009 and references therein). For the above reasons, this chapter is mainly focused on the length-based PBE.

6.2.2 Standard method of moments

The standard method of moments (SMM) is one of the oldest methods for solving the PBE. The moments of the particle size distribution are defined as follows:

$$m_k = \int_0^{+\infty} n(L; \mathbf{x}, t)L^k \, dL \tag{6.17}$$

Applying the definition of moments to Eq. (6.12) results in

$$\begin{aligned}
\frac{\partial m_k}{\partial t} + \nabla \cdot [\mathbf{u}m_k] - \nabla \cdot [\Gamma_{\text{eff}}\nabla m_k] &= (0)^k B(\mathbf{x}, t) \\
+ \int_0^{+\infty} kL^{k-1}G(L)n(L)\,dL + \overline{B_k^a} - \overline{D_k^a} + \overline{B_k^b} - \overline{D_k^b}
\end{aligned} \tag{6.18}$$

where $B(\mathbf{x},t)$ is the nucleation rate, and

$$\overline{B_k^a} = \frac{1}{2}\int_0^{+\infty} n(\lambda)\int_0^{+\infty} \beta(u,\lambda)(u^3 + \lambda^3)^{k/3}n(u)\mathrm{d}u\mathrm{d}\lambda \tag{6.19}$$

$$\overline{D_k^a} = \int_0^{+\infty} L^k n(L)\int_0^{+\infty} \beta(L,\lambda)n(\lambda)\mathrm{d}\lambda\mathrm{d}L \tag{6.20}$$

$$\overline{B_k^b} = \int_0^{+\infty} L^k \int_0^{+\infty} \psi(\lambda)b(L|\lambda)n(\lambda)\mathrm{d}\lambda\mathrm{d}L \tag{6.21}$$

$$\overline{D_k^b} = \int_0^{+\infty} L^k \psi(L)n(L)\mathrm{d}L \tag{6.22}$$

Equation (6.18) cannot be solved directly since the growth, aggregation, and breakage related terms must be expressed explicitly as the functions of moments. In actual crystallization processes, the aggregation rate $\beta(L,\lambda)$ and the breakage rate $\psi(L)$ are functions of the crystal size and the size-dependent growth rate, i.e., $G = G_0(1 + \gamma L)^b$ with $b < 1$, and/or the growth dispersion is more commonly encountered. Thus, for applying the SMM, some assumptions must be made, i.e., size-independent growth and neglecting aggregation and breakage. Then the set of moment equations, for example from the 0th to the 4th, is expressed as

$$\frac{\partial m_k}{\partial t} + \nabla \cdot [\mathbf{u}m_k - \Gamma_{\mathrm{eff}}\nabla m_k] = 0^k B(\mathbf{x}) + jm_{k-1}G \quad (k = 0 - 4) \tag{6.23}$$

The SMM has the advantages of easy manipulation and being computationally time-saving. These are important for CFD–PBE simulations since computational costs, complexity and manipulability are important factors when choosing an appropriate technique. Thus, the SMM has been widely used for the CFD–PBE modeling in both reacting and nonreacting systems. Some successful CFD–PBE simulations with the SMM on precipitation processes in relatively simple and complex geometries have been reported in the past decade, e.g., in a coaxial pipe mixer (Öncül et al., 2005), a tubular reactor (Marchisio et al., 2002), a semi-batch stirred tank (Baldyga et al., 2007; Vicum and Mazzotti, 2007) and a continuous stirred tank reactor (CSTR; Jaworski and Nienow, 2003; Wang et al., 2007). Generally, $BaSO_4$ has been used as a model product to study the influence of activity coefficient and operating conditions, etc. Size-independent growth was adopted with aggregation ignored or just a constant empirical kernel used (Marchisio et al., 2002).

The most distinct drawback of the SMM is that size-dependent (typically nonlinear) growth, aggregation, and breakage cannot be handled, which greatly limits its applications. To overcome these weak points, the QMOM is a good choice (McGraw, 1997; Marchisio et al., 2003b).

6.2.3 Quadrature method of moments

The QMOM was proposed by McGraw (1997) to solve the aerosol dynamics, and further developed by Marchisio et al. (2003c). The QMOM is based on the following Gaussian quadrature approximation:

$$m_k = \int_0^{+\infty} n(L; \mathbf{x}, t) L^k \, \mathrm{d}L \approx \sum_{i=1}^{N_d} w_i L_i^k \tag{6.24}$$

where L_i (abscissas) and w_i (weights) are determined through the product-difference (PD) algorithm (McGraw, 1997). Applying Eq. (6.24) to Eq. (6.18), the set of moment equations becomes

$$\frac{\partial m_k}{\partial t} + \nabla \cdot [\mathbf{u} m_k] - \nabla \cdot [\Gamma_{\mathrm{eff}} \nabla m_k] = (0)^k B(\mathbf{x}) + k \sum_i L_i^{k-1} G(L_i) w_i$$

$$+ \frac{1}{2} \sum_i w_i \sum_j w_j \beta_{ij} (L_i^3 + L_j^3)^{k/3} - \sum_i L_i^k w_i \sum_j w_j \beta_{ij} \tag{6.25}$$

$$+ \sum_i \psi_i \overline{b}_i^{(k)} w_i - \sum_i L_i^k \psi_i w_i$$

where $\beta_{ij} = \beta(L_i, L_j)$, $\psi_i = \psi(L_i)$, and

$$\overline{b}_i^{(k)} = \int_0^{+\infty} L^k b(L|L_i) \mathrm{d}L \tag{6.26}$$

The QMOM has been tested and compared with other techniques, showing its great potential in solving the PBE with good accuracy by using a very small number of equations (Marchisio et al., 2006). Its coupling with CFD in non-reacting systems has been reported (Marchisio et al., 2003b; Gimbun et al., 2009; Petitti et al., 2010). Recently, the QMOM coupled with CFD has been employed to simulate the precipitation processes in confined impinging jet reactors (Gavi et al., 2007) and in stirred tanks (Cheng et al., 2009). The QMOM can handle aggregation and breakage terms and size-dependent growth with relative ease. Besides its applicability, the reported work also indicates that it is a strong and promising closure method for the PBE in terms of computational cost and efficiency, especially when multiphase-based CFD–PBE modeling must be used.

Neither the SMM nor the QMOM can directly obtain the crystal size distribution (CSD). Generally, low-order moments of the CSD are sufficient to characterize bubbles, drops, or crystal quality. However, in some processes, like polymerization, crystallization, and aerosol handling, the control of a full CSD may be necessary, due to the strong dependence of the physicochemical and mechanical properties (e.g., filtration rate and fluidization properties) of products on the characteristics of corresponding CSDs (Kalani and Christofides, 2002). In that case, the CSD can be the major decisive factor in the ultimate use of precipitated products. For the moment methods, the full CSD is commonly reconstructed from low-order moments using relevant numerical techniques. However, this is numerically unstable and no universally applicable method exists (Rigopoulos and Jones, 2003; John et al., 2007), which represents a serious problem in certain applications. Another approach is to first give a simple priori shape for a CSD (Gaussian, log-normal, etc.), and then indirectly to retrieve the CSD from two or three low-order moments (John et al., 2007). This may be practical for a simple and ideal distribution, but very difficult for some complicated and especially multimodal distributions (Diemer and Olson, 2002). Thus, the CFD–PBE simulation of some systems such as precipitation or drops coalescence and breakage with the MCM or the discretized method will become more important and useful in the future.

6.2.4 **Multi-class method or discretized method**

In the MCM techniques, the growth term and the secondary term are treated separately. For pure aggregation and/or breakage problems, the multi-class methods proposed before 1996 had several distinct drawbacks as reviewed by Kostoglou (2007): many of them conserved only one integral property of the particles, others were designed only for a specified grid type, or they were very complicated. The first general formulation conserving two selected properties (moments) is the fixed pivot technique developed by Kumar and Ramkrishna (1996a). This technique is the most extensively used discretized method since then. Although the numerically calculated moments are fairly accurate, this technique always has problems with overprediction of the number density. Afterwards, a moving pivot technique was presented by Kumar and Ramkrishna (1996b) to overcome the overprediction. However, it is more complex and gives difficulties to the solution of resulted set of ordinary stiff differential equations. Moreover, by allowing the representative mass point to move, coupling with CFD becomes intractable.

Based on the work of Kumar and Ramkrishna (1996a), Kumar et al. (2006) presented a cell average technique, which has all the advantages of the fixed pivot technique but alleviates the problem of overprediction to a great extent. As investigated by Kumar et al. (2008), the computation time of the cell average technique is comparable to and in some cases even less than that of the fixed pivot technique. Finite volume methods have been employed to solve aggregation and/or breakage problems in recent years. By reformulating the PBE in a mass conservation form, Filbet and Laurencot (2004) proposed a finite volume scheme (FVS) for solving aggregation problems. This FVS has been employed by several authors to deal with aggregation and/or breakage terms (Qamar and Warnecke, 2007; Qamar et al., 2009). Following Filbet and Laurencot (2004), Kumar (2006) developed a similar approach for solving aggregation and/or breakage PBEs. He found that the consistency of the cell average technique was superior to that of the FVS. Although the FVS is a good alternative to solve aggregation and breakage problems, the cell average technique was recommended by Kumar (2006) since it is easy to implement and faster than the FVS. In this chapter we will introduce primarily fixed pivot techniques for treating secondary terms, and the techniques for dealing with the growth term will be discussed later in this section.

The MCM partitions the entire continuous size range into a number of small contiguous subclasses and converts the PBE into a certain number of discretized

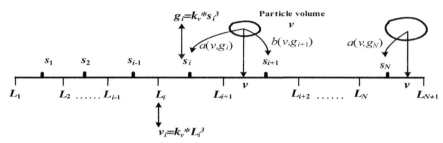

FIGURE 6.1 Sketch of the particle grid.

equations for each size bin. It tracks directly the discretized CSD, and the incorporation of aggregation, breakage and size-dependent growth becomes straightforward (Kumar and Ramkrishna, 1996a; Kumar et al., 2006). Figure 6.1 shows a typical grid, in which the entire size range is divided into small bins, and the interval between two sizes L_i and L_{i+1} is called the ith size bin. The particle population and/or other related properties for the ith size bin are represented at the pivot size s_i (not necessarily the arithmetic mean of L_i and L_{i+1}). Every particle size L_i corresponds to a unique particle volume $v_i = k_v L_i^3$. Likewise, s_i corresponds to the unique pivot volume $g_i = k_v s_i^3$. During aggregation, the most probable case is that a newly produced particle does not match the pivot size of the ith bin $s_i(g_i)$. Thus, this particle needs to be assigned to its adjacent pivot sizes, and in the process of reassignment, all the integral properties of the particles (such as number, mass, and specific area) should be well preserved. Generally, mass and number are the most important of the two integral properties. As illustrated in Figure 6.1, a newly aggregated particle from g_j and g_k, $v = g_j + g_k = k_v s_j^3 + k_v s_k^3$, which falls between the pivot sizes g_i and g_{i+1}, is assigned to g_i and g_{i+1} with respective assigning fractions, $a(v, g_i)$ and $b(v, g_{i+1})$ (Kumar and Ramkrishna, 1996a; Cheng et al., 2012a).

Integrating Eq. (6.1) over L_i and L_{i+1} and then multiplying both sides by the representative volume, g_i, we have

$$\frac{\partial}{\partial t}[\rho \alpha_i] + \nabla \cdot [\rho \mathbf{u} \alpha_i] - \nabla \cdot [\Gamma_{\text{eff}} \nabla \alpha_i] = 0^{i-1} \rho g_i (v_1 / g_1) B(t) +$$
$$\left[-\rho g_i \int_{L_i}^{L_{i+1}} \frac{\partial [G(L)n(L,t)]}{\partial L} \, dL \right] + \rho \left[\sum_{k=i}^{N} n_{i,k} \psi_k \alpha_k g_i / g_k - \psi_i \alpha_i \right] +$$
$$\rho \left[\sum_{\substack{j,k \\ g_{i-1} \leq (g_j + g_k) \leq g_{i+1}}}^{j \geq k} (1 - \tfrac{1}{2}\delta_{j,k}) \eta \beta_{j,k} \alpha_j \alpha_k g_i / g_j / g_k - \alpha_i \sum_{k=1}^{N} \beta_{i,k} \alpha_k / g_k \right] \tag{6.27}$$

where $\alpha_i = g_i N_i$ and N_i are the volume fraction and the number of crystals in the ith size bin per unit volume of suspension respectively, $\beta_{j,k} = \beta(g_j, g_k) = \beta(s_j, s_k)$, and η takes the following form:

$$\eta = \begin{cases} a(v, g_i) = \dfrac{g_{i+1} - v}{g_{i+1} - g_i} \, , & g_i \le v \le g_{i+1} \\[3mm] b(v, g_i) = \dfrac{v - g_{i-1}}{g_i - g_{i-1}} \, , & g_{i-1} \le v \le g_i \end{cases} \tag{6.28}$$

The growth term in Eq. (6.27) still needs further discretization. Because of its hyperbolic nature, the presence of sharp moving fronts or discontinuity is unavoidable (Kumar and Ramkrishna, 1997), and the growth term, which looks simple, poses great difficulty in solving the combined processes. The approximation of the advective growth term by finite-difference-type schemes (e.g., backward, central, or upwinding) introduces an artificial viscosity in the solution, which leads to smearing of the solution around discontinuous or steep fronts (Qamar and Warnecke, 2007). Schemes like the weighted essentially non-oscillator (WENO) have also been used for discretization of the growth term (Lim et al., 2002). Although the WENO scheme shows improvements of accuracy and stability over the conventional finite-difference ones, it is computationally expensive and still shows numerical diffusion errors to some extent near discontinuous or steep fronts.

The method of characteristics (MOC) is considered to be the best way to deal with the hyperbolic growth term (Kumar and Ramkrishna, 1997; Qamar and Warnecke, 2007). With the MOC, the solution moves along the propagation pathline, which simply eliminates the advective growth term from the PBE. Therefore, a highly accurate solution even at discontinuous points can be obtained. However, in the case of stiff nucleation, it is necessary to adaptively determine the time step to add a nucleus size bin. Furthermore, for solving the PBEs involving agglomeration and breakage terms, the MOC can take longer computational time than other spatial discretization methods (Lim et al., 2002). Thus, the MOC scheme is applicable only in homogeneous systems for the time being. For CFD–PBE modeling, each physical cell in the computational grid is well mixed and has its own growth rate. The combination with the MOC is thus intractable.

High-resolution schemes have been used primarily for numerical solution of hyperbolic systems such as astrophysical flows and gas dynamics. These schemes provide high accuracy for simulating hyperbolic conservation laws while reducing numerical diffusion and eliminating nonphysical oscillations that can occur with classical methods. Thus, much attention has been paid to employing these schemes for solving the growth term in recent years (see Ma et al., 2002; Woo et al., 2006; Gunawan et al., 2008; Qamar et al., 2009 and references therein). Several high-resolution schemes have been used in the literature. The high-resolution semi-discrete FVS of Koren (1993) has been used by Qamar and coworkers (Qamar et al., 2006, 2009; Qamar and Warnecke, 2007). Woo et al. (2006) have coupled a new family of high-resolution central schemes developed by Kurganov and Tadmor (2000) with CFD to simulate the antisolvent crystallization (nucleation and growth) in a 2D stirred tank with high accuracy and good application flexibility. Thus, the high-resolution schemes can be a good choice for dealing with the growth term. However, these schemes are generally based on a uniform grid and their coupling with CFD has rarely been investigated. A straightforward first-order upwind scheme, which has been used by some researchers in CFD–PBE modeling (Muhr et al., 1996; Rigopoulos and Jones, 2003; Rigopoulos, 2007; Cheng et al., 2012a), is given as follows:

$$-\int_{L_i}^{L_{i+1}} \frac{g_i \, \partial[G(L)n(L)]}{\partial L} \, \mathrm{d}L = \begin{cases} -\dfrac{G(s_1)\alpha_1}{L_2 - L_1}, & i = 1 \\[2ex] \dfrac{G(s_{i-1})\alpha_{i-1}g_i/g_{i-1}}{L_i - L_{i-1}} - \dfrac{G(s_i)\alpha_i}{L_{i+1} - L_i}, & i = 2,3,\dots,N-1 \\[2ex] \dfrac{G(s_{N-1})\alpha_{N-1}g_N/g_{N-1}}{L_N - L_{N-1}}, & i = N \end{cases} \tag{6.29}$$

6.3 CRYSTALLIZER MODELING PROCEDURES

In this section, we will show how to model a precipitation process in a 3D stirred tank, using BaSO$_4$ precipitation as the model reaction and the techniques for solving the PBE as discussed in previous sections. The stirred tank is assumed to be operated in a continuous mode. Figure 6.2 illustrates a commonly used stirred tank for modeling of BaSO$_4$ precipitation, as in Jaworski and Nienow (2003), Wang et al. (2006a, 2007), and Cheng et al. (2009). Generally, a standard six-bladed Rushton turbine is used. Two inlet tubes are simulated and through each a solution containing either BaCl$_2$ or Na$_2$SO$_4$ is fed.

Apart from solving the PBE, e.g., the moment equations from the SMM or the QMOM and the discretized equations from the MCM, we first need to get the three-dimensional flow field, for example by solving the Reynolds-averaged Navier–Stokes

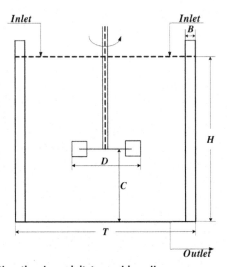

FIGURE 6.2 Sketch of the stirred precipitator and impeller.

equations (RANS). Since the precipitated particles are generally smaller than 20 μm and the solid concentration is very low, the particles closely follow the liquid and the influence of the dilute solid phase on liquid flow can be neglected. The single-phase RANS can thus be applied (Jaworski and Nienow, 2003; Wang et al., 2007). Since the supersaturation is produced by the reaction of barium chloride and sodium sulfate, the species transport equations must also be solved to get the spatial distribution of concentrations and thus the distribution of supersaturation. To solve the PBE, some models must be known, such as the nucleation and growth kinetics, the aggregation rate model, the breakage rate model, and the particle fragmentation distribution expression.

6.3.1 Species transport equations

The transport equations for different chemical species, described in terms of molar concentrations c_i ($i = Ba^{2+}$, Cl^-, Na^{2+} and SO_4^{2-}), are modeled as

$$\frac{\partial(\rho c_i)}{\partial t} + \nabla \cdot [\rho \mathbf{u} c_i - \Gamma_{eff} \nabla c_i] = S_i \tag{6.30}$$

The effective diffusion coefficient, Γ_{eff}, is computed as the turbulent viscosity μ_t divided by the turbulent Schmidt number Sc_t, i.e., $\Gamma_{eff} = \mu_t/Sc_t$. The source term S_i is related to the specific crystal growth rate S_g and the second moment m_2 by

$$S_i = \pm \rho S_g = \pm \rho (3m_2 G) k_v \frac{\rho_{BaSO_4}}{M_{BaSO_4}} \tag{6.31}$$

with a minus sign for Ba^{2+} and SO_4^{2-}. For nonreacting ions Na^+ and Cl^-, S_i is set to zero.

6.3.2 Nucleation and growth kinetics

Using Nielsen's (1964) experimental data of $BaSO_4$ precipitation, Baldyga et al. (1995) developed an expression for the nucleation rate considering both heterogeneous and homogeneous nucleation. This expression was corrected by Cheng et al. (2012a) to make this expression continuous mathematically at $\Delta c = 10 \text{ mol/m}^3$, which is the very point distinguishing between heterogeneous and homogeneous nucleation. Figure 6.3 shows Nielsen's original experimental data and the new expression. The nucleation rate B is then given as

$$B = \begin{cases} 2.83 \times 10^{10} \Delta c^{1.775}, & \text{if } \Delta c \leq 10 \text{ mol/m}^3 \text{ (heterogeneous)} \\ 2.33 \times 10^{-2} \Delta c^{13.86}, & \text{if } \Delta c > 10 \text{ mol/m}^3 \text{ (homogeneous)} \end{cases} \tag{6.32}$$

where $\Delta c = (S-1)\sqrt{K_{sp}}$, K_{sp} is the solubility product, which is $1.14 \times 10^{-4} \text{ mol}^2/\text{m}^6$ at room temperature for $BaSO_4$, and S is the supersaturation ratio, defined as $S = \sqrt{c_A c_B / K_{sp}}$.

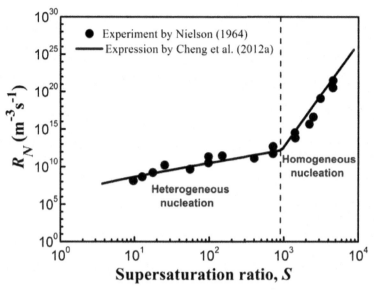

FIGURE 6.3 Nucleation rate expression (Cheng et al., 2012a).

Several growth rate expressions for $BaSO_4$ have been used in the literature. The local values of growth rate G can be computed from the classical relation:

$$G = k_g (S_a - 1)^2 \qquad (6.33)$$

where $k_g = 4.0 \times 10^{-11}$ m/s (Wei and Garside, 1997), $S_a = \gamma_{ac}S$, and γ_{ac} is the activity coefficient computed using Bromley's method (Bromley, 1973), which provided accurate data up to an ionic strength of 6 M. The two-step model including the surface integration step and the molecular diffusion step has also been employed in some studies (Marchisio et al., 2002; Wang et al., 2006a, 2007; Baldyga et al., 2007).

$$G = k_r \left(\sqrt{c_{As}c_{Bs}} - \sqrt{K_{sp}} \right)^2 = k_d(c_A - c_{As}) = k_d(c_B - c_{Bs}) \qquad (6.34)$$

where c_{As} and c_{Bs} are the reactant concentrations on the crystal surface, k_r is a kinetic constant equal to 5.8×10^{-8} (m/s)/(m³/mol)² (Nielsen and Toft, 1984; Marchisio et al., 2002), and k_d is the mass transfer coefficient with a constant value of 10^{-7} (m/s)/(m³/mol) (Nagata and Nishikawa, 1972; Wang et al., 2007). For given values of c_A and c_B, the growth rate G can be found by solving Eq. (6.34) using the Newton–Raphson method.

6.3.3 Aggregation and breakage kernels

Particles must be brought into close proximity by a transport mechanism, to produce collision, and then the aggregation can occur. Brownian motion (usually for particles smaller than 1 μm) and flow shear (for particles in the range 1–50 μm) are the two main controlling mechanisms, which induce collision for precipitated particles. The collision kernel functions for Brownian and shear-induced collisions are expressed as (Saffman and Turner, 1956)

$$Q_{Br}(L_i, L_j) = \frac{2k_B T}{3\mu} \frac{(R_i + R_j)^2}{R_i R_j} \tag{6.35}$$

and

$$Q_{Fl}(L_i, L_j) = 1.29 G_{sh}(R_i + R_j)^3 \tag{6.36}$$

respectively, where L_i and L_j are the characteristic particle sizes, R_i and R_j are the corresponding collision radiuses, k_B is the Boltzmann constant, and $G_{sh} = (\varepsilon/v)^{1/2}$ is the characteristic velocity gradient (shear rate) of the flow field. It is assumed that the two collision mechanisms are linearly additive to give the overall collision rate as

$$Q(L_i, L_j) = Q_{Br}(L_i, L_j) + Q_{Fl}(L_i, L_j) \tag{6.37}$$

Not all collisions result in successful aggregation due to hydrodynamic interaction, resistance forces from the viscous fluid layer, and/or insufficient time for restructuring after collision. All these are reflected in the collision efficiency $\alpha(L,\lambda)$. Several models have been developed in the literature to calculate the collision efficiency of porous aggregates (permeable floc models; Bäbler, 2008). In stirred precipitation systems, aggregates are compact because they are subject to strong flow shear. Thus, the collision efficiency calculated from the impermeable flocs model can be adopted (Cheng et al., 2009). The aggregation rate can be calculated as $\beta(L,\lambda) = \alpha(L,\lambda)Q(L,\lambda)$.

The hydrodynamic stress-induced fragmentation is most commonly encountered in precipitation systems. Several expressions for the breakage rate kernel have been developed by different authors and are summarized by Marchisio et al. (2003b). A semi-theoretical expression that has been found to apply to a wide variety of fragmentation phenomena is the power-law breakage kernel (Wójcik and Jones, 1998; Kramer and Clark, 1999):

$$\psi(L) = c_1 v^x \varepsilon^y L^\gamma \tag{6.38}$$

where c_1 is a dimensionless empirical constant. By fitting the model to experimental data, Peng and Williams (1994) found that the exponent γ can be assumed to be between 1 and 3, with $\gamma = 2$ usually being taken. For the fragment distribution functions, symmetric fragmentation, erosion and uniform distribution, etc. have been used in the literature (Marchisio et al., 2003a). The commonly used uniform distribution is given as

$$b(L|\lambda) = \begin{cases} 6L^2/\lambda^3 & \text{if } 0 < L < \lambda \\ 0 & \text{otherwise} \end{cases} \tag{6.39}$$

$$\overline{b}_i^{(k)} = L_i^k \frac{6}{k+3} \tag{6.40}$$

6.3.4 Computational details

The simulations undertaken in the literature were programmed with FORTRAN, as in Wang et al. (2006a, 2007), Cheng et al. (2009), and Zhang et al. (2009), or by means of some commercial codes like Fluent (Ansys, Inc.), as in Marchisio et al. (2002), Jaworski and Nienow (2003), Öncül et al. (2005), Gavi et al. (2007), Guo et al. (2009), Cheng et al. (2012a) and references therein. The three-dimensional flow field in the stirred tank is first obtained by solving the single-phase Reynolds-averaged Navier–Stokes equations. The correlation between the rotating impeller and the stationary baffles is treated with the multi-reference frame (MRF). Generally, the single-phase standard k–ε or RNG k–ε turbulence model is applied. The coupling of velocity and pressure is resolved using SIMPLE or SIMPLEC. A grid independence check can be conducted by comparing the flow fields from different numbers of grids or by means of boundary and gradient adaptation. After reaching convergence with the residuals of the flow equations being well below 10^{-4}, the velocity and turbulence fields are saved and kept unchanged for subsequent simulations of precipitation. Figure 6.4 illustrates the flow field and the distribution of turbulent energy dissipation rates (Cheng et al., 2009).

On the basis of the already known flow field, the species transport equation (6.30) and the PBE equations will be solved. For the SMM, Eq. (6.23) is used and generally five moment equations are solved to get the zeroth- to fourth-order moments of the CSD. For the QMOM, Eq. (6.25) is used and the number of moment equations being solved is dependent on the nodes used, generally two nodes ($N_d = 2$) or three nodes ($N_d = 3$). By applying the QMOM with N_d nodes, the first $2N_d$ moments will be tracked. For example, if $N_d = 3$, the first six moments, from m_0 to m_5, are calculated.

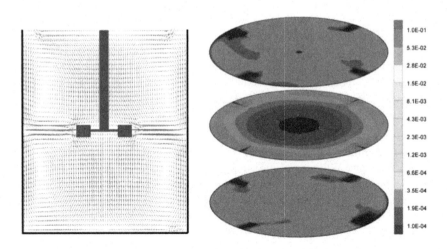

(a) Flow field   (b) Turbulent energy dissipation rate

FIGURE 6.4 (a) Flow field in a vertical cross-section. (b) Distribution of turbulent energy dissipation rate in horizontal cross-sections ($T = 0.27$ m, $H = 0.27$ m, standard Rushton turbine with $D = T/3$, $N = 120$ rpm, $\tau = 430$ s) (Cheng et al., 2009).

For the MCM, Eq. (6.27) is used. The number of discretized equations is dependent on the number of size bins, e.g., if 36 size bins are adopted, then 40 equations need to be solved simultaneously (36 size bin equations plus four species transport equations). Because of the stiffness of nucleation, smaller relaxation factors are used for the species equations of Ba^{2+} and Cl^- and the first size bin equation. When using Fluent, the PBE and species transport equations are incorporated into Fluent through user-defined scalars (UDSs) and user-defined functions (UDFs). The convergence criterion is that the residuals of all PBE and species transport equations are below 10^{-6}. Aggregation and breakage processes can be considered with the QMOM and the MCM. The MCM is more time-consuming compared to the SMM and the QMOM, especially when secondary terms such as aggregation and breakage are considered, since the discretized expression of the aggregation term in Eq. (6.27) in the MCM is involved with a triple loop of aggregation (see Cheng et al., 2012a).

6.3.5 Simulated results of precipitation processes

In this section, some typical simulated results of $BaSO_4$ precipitation in published work are given. By solving the species transport equations and the PBE equations, the spatial distributions of supersaturation, nucleation, and growth can be obtained. Figure 6.5 illustrates the distribution of local supersaturation ratio s_a at different

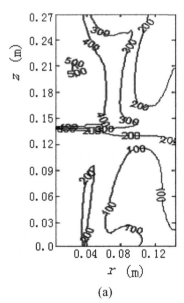
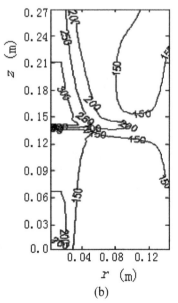

(a)   (b)

FIGURE 6.5 Local supersaturation ratio s_a distribution at different impeller speeds ($C = T/2$, $\tau = 430$ s, semi-batch, feed point near the surface, $C = 0.10$ kmol/m³) (Wang et al., 2006a).

(a) $N = 200$ rpm. (b) $N = 500$ rpm.

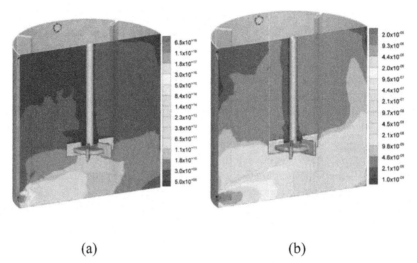

(a)   (b)

FIGURE 6.6 Distributions of nucleation rate and growth rate (pre-mixed feed, $C = 30$ mol/m^3, $\tau = 63$ s, $N = 372$ rpm) (Cheng et al., 2012a).

(a) Nucleation rate. (b) Growth rate.

impeller speeds. It is seen that s_a becomes more uniform and its mean value decreases as the impeller speed increases, due to the better macroscopic mixing in the reactor induced by stronger liquid circulation and intensified turbulence (Wang et al., 2006a).

Figure 6.6 gives the distributions of nucleation rate and growth rate across a stirred tank (Cheng et al., 2012a). Extensive nucleation and growth occur in the near vicinity of the feed inlet near the bottom in the case of premixed precipitation.

By implementing the SMM, Wang et al. (2006a) simulated the precipitation of BaSO$_4$, and the effects of feeding location, feed concentration, impeller speed and residence time were investigated. Figure 6.7 shows the effect of impeller speed on the mean particle size, d_{32}, and on the coefficient of variation (C.V.). It is seen that d_{32} increases with impeller speed. Other work has shown that the mean size increases with the agitation speed and reaches a plateau at large agitation speeds (Pohorecki and Baldyga, 1988; Vicum and Mazzotti, 2007). However, in the premixed precipitation cases, the agitation speed has very limited influence on d_{32}, which even decreases a little with the impeller speed when the speed is large (Cheng et al., 2012a).

Some other researchers have successfully simulated the precipitation of BaSO$_4$ using the SMM (Baldyga and Orciuch, 1997; Jaworski and Nienow, 2003; Öncül et al., 2005). The residence time and the crystal shape factor significantly affect the local supersaturation and the volume-averaged crystal size. Turbulent mixing is another important factor, as shown by Marchisio et al. (2002) and Vicum and Mazzotti (2007).

Generally, in the above works discrepancies between predicted results and experimental data become larger particularly at higher concentrations, due partly to the neglect of aggregation. The aggregation was clearly seen in BaSO$_4$ precipitation processes (Wong et al., 2001; Marchisio et al., 2002; Kucher et al., 2006). Using the

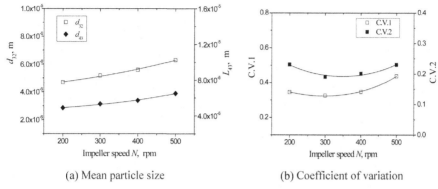

(a) Mean particle size   (b) Coefficient of variation

FIGURE 6.7 Effects of impeller speed on mean particle size (a) and coefficient of variation (b) (Wang et al., 2006a).

QMOM, Gavi et al. (2007) studied the precipitation of $BaSO_4$ in a confined imping-ing jet reactor. Predictions were found to be in good agreement with experimen-tal data considering only the Brownian motion-induced aggregation. Cheng et al. (2012a) also showed that the predicted results can be considerably improved when choosing appropriate aggregation and breakage kernels.

It has been shown that the QMOM with two nodes can work with acceptable ac-curacy for aggregation and breakage problems (Marchisio et al., 2003b). Also, it can be seen in Figure 6.8 that when neglecting secondary terms, the predictions from the

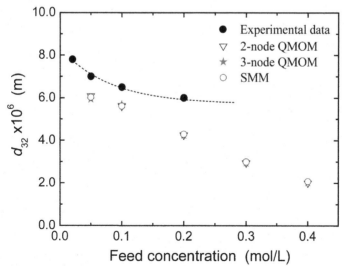

FIGURE 6.8 Comparison between predicted d_{32} at different feed concentrations using SMM and two- and three-node QMOM and experimental data ($T = 0.27$ m, $H = 0.27$ m, standard Rushton turbine with $D = T/3$, $N = 120$ rpm, $\tau = 430$ s) (Cheng et al., 2009).

SMM and two-node and three-node QMOM are very close (Cheng et al., 2009). The predictions from two-node and three-node QMOM are also very close (the error for m_0 is below 0.4%).

The published work on CFD–PBE simulation with the discretized method in reacting systems (e.g., precipitation) is limited, but more comprehensive and in-depth work in this aspect is needed. Mühlenweg et al. (2002) simulated the gas-phase synthesis of nanoparticles in an ideal plug flow reactor. Their work shows that CFD coupled with the MCM is principally feasible. Woo et al. (2006) coupled the high-resolution FVS with CFD to simulate the antisolvent crystallization in a 2D stirred tank. The effects of different operating conditions and scale-up rules on the full CSD have been well analyzed qualitatively. Veroli and Rigopoulos (2010) presented a framework for modeling turbulent precipitation using a coupled discretized PBE–transported PDF (probability density function) method. The authors applied this method to the precipitation of $BaSO_4$ in a 2D turbulent pipe flow and made a comparison with the published experimentally measured full CSDs, showing good agreement both in size and in shape.

The above work of the CFD–PBE simulations with the MCM is limited to 2D flow field with the important aspect of aggregation neglected. Thus, using the first-order upwind scheme for the growth term and the fixed pivot for the aggregation, Cheng et al. (2012a) simulated the precipitation process of $BaSO_4$ in a 3D stirred tank. Figure 6.9 illustrates the normalized volumetric size distributions for different-sized bin numbers in the cases with and without aggregation (Cheng et al., 2012a). It can be found that for the MCM, the finer the grid size, the more accurate the solution, but the greater computational cost it demands. However, it is necessary to find a tradeoff between accuracy and computational cost. It is also seen in Figure 6.9 that the finer the grid, the larger the peak value of the CSD and the smaller and shorter the tail of the CSD is.

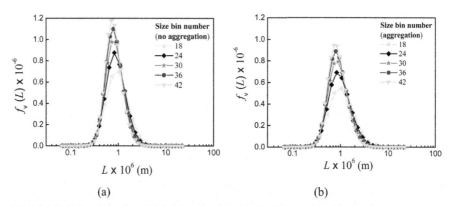

(a)   (b)

FIGURE 6.9 Volumetric size distribution simulated with different-size bin numbers (Cheng et al., 2012a).

(a) Without aggregation. (b) Aggregation considered.

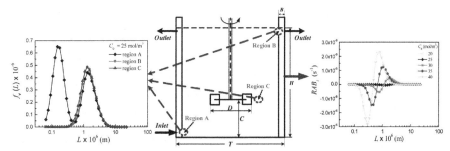

FIGURE 6.10 CSDs of three investigated regions (A, B, and C) and the net birth rate of the *i*th size bin due to aggregation at different concentrations (Cheng et al., 2012a).

Figure 6.10 gives the CSDs at different spatial locations and the net birth rate of the *i*th size bin due to aggregation at different concentrations (Cheng et al., 2012a). Bimodal distributions are found at some locations, which cannot be achieved through the SMM or the QMOM. With the MCM, not only is the integral effect of aggregation on the CSD seen, but also the aggregation rate for each size bin is calculated.

The MCM with aggregation incorporated is highly computation-intensive. Moreover, the lower peak and widening/lengthening of the tail of the CSD are still problems, due partly to the treatment of the growth term. Although detailed information cannot be obtained using moment methods, they are useful for sophisticated control of some processes like nanoparticle production through precipitation.

6.4 MACROMIXING AND MICROMIXING

Precipitation is a fast reaction-related process and the mixing in precipitators may be the rate-controlling step. Mixing problems, which are a key issue in deciding the conversion and selectivity in complex chemical systems, are unavoidable for these systems, especially when scaling up. The relative importance of chemical reactions to mixing is generally represented by comparing their respective characteristic time scales, as in Fox (2003). The reaction crystallization such as the barium sulfate precipitation is usually of the order of 10^{-8} to 10^{-9} s, which is several orders of magnitude smaller than the fastest mixing time. Thus mixing, especially micromixing, plays a key role in the precipitation process. Micromixing affects the rates of chemical reactions, and the subsequent nucleation and crystal growth processes.

Macromixing performance is commonly characterized by blending time both in single-phase and two-phase systems, on which tracer experiments have extensively been conducted (see, for example, Wang et al., 2006b). In a tracer experiment, a certain quantity of tracer is injected at some locations in a reactor and the tracer concentration as a function of time is monitored to obtain the mixing time. CFD simulations of macromixing have been conducted by solving the flow field together with the tracer transport equation based on the already resolved flow field (Ranade and Bourne, 1991; Jaworski et al., 2000; Wang et al., 2006b), which can be easily implemented using available CFD codes. However, introduction of a dispersed phase will pose severe difficulties in modeling multiphase flow fields. Thus, predicting multiphase flow fields with reasonable accuracy is still challenging. Micromixing experiments have been commonly carried out on selected chemical test reactions (Fournier et al., 1996), such as consecutive competing and parallel competing reactions. Unlike single-phase micromixing experiments, where several test reaction systems have been employed, two-phase micromixing experiments basically use the iodide/iodate and diazo-coupling methods, as in Villermaux and Fournier (1994), Hofinge et al. (2011), Yang et al. (2013) and references therein, since for other systems new problems can arise (Cheng et al., 2012b). The micromixing efficiency is commonly characterized using the segregation index X_Q. If micromixing is perfect, X_Q is 0; if it is total segregation, X_Q is 1. Any value between the two extremities represents partial segregation. Figure 6.11 shows experimental results in a solid–liquid stirred tank (Yang et al., 2013). Generally, the higher the stirrer speed to enhance turbulence, the smaller the value of X_Q. The influence of the existence of particles on micromixing is also illustrated in this figure.

To describe the micromixing process numerically, a suitable micromixing model must be incorporated. Several micromixing models have been proposed and applied in a single-phase system, e.g., the exchange-with-the-mean (IEM) model (David and Villermaux, 1987), the multi-environment model (Villermaux and Zoulalian,1969), the engulfment–deformation–diffusion (EDD) model (Baldyga and Bourne, 1984), the engulfment model (E-model) (Baldyga and Bourne, 1989), and the PDF (probability density function)-related models (the presumed PDF model and the transported PDF model) (Villermaux and Falk, 1994; Fox, 1998; Marchisio et al., 2001). However, studies on two-phase micromixing from modeling and numerical perspectives are still lacking.

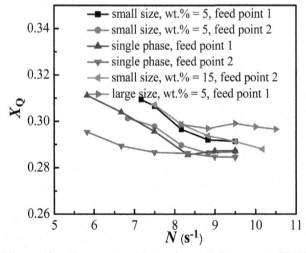

FIGURE 6.11 Effects of impeller speed and feed point on X_Q (Yang et al., 2013).

The IEM model has been used to study the effect of micromixing on crystal sizes in precipitation by Pohorecki and Baldyga (1979) and Garside and Tavare (1985). The environment model is another commonly used model for micromixing in the precipitation process, where it is assumed that the vessel volume consists of two or more separate environments having extreme states of micromixing: completely segregated or well mixed. The EDD model has been widely employed to study the mixing in the barium sulfate precipitation process in recent years (Phillips et al., 1999). In the E-model, the vessel volume is divided into two zones, i.e. the mixing-precipitation zone and the environment zone. All reactants take part in reactions in both zones. For the presumed PDF methods, the β-PDF and finite-mode PDF (FM-PDF) models have been widely used in single and complex reactions and in precipitation processes (Marchisio et al., 2001, 2002; Woo et al., 2006; Gavi et al., 2007; Wang et al., 2007). The presumed PDF methods are implemented and solved with relative ease, and their incorporation into existing popular CFD models is straightforward. To describe the micromixing-sensitive precipitation reaction process, a complete CFD–PBE–PDF model is commonly required.

The FM-PDF model is widely chosen to model micromixing. Every cell in the computational domain contains N different modes or environments, which correspond to the discretization of the composition PDF in a finite set of delta function (Fox, 1998):

$$f_{\varphi}(\psi;x,t) = \sum_{n=1}^{N} p_n(x,t) \prod_{\alpha=1}^{m} \delta\left[\psi_\alpha - \langle\varphi_\alpha\rangle_n(x,t)\right] \tag{6.41}$$

where $f_{\varphi}(\psi; \mathbf{x}, t)$ is the joint PDF of all scalars, concentrations and moments, etc. appearing in the precipitation model; N is the number of modes; $p_n(x,t)$ is the probability of mode n; $\langle\phi_\alpha\rangle_n(x,t)$ is the value of scalar α corresponding to mode n; and m is the total number of scalars. By definition, the probabilities, p_n, sum to unity, and the average value of any scalar is defined by integration with respect to ψ. For the system under consideration, knowledge of the mixture fraction and a reaction progress variable suffices to predict the reactant concentration in each environment. Thus, the first scalar is specified to the mixture fraction $\phi_1(x,t) \equiv \zeta(x,t)$. From Eq. (6.41), the average value of the mixture fraction is given by

$$\langle\xi\rangle = \sum_{n=1}^{N} p_n \langle\xi\rangle_n \tag{6.42}$$

Previous results (Piton et al., 2000) showed that three modes were sufficient to work with good accuracy. The three-mode PDF model is taken for discretization of the reacting system in three environments, where environments 1 and 2 correspond to unmixed reactants A and B respectively, and environment 3 corresponds to the environment in which reactants A and B are mixed well and react.

At steady state, the scalar transport equations for the probabilities of modes 1 and 2, and for the weighted mixture fraction in environment 3, $s_3 \equiv p_3\langle\xi\rangle_3$, are

$$\langle u_j\rangle \frac{\partial p_1}{\partial x_j} = \frac{\partial}{\partial x_j}\left(\Gamma_{\text{eff}} \frac{\partial p_1}{\partial x_j}\right) + \gamma_s p_3 - \gamma p_1(1-p_1) \tag{6.43}$$

$$\langle u_j \rangle \frac{\partial p_2}{\partial x_j} = \frac{\partial}{\partial x_j}\left(\Gamma_{\text{eff}} \frac{\partial p_2}{\partial x_j}\right) + \gamma_s p_3 - \gamma p_2 (1 - p_2) \tag{6.44}$$

$$\langle u_j \rangle \frac{\partial s_3}{\partial x_j} = \frac{\partial}{\partial x_j}\left(\Gamma_{\text{eff}} \frac{\partial s_3}{\partial x_j}\right) - \gamma_s p_3 (\langle \xi \rangle_1 + \langle \xi \rangle_2) + \tag{6.45}$$
$$\gamma p_1 (1 - p_1)\langle \xi \rangle_1 + \gamma p_2 (1 - p_2)\langle \xi \rangle_2$$

where $p_3 = 1 - p_1 - p_2$, Γ_{eff} is the turbulent diffusivity, $\langle u_j \rangle$ is the mean velocity in the j direction, and γ and γ_s are the micromixing rate and the spurious dissipation rate respectively:

$$\gamma = C_\phi \frac{\varepsilon}{k} \frac{\langle \xi'^2 \rangle}{[p_1 (1 - p_1)(1 - \langle \xi \rangle_3)^2 + p_2 (1 - p_2)\langle \xi \rangle_3^2]} \tag{6.46}$$

$$\gamma_s = \frac{2\Gamma_{\text{eff}}}{1 - 2\langle \xi \rangle_3 (1 - \langle \xi \rangle_3)} \frac{\partial \langle \xi \rangle_3}{\partial x_j} \frac{\partial \langle \xi \rangle_3}{\partial x_j} \tag{6.47}$$

where C_ϕ is a constant of the order of unity and $\langle \xi'^2 \rangle$ is the mixture fraction variance, defined as

$$\langle \xi'^2 \rangle = \langle \xi^2 \rangle - \langle \xi \rangle^2 \tag{6.48}$$

In Eq. (6.48), $\langle \xi^2 \rangle$ is the second moment of the mixture fraction defined by

$$\langle \xi^2 \rangle = \sum_{n=1}^{N} p_n \langle \xi^2 \rangle_n \tag{6.49}$$

As environments 1 and 2 contain only reactant A or B respectively, the mixture fractions are $\langle \xi \rangle_1 = 1$ and $\langle \xi \rangle_2 = 0$. Thus, Eq. (6.42) can be written as

$$\langle \xi \rangle = p_1 + s_3 \tag{6.50}$$

whereas the mixture fraction variance can be given by

$$\langle \xi'^2 \rangle = p_1 + \frac{s_3^2}{p_3} - \langle \xi \rangle^2 \tag{6.51}$$

To obtain a simpler expression, the population balance can be expressed in terms of the SMM method. It should be noted that nucleation and growth occur only in environment 3. Using this approach, the governing equations for the CSD moments at steady state are

$$\frac{\partial}{\partial x_j}\left(\langle u_j \rangle \overline{m_0}\right) = \frac{\partial}{\partial x_j}\left(\Gamma_{\text{eff}} \frac{\partial \overline{m_0}}{\partial x_j}\right) + B\left(\langle c_A \rangle_3, \langle c_B \rangle_3\right) p_3 \tag{6.52}$$

$$\frac{\partial}{\partial x_j}\left(\langle u_j \rangle \overline{m_k}\right) = \frac{\partial}{\partial x_j}\left(\Gamma_{\text{eff}} \frac{\partial \overline{m_k}}{\partial x_j}\right) + kG\left(\langle c_A \rangle_3, \langle c_B \rangle_3\right)\overline{m_{k-1}}, \quad k = 1-4 \tag{6.53}$$

where $\overline{m}_j$ is the mean value of the jth moment of the particle number density function $\overline{m}_j = p_3 \langle m_j \rangle_3$, and $\langle c_A \rangle_3$ and $\langle c_B \rangle_3$ are the local reactant concentrations in environment 3 that can be calculated by introducing the reaction progress variable Y:

$$\frac{c_A}{c_{A0}} = \xi - \xi_s Y, \quad \frac{c_B}{c_{B0}} = (1 - \xi) - (1 - \xi_s)Y \qquad (6.54)$$

$$\frac{\partial}{\partial x_j}\left(\langle u_j \rangle \langle Y \rangle\right) = \frac{\partial}{\partial x_j}\left(\Gamma_{\text{eff}}\frac{\partial \langle Y \rangle}{\partial x_j}\right) + \frac{\rho_3 k_V G \overline{m}_2}{M \xi_s c_{A0}} \qquad (6.55)$$

where c_{A0} and c_{B0} are the inlet concentrations of the two reactants in their separate feed streams, ρ is the crystal density, k_V is the crystal shape factor, and M is the crystal molecular weight.

The presumed PDF methods are implemented and solved with relative ease, and their incorporation into existing popular CFD models is straightforward. Therefore, CFD–PBE–PDF has been widely used to simulate the precipitation of BaSO$_4$ in tubular or/and stirred reactors (Marchisio et al., 2001; Marchisio and Barresi, 2003; Gavi et al., 2007; Wang et al., 2007; Zhang et al., 2009). Impeller speed is a major parameter affecting the mixing intensity and thus the mean crystal size in a stirred tank. However, the literature reports controversially that an increase in impeller speed may increase, decrease, produce a minimum, or not affect the mean crystal size at all (Wang et al., 2007). Figure 6.12 gives some experimental and predicted results of BaSO$_4$ precipitation in a continuous stirred tank at different impeller speeds

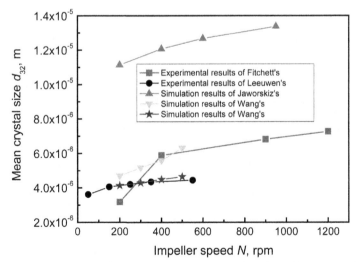

FIGURE 6.12 Two sets of experimental data and corresponding simulation results of BaSO$_4$ precipitation at different impeller speeds by some researchers ($T = H = 0.27$ m, Rushton turbine, $D = T/3$, $\tau = 430$ s, $C = 0.10$ kmol/m^3) (Wang et al., 2007).

in the literature, as summarized by Wang et al. (2007). It is seen that the crystal size increases with the impeller speed. The reason might be that when the impeller speed is increased, the local supersaturation level is lowered and the growth is relatively favored at lower supersaturations, while nucleation is damped more seriously.

6.5 SUMMARY AND PERSPECTIVE

Studies of crystallization and crystallizers cover a vast range of aspects, thus in this chapter we focus particularly on the modeling of crystallizers, especially the reactive crystallization processes. In Section 6.1, the commonly used numerical methods for the PBE are described. In Section 6.2, the general PBE and some schemes like the standard method of moments, the quadrature method of moments and the multi-class method are described in detail. The strong and weak points of each type of scheme are discussed. Since precipitation is a fast reaction process and the mixing problem is unavoidable, in Section 6.4 we discussed macromixing and micromixing: experimental and numerical studies on macromixing, and experimental work on micromixing in single- and two-phase systems are considered. The moment method, especially the QMOM and its extensions, is the most widely used in CFD–PBE modeling, particularly in reacting systems. CFD–PBE modeling with the multi-class method has been used mainly in nonreacting systems, such as bubble coalescence and breakage in columns and stirred tanks, but its application in reacting systems is limited to simplified cases, though its potential is considerable. Some relatively simplified models, such as Brownian and shear-induced aggregation kernels and empirical breakage kernels, have been employed to describe the secondary processes in crystallization modeling. In many cases, the accuracy of simulation can be improved; however, selection and usage of these models must be done with care since the theoretical background and validations of these models are still limited.

Studies of crystallization and crystallizers in China have achieved great progress. Fundamental research on crystallization chemistry and other aspects of molecular and microcrystal scales provides a sound basis for development of specific crystallization processes. Much attention has been paid to the determination of crystallization kinetics of pharmaceutical materials, mainly the nucleation and growth rates. For kinetic studies, methods based on measuring the induction time are widely used on the kinetics of dexametasone sodium phosphate in the ethanol–acetone system (Hao et al., 2005), cloxacillin sodium in the methanol–butyl acetate system (Zhi et al., 2011), and L-tryptophan in alcohols–water system (Chen et al., 2012), etc. Methods based on combining PBE (mainly the moment methods) and CSD measurement have also been employed to determine the kinetics, such as the determination of growth and breakage kinetics of L-threonine crystals by Bao et al. (2006). Another focus of fundamental research is measurement of thermodynamical data such as the solubility and metastable zone (Chen et al., 2012), and studying the influence of impurities on the crystallization of pharmaceuticals (Dang and Wei, 2010). Also, study of the control of crystal habit (or morphology) laid the scientific basis for developing the technologies for products of special functionality. Experimental work has been done to investigate the effects of solvent and impurity on the crystal habit of pharmaceutical products, and to provide basic information for solvent selection to obtain desired crystal habits (Nie et al., 2007). Recently, molecular modeling has been employed to predict the influence of ionic impurities and solvent molecules on crystal morphology modification (Dang and Wei, 2010; Gu et al., 2013). However, study of the mechanism of control of crystalline morphology and crystal habit is still inadequate, especially the chemical engineering manipulation for obtaining the desired morphology.

The purpose of designing a crystallizer is to realize the chemical engineering conditions designated by crystallochemical research in laboratories through the measures of macroscopic operation and control. Therefore, macro-scale multiphase hydrodynamic environments that favor the controlled crystallization process should be implemented. At present, correct and accurate description of the micro-scale phenomenon in macroscopic environments and macro-models is still a great challenge facing chemists and chemical engineers. The research of micromixing could be the way to bridge the gap between the micro-phenomena and the macro-models, but this is based on consummate CFD–PBE modeling. Generally, when scaling up the crystallization process from lab to pilot and to industrialization, PBE modeling, mass balance plus energy balance on macro-scale and some experience are still the main ways despite inevitable uncertainties or risks. It is expected that CFD–PBE modeling will find vast applications in research and development efforts. Since size and shape are both key parameters for pharmaceutical crystals, two-dimensional CFD–PBE modeling in this respect is needed.

Numerical implementation of CFD–PBE modeling demands continuing efforts. Due to the limitations of the SMM, the QMOM will be one of the major moment methods, and improvements in the QMOM are being made, such as the conditional quadrature method of moments (CQMOM) and the sectional quadrature method of moments (SQMOM) (Attarakih et al., 2009). The MCM coupled with CFD is computationally intensive. However, with the development of computer technology, it will become more important and useful in the future due to its capabilities of precisely predicting the full CSD, which is very important in the pharmaceutical crystallization and precipitation processes. For crystallization modeling, the aggregation and breakage models are important because, for large crystals, attrition and breakage in dense crystal slurries are unavoidable, and for precipitated crystallites, aggregation is commonly the key factor determining the final quality of the product. Available aggregation and breakage models contain many uncertainties and empirical parameters, so that they cannot be used generally. Therefore, in-depth studies on the collision frequency, collision efficiency, breakage rate, and fragmentation distribution will continue in the future. Moreover, more efficient formulations for the growth term in the MCM need further attention. In this chapter, only one-dimensional PBE modeling is discussed; however, multi-dimensional PBE modeling seems to be more valuable and it will receive more attention, especially in the pharmaceutical crystallization processes, because size and shape are both key factors in determining the bioavailability of the final capsules. Thus, extending one-dimensional PBE to multi-dimensional PBE and development of corresponding algorithms are necessary. Actually, some work has been done in this respect, such as extensions of the fixed pivot to the two-dimensional case (Nandanwar and Kumar, 2008; Chauhan et al., 2010) and the extension of the cell average technique to two dimensions by Kumar et al. (2011), with focus on homogeneous systems.

NOMENCLATURE

$a(v,g_l)$	particle assignment coefficient	–
$b(v,g_l)$	particle assignment coefficient	–
$b(L\|\lambda)$	fragmentation distribution function	m^{-1}
$B(\mathbf{x},t)$	nucleation rate	$\#/(m^3 \cdot s)$
$B^a(L)$	birth rate due to aggregation	$\#/(m^4 \cdot s)$
$B^b(L)$	birth rate due to breakage	$\#/(m^4 \cdot s)$
$\overline{B}^a_k$	kth moment transformation of birth term due to aggregation	$\# \cdot m^k/(m^3 \cdot s)$
$\overline{B}^b_k$	kth moment transformation of birth term due to breakage	$\# \cdot m^k/(m^3 \cdot s)$
c_i	molar concentration of chemical entity i	mol/m^3
D	diameter of Rushton turbine impeller	m
$D^a(L)$	death rate due to aggregation	$\#/(m^4 \cdot s)$
$D^b(L)$	death rate due to breakage	$\#/(m^4 \cdot s)$
$\overline{D}^a_k$	kth moment transformation of death term due to aggregation	$\# \cdot m^k/(m^3 \cdot s)$
$\overline{D}^b_k$	kth moment transformation of death term due to breakage	$\# \cdot m^k/(m^3 \cdot s)$
$f_\phi(\psi;x,t)$	joint PDF of all scalars	–
$G(L)$	linear growth rate	m/s
G_{sh}	fluid shear rate	s^{-1}
H	height of the tank	m
k_B	Boltzmann constant	J/K
k_d	mass transfer coefficient	$mol/(m^2 \cdot s)$
K_{sp}	solubility product	mol^2/m^6
L	particle size	m
L_i	abscissa (or node) of quadrature approximation	m
m_k	kth moment of crystal size distribution	$\# \cdot m^k/m^3$
$n(L;\mathbf{x},t)$	number density function	$\#/m^4$
$p_n(x,t)$	probability of model n	–
$Q(L,\lambda)$	collision rate	m^3/s
S	supersaturation ration, dimensionless	
S_g	specific crystal growth rate	$1/(m^3 \cdot s)$
Sc_t	turbulent Schmidt number	–
t	time	s
$\mathbf{u}$	Reynolds-averaged velocity vector	m
w_i	weight of quadrature approximation	–

x	space vector	m
Greek letters		
α_i	volume fraction of the ith size bin per unit volume	–
$\alpha(L,\lambda)$	collision efficiency	–
$\beta(L,\lambda)$	aggregation rate	m^3/s
γ_{ac}	activity coefficient	–
γ_s	spurious dissipation rate	–
$\langle \phi_\alpha \rangle_n (x,t)$	value of scalar corresponding to mode n	
Γ_{eff}	effective diffusivity	m^2/s
ε	turbulent energy dissipation rate	m^2/s^3
μ	dynamic viscosity of the fluid	Pa·s
ν	kinematic viscosity	m^2/s
ρ	density	kg/m^3
$\psi(L)$	breakage kernel	$1/s$
ξ	property vector specifying the state of particle	–
ζ_i	ith component of property vector	–
ζ_j	jth component of flux in ξ-space	–
Subscripts		
Br	Brownian	
eff	effective	
Fl	flow	
sh	shear	

REFERENCES

Attarakih, M. M., Drumm, C., & Bart, H. -J. (2009). Solution of the population balance equation using the sectional quadrature method of moments (SQMOM). *Chem. Eng. Sci.*, *64*, 742–752.

Bäbler, M. U. (2008). A collision efficiency model for flow-induced coagulation of fractal aggregates. *AIChE J.*, *54*(7), 1748–1760.

Baldyga, J., & Bourne, J. R. (1984). A fluid-mechanical approach to rubulent mixing and chemical reaction. Part 1. Inadequacies of available methods. *Chem. Eng. Commun.*, *28*, 231–241.

Baldyga, J., & Bourne, J. R. (1989). Simplification of micromixing calculations. I. Derivation and application of new model. *Chem. Eng. J.*, *42*, 83–92.

Baldyga, J., & Orciuch, W. (1997). Closure problem for precipitation. *Chem. Eng. Res. Des.*, *75*(A2), 160–170.

Baldyga, J., Podgorska, W., & Pohorecki, R. (1995). Mixing-precipitation model with application to double feed semibatch precipitation. *Chem. Eng. Sci.*, *50*(8), 1281–1300.

Baldyga, J., Makowski, Ł., & Orciuch, W. (2007). Double-feed semibatch precipitation effects of mixing. *Chem. Eng. Res. Des.*, *85*(5), 745–752.

Bao, Y., Zhang, J., Yin, Q., & Wang, J. (2006). Determination of growth and breakage kinetics of L-threonine crystals. *J. Cryst. Growth*, *289*, 317–323.

Bromley, L. A. (1973). Thermodynamic properties of strong electrolytes in aqueous solutions. *AIChE J.*, *19*, 313–320.

Chauhan, S. S., Chakraborty, J., & Kumar, S. (2010). On the solution and applicability of bivariate population balance equations for mixing in particle phase. *Chem. Eng. Sci.*, *65*, 3914–3927.

Chen, Q., Wang, J., & Bao, Y. (2012). Determination of the crystallization thermodynamics and kinetics of L-tryptophan in alcohols–water system. *Fluid Phase Equilib.*, *313*, 182–189.

Cheng, J., Yang, C., & Mao, Z. -S. (2012a). CFD–PBE simulation of premixed continuous precipitation incorporating nucleation, growth and aggregation in a stirred tank with multi-class method. *Chem. Eng. Sci.*, *68*(1), 469–480.

Cheng, J., Feng, X., Cheng, D., & Yang, C. (2012b). Retrospect and perspective of micro-mixing studies in stirred tanks. *Chinese J. Chem. Eng.*, *20*(1), 178–190.

Cheng, J. C., Yang, C., Mao, Z. -S., & Zhao, C. J. (2009). CFD modeling of nucleation, growth, aggregation, and breakage in continuous precipitation of barium sulfate in a stirred tank. *Ind. Eng. Chem. Res.*, *48*(15), 6992–7003.

Costa, C. B. B., Maciel, M. R. W., & Filho, R. M. (2007). Considerations on the crystallization modeling: Population balance solution. *Comput. Chem. Eng.*, *31*, 206–218.

Dang, L., & Wei, H. (2010). Effects of ionic impurities on the crystal morphology of phosphoric acid hemihydrate. *Chem. Eng. Res. Des.*, *88*, 1372–1376.

David, R., & Villermaux, J. (1987). Interpretation of micromixing effects on fast consecutive competing reactions in semibatch stirred tanks by a simple interaction-model. *Chem. Eng. Commun.*, *54*, 333–352.

Diemer, R. B., & Olson, J. H. (2002). A moment methodology for coagulation and breakage problems: part 2 – Moment models and distribution reconstruction. *Chem. Eng. Sci.*, *57*, 2211–2228.

Falope, G. O., Jones, A. G., & Zauner, R. (2001). On modelling continuous agglomerative crystal precipitation via Monte Carlo simulation. *Chem. Eng. Sci.*, *56*, 2567–2574.

Filbet, F., & Laurencot, P. (2004). Numerical simulation of the Smoluchowski coagulation equation. *SIAM J. Scientific Computing.*, *25*, 2004–2028.

Fournier, M. C., Falk, L., & Villermaux, J. (1996). A new parallel competing reaction system for assessing micromixing efficiency. Experimental approach. *Chem. Eng. Sci.*, *51*(22), 5053–5064.

Fox, R. O. (1998). On the relationship between Lagrangian micromixing models and computational fluid dynamics. *Chem. Eng. Process: Process Intensif.*, *37*, 521–535.

Fox, R. O. (2003). *Computational models for turbulent reacting flows*. Cambridge: Cambridge University Press.

Garside, J., & Tavare, N. (1985). Mixing, reaction and precipitation: Limits of micromixing in an MSMPR crystallizer. *Chem. Eng. Sci.*, *40*(8), 1485–1493.

Gavi, E., Rivautella, L., Marchisio, D. L., Vanni, M., Barresi, A. A., & Baldi, G. (2007). CFD modelling of nano-particle precipitation in confined impinging jet reactors. *Chem. Eng. Res. Des.*, *85*(5), 735–744.

Gimbun, J., Rielly, C. D., & Nagy, Z. K. (2009). Modelling of mass transfer in gas–liquid stirred tanks agitated by Rushton turbine and CD-6 impeller: A scale-up study. *Chem. Eng. Res. Des.*, *87*, 437–451.

Gu, H., Li, R., Sun, Y., Li, S., Dong, W., & Gong, J. (2013). Molecular modeling of crystal morphology of ginsenoside compound K solvates and its crystal habit modification by solvent molecules. *J. Cryst. Growth*, *373*, 146–150.

Gunawan, R., Fusman, I., & Braatz, R. D. (2008). Parallel high-resolution finite volume simulation of particulate processes. *AIChE J.*, *54*, 1449–1458.

Guo, S., Evans, D. G., Li, D., & Duan, X. (2009). Experimental and numerical investigation of the precipitation of barium sulfate in a rotating liquid film reactor. *AIChE J.*, *55*(8), 2024–2034.

Hao, H., Wang, J., & Wang, Y. (2005). Determination of induction period and crystal growth mechanism of dexamethasone sodium phosphate in methanol–acetone system. *J. Cryst. Growth*, *274*, 545–549.

Hofinge, J., Sharpe, R. W., Bujalski, W., Bakalis, S., Assirelli, M., Eaglesham, A., & Nienow, A. W. (2011). Micromixing in two-phase (g-l and s-l) systems in a stirred tank. *Can. J. Chem. Eng.*, *89*, 1029–1039.

Hounslow, M. J., Ryall, R. L., & Marshall, V. R. (1988). Discretized population balance for nucleation, growth, and aggregation. *AIChE J.*, *34*(11), 1821–1832.

Hulburt, H. M., & Katz, S. (1964). Some problems in particle technology. *Chem. Eng. Sci.*, *19*, 555–574.

Jaworski, Z., Bujalski, W., Otomo, N., & Nienow, A. (2000). CFD study of homogenization with dual Rushton turbines – comparison with experimental results. Part I: Initial studies. *Chem. Eng. Res. Des.*, *78*(3), 327–333.

Jaworski, Z., & Nienow, A. W. (2003). CFD modelling of continuous precipitation of barium sulphate in a stirred tank. *Chem. Eng. J.*, *91*, 167–174.

John, V., Angelov, I., Öncül, A. A., & Thévenin, D. (2007). Techniques for the reconstruction of a distribution from a finite number of its moments. *Chem. Eng. Sci.*, *62*, 2890–2904.

Kalani, A., & Christofides, P. D. (2002). Simulation, estimation and control of size distribution in aerosol processes with simultaneous reaction, nucleation, condensation and coagulation. *Comput. Chem. Eng.*, *26*(7–8), 1153–1169.

Koren, B. (1993). A robust upwind discretization method for advection, diffusion and source terms. In C. B. Vreugdenhill, & B. Koren (Eds.), *Numerical methods for advection–diffusion problems (Notes on Numerical Fluid Mechanics, Chap 5, Vol. 45, pp. 117–138)*. Braunschweig: Vieweg Verlag.

Kostoglou, M. (2007). Extended cell average technique for the solution of coagulation equation. *J. Colloid Interface Sci.*, *306*, 72–81.

Kramer, T. A., & Clark, M. M. (1999). Incorporation of aggregate breakup in the simulation of orthokinetic coagulation. *J. Colloid Interface Sci.*, *216*, 116–126.

Kucher, M., Babic, D., & Kind, M. (2006). Precipitation of barium sulfate: Experimental investigation about the influence of supersaturation and free lattice ion ratio on particle formation. *Chem. Eng. Process*, *45*, 900–907.

Kumar, J. (2006). *Numerical approximations of population balance equations in particulate systems*. Ph. D. thesis. Germany: Otto-von-Guericke University.

Kumar, J., Peglow, M., Warnecke, G., Heinrich, S., & Morl, L. (2006). Improved accuracy and convergence of discretized population balance for aggregation: The cell average technique. *Chem. Eng. Sci.*, *61*(10), 3327–3342.

Kumar, J., Peglow, M., Warnecke, G., & Heinrich, S. (2008). An efficient numerical technique for solving population balance equation involving aggregation, breakage, growth and nucleation. *Powder Technol.*, *182*(1), 81–104.

Kumar, R., Kumar, J., & Warnecke, G. (2011). Numerical methods for solving two-dimensional aggregation population balance equations. *Comput. Chem. Eng.*, *35*, 999–1009.

Kumar, S., & Ramkrishna, D. (1996a). On the solution of population balance by discretization I A fixed pivot technique. *Chem. Eng. Sci.*, *51*, 1311–1332.

Kumar, S., & Ramkrishna, D. (1996b). On the solution of population balance by discretization II. A moving pivot technique. *Chem. Eng. Sci.*, *51*, 1333–1342.

Kumar, S., & Ramkrishna, D. (1997). On the solution of population balance equations by discretization – III. Nucleation, growth and aggregation of particles. *Chem. Eng. Sci.*, *52*, 4659–4679.

Kurganov, A., & Tadmor, E. (2000). New high-resolution central schemes for nonlinear conservation laws and convection–diffusion equations. *J. Comput. Phys.*, *160*, 241–282.

Lim, Y. I., Le Lann, J. -M., Meyer, X. M., Joulia, X., Lee, G., & Yoon, E. S. (2002). On the solution of population balance equations (PBE) with accurate front tracking methods in practical crystallization processes. *Chem. Eng. Sci.*, *57*(17), 3715–3732.

Ma, D. L., Tafti, D. K., & Braatz, R. D. (2002). High-resolution simulation of multidimensional crystal growth. *Ind. Eng. Chem. Res.*, *41*, 6217–6223.

Mahoney, A. W., & Ramkrishna, D. (2002). Efficient solution of population balance equations with discontinuities by finite elements. *Chem. Eng. Sci.*, *57*(7), 1107–1119.

Marchal, P., David, R., Klein, J. P., & Villermaux, J. (1988). Crystallization and precipitation engineering – I. An efficient method for solving population balance in crystallization with agglomeration. *Chem. Eng. Sci.*, *43*(1), 59–67.

Marchisio, D. L., & Barresi, A. A. (2003). CFD simulation of mixing and reaction: The relevance of the micro-mixing model. *Chem. Eng. Sci.*, *58*, 3579–3587.

Marchisio, D. L., & Fox, R. O. (2005). Solution of population balance equations using the direct quadrature method of moments. *J. Aerosol Sci.*, *36*, 43–73.

Marchisio, D. L., Barresi, A. A., & Fox, R. O. (2001). Simulation of turbulent precipitation in a semi-batch taylor-couette reactor using CFD. *AIChE J.*, *47*(3), 664–676.

Marchisio, D. L., Barresi, A. A., & Garbero, M. (2002). Nucleation, growth, and agglomeration in barium sulfate turbulent precipitation. *AIChE J.*, *48*(9), 2039–2050.

Marchisio, D. L., Vigil, R. D., & Fox, R. O. (2003a). Quadrature method of moments for aggregation–breakage processes. *J. Colloid Interface Sci.*, *258*, 322–334.

Marchisio, D. L., Vigil, R. D., & Fox, R. O. (2003b). Implementation of the quadrature method of moments in CFD codes for aggregation–breakage problems. *Chem. Eng. Sci.*, *58*, 3337–3351.

Marchisio, D. L., Pikturna, J. T., Fox, R. O., & Vigil, R. D. (2003c). Quadrature method of moments for population-balance equations. *AIChE J.*, *49*(5), 1266–1276.

Marchisio, D. L., Soos, M., Sefcik, J., & Morbidelli, M. (2006). Role of turbulent shear rate distribution in aggregation and breakage processes. *AIChE J.*, *52*(1), 158–173.

McGraw, R. (1997). Description of aerosol dynamics by the quadrature method of moments. *Aerosol Sci. Technol.*, *27*(2), 255–265.

Mühlenweg, H., Gutsch, A., Schild, A., & Pratsinis, S. E. (2002). Process simulation of gas-to-particle-synthesis via population balances: Investigation of three models. *Chem. Eng. Sci.*, *57*, 2305–2322.

Muhr, H., David, J., Villermaux, J., & Jezequel, P. H. (1996). Crystallization and precipitation engineering VI: Solving population balance in the case of the precipitation of silver bromide crystals with high primary nucleation rate by using first order upwind differentiation. *Chem. Eng. Sci.*, *51*, 309–319.

Nagata, S., & Nishikawa, M. (1972). Mass transfer from suspended microparticles in agitated liquids. *Proceedings of the first Pacific chemical engineering congress* (pp. 301–320).

Nandanwar, M. N., & Kumar, S. (2008). A new discretization of space for the solution of multi-dimensional population balance equations: Simultaneous breakup and aggregation of particles. *Chem. Eng. Sci.*, *63*, 3988–3997.

Nicmanis, M., & Hounslow, M. J. (1996). A finite element analysis of the steady state population balance equation for particulate systems: Aggregation and growth. *Comput. Chem. Eng.*, *20*, S261–S266.

Nie, Q., Wang, J., Wang, Y., & Bao, Y. (2007). Effects of solvent and impurity on crystal habit modification of 11α-hydroxy-16α,17α-epoxyprogesterone. *Chinese J. Chem. Eng.*, *15*(5), 648–653.

Nielsen, A. E. (1964). *Kinetics of precipitation*. London: Pergamon Press.

Nielsen, A. E., & Toft, J. M. (1984). Electrolyte crystal growth kinetics. *J. Cryst. Growth*, *67*, 278–288.

Öncül, A. A., Sundmacher, K., & Thévenin, D. (2005). Numerical investigation of the influence of the activity coefficient on barium sulphate crystallization. *Chem. Eng. Sci.*, *60*, 5395–5405.

Peng, S. J., & Williams, R. A. (1994). Direct measurement of floc breakage in flowing suspension. *J. Colloid Interface Sci.*, *166*, 321–332.

Petitti, M., Nasuti, A., Marchisio, D. L., Vanni, M., Baldi, G., Mancini, N., & Podenzani, F. (2010). Bubble size distribution modeling in stirred gas–liquid reactors with QMOM augmented by a new correction algorithm. *AIChE J.*, *56*(1), 36–53.

Phillips, R., Rohani, S., & Baldyga, J. (1999). Micromixing in a single-feed semi-batch precipitation process. *AIChE J.*, *45*(1), 82–92.

Piton, D., Fox, R., & Marcant, B. (2000). Simulation of fine particle formation by precipitation using computational fluid dynamics. *Can. J. Chem. Eng.*, *78*(5), 983–993.

Pohorecki, R., & Baldyga, J. (1979). *Processes of industrial crystallization*, Amsterdam: North-Holland.

Pohorecki, R., & Baldyga, J. (1988). The effects of micromixing and the manner of reactor feeding on precipitation in stirred tank reactors. *Chem. Eng. Sci.*, *43*(8), 1949–1954.

Qamar, S., & Warnecke, G. (2007). Numerical solution of population balance equations for nucleation, growth and aggregation processes. *Comput. Chem. Eng.*, *31*, 1576–1589.

Qamar, S., Elsner, M. P., Angelov, I., Warnecke, G., & Seidel-Morgenstern, A. (2006). A comparative study of high resolution schemes for solving population balances in crystallization. *Comput. Chem. Eng.*, *30*, 1119–1131.

Qamar, S., Warnecke, G., & Elsner, M. P. (2009). On the solution of population balances for nucleation, growth, aggregation and breakage processes. *Chem Eng. Sci.*, *64*, 2088–2095.

Ranade, V. V., & Bourne, J. R. (1991). Reactive mixing in agitated tanks. *Chem. Eng. Commun.*, *99*, 33–53.

Randolph, A. D., & Larson, M. A. (1988). *Theory of particulate processes* (2nd ed.). San Diego, CA: Academic Press.

Rigopoulos, S. (2007). PDF method for population balance in turbulent reactive flow. *Chem. Eng. Sci.*, *62*, 6865–6878.

Rigopoulos, S., & Jones, A. G. (2003). Finite-element scheme for solution of the dynamic population balance equation. *AIChE J.*, *49*, 1127–1139.

Saffman, P. G., & Turner, J. S. (1956). On the collision of drops in turbulent clouds. *J. Fluid Mech.*, *1*, 16–30.

Verkoeijen, D., Pouw, G. A., Meesters, G. M. H., & Scarlett, B. (2002). Population balances for particulate processes – A volume approach. *Chem. Eng. Sci.*, *57*(12), 2287–2303.

Veroli, G. D., & Rigopoulos, S. (2010). Modeling of turbulent precipitation: A transported population balance–PDF method. *AIChE J.*, *56*(4), 878–892.

Vicum, L., & Mazzotti, M. (2007). Multi-scale modeling of a mixing–precipitation process in a semibatch stirred tank. *Chem. Eng. Sci.*, *62*(13), 3513–3527.

Villermaux, J., & Falk, L. (1994). A generalizing mixing model for initial contacting of reactive fluids. *Chem. Eng. Sci.*, *49*, 5127–5140.

Villermaux, J., & Fournier, M. C. (1994). Potential use of a new parallel reaction systems to characterize micromixing in stirred tank. *AIChE Symp. Ser.*, *299*(90), 50–54.

Villermaux, J., & Zoulalian, A. (1969). Etat de mélange du fluide dans un réacteur continu. A propos d'un modèle de Weinstein et Adler. *Chem. Eng. Sci.*, *24*(9), 1513–1517.

Wang, Z., Mao, Z. -S., Yang, C., & Shen, X. Q. (2006a). Computational fluid dynamics approach to the effect of mixing and draft tube on the precipitation of barium sulfate in a continuous stirred tank. *Chinese J. Chem. Eng.*, *14*(6), 713–722.

Wang, Z., Mao, Z. -S., & Shen, X. Q. (2006b). Numerical simulation of macroscopic mixing in a rushton impeller stirred tank. *Chinese J. Process Eng.*, *6*, 857–863.

Wang, Z., Zhang, Q. H., Yang, C., Mao, Z. -S., & Shen, X. Q. (2007). Simulation of barium sulfate precipitation using CFD and FM-PDF modeling in a continuous stirred tank. *Chem. Eng. Technol.*, *30*(12), 1642–1649.

Wei, H., & Garside, J. (1997). Application of CFD modelling to precipitation systems. *Chem. Eng. Res. Des.*, *75*(A2), 219–227.

Wójcik, J. A., & Jones, A. G. (1998). Particle disruption of precipitated $CaCO_3$ crystal agglomerates in turbulently agitated suspensions. *Chem. Eng. Sci.*, *53*(5), 1097–1101.

Wong, D. C. Y., Jaworski, Z., & Nienow, A. W. (2001). Effect of ion excess on particle size and morphology during barium sulphate precipitation: An experimental study. *Chem. Eng. Sci.*, *56*, 727–734.

Woo, X. Y., Tan, R. B. H., Chow, P. S., & Braatz, R. D. (2006). Simulation of mixing effects in antisolvent crystallization using a coupled CFD–PDF–PBE approach. *Cryst. Growth Des.*, *6*(6), 1291–1303.

Yang, L., Cheng, J., Fan, P., Yang, C., & Mao, Z. -S. (2013). Micromixing of solid–liquid systems in a stirred tank with double impellers. *Chem. Eng. Technol.*, *36*(3), 1–8.

Zhang, Q., Mao, Z. -S., Yang, C., & Zhao, C. (2009). Numerical simulation of barium sulphate precipitation process in a continuous stirred tank with multiple-time-scale turbulent mixer model. *Ind. Eng. Chem. Res.*, *48*(1), 424–429.

Zhi, M., Wang, Y., & Wang, J. (2011). Determining the primary nucleation and growth mechanism of cloxacillin sodium in methanol–butyl acetate system. *J. Cryst. Growth*, *314*, 213–219.

Index

Printed in the United States
By Bookmasters